Dynamics of Offshore Structures

Cover photograph. This historical Argus Island Tower was a U.S. Navy facility, located 39 km off the southwest coast of Bermuda in a water depth of 58 m. Built in 1960, the Tower was used for about 10 years for underwater acoustic research and for submarine detection. The two enclosed levels on top of this four legged jacket structure had space for diesel generators, living quarters, and laboratories. During the first few years of the Tower's existence, it was subjected to storm-generated waves approaching 21 m, which was also the wave height upon which the Tower design was based. The 1969 inspections of the Tower revealed storm damage to many of its subsurface welded brace connections, damage that was deemed too closely to repair and subsequently maintain. Thus, demolition using shaped charges toppled the Tower in 1976, and its remains now rest on the coral floor of the sea.

Dynamics of Offshore Structures

James F. Wilson, Editor

John Wiley & Sons, Inc.

Published by John Wiley & Sons, Inc., Hoboken, New Jersey
Published simultaneously in Canada

For general information on our other products and services or for technical support, please contact our Customer Care Department within the United States at
(800) 762-2974, outside the United States at (317) 572-3993 or fax (317) 572-4002.

Wiley also publishes its books in a variety of electronic formats. Some content that
appears in print may not be available in electronic books.

Library of Congress Cataloging-in-Publication Data:

Wilson, James F. (James Franklin), 1933–
    Dynamics of offshore structures / James F. Wilson, Bruce J. Muga,
Lymon C. Reese.—2nd ed.
      p. cm.
    Includes index.
    New ed. of: Dynamics of offshore structures / James F. Wilson, editor. 1984.
    ISBN 0-471-26467-9 (cloth: alk. paper)
    1. Offshore structures.   I. Muga, Bruce J. (Bruce Jennings)
II. Reese, Lymon C., 1917–   III. Dynamics of offshore structures. IV. Title.

TC1665.W55 2002
627′.98—dc21   2002028858

10 9 8 7 6 5 4 3 2 1

Contents

Preface

This book is intended for three groups: (1) students and professors of structural and ocean engineering; (2) engineers and scientists in academic institutions, government laboratories, and industries involved in research on offshore installations, especially fluid-structure-soil interactions; and (3) practicing professional engineers who consider conceptual designs and need to employ dynamic analysis to evaluate facilities constructed offshore. The material herein was originally prepared by the three contributors for short courses attended by engineering practitioners, and for university courses taken by engineering seniors and graduate students.

Compared to the first edition, this second edition includes more example problems to illustrate the dynamic modeling, analysis, and solution of deterministic and stochastic responses for a wide variety of structures offshore, which include buoys, moored ships, and platforms of the fixed-bottom, cable-stayed, and gravity-type designs. Also, the extensive references of the first edition are updated, especially source material involving offshore waves, structural modal damping, and fluid-structure-soil interactions.

As in the first edition, this second edition addresses the basic physical ideas, structural modeling, and mathematical methods needed to analyze the dynamic behavior of structures offshore. Chapter 1 summarizes existing installations and points out future challenges. In subsequent chapters, careful attention is given to the many and sometimes subtle assumptions involved in formulating both the structural model and the natural forces imposed by the often hostile environment. The analyses in these chapters focus on plane motions of elastic structures with linear and nonlinear restraints, motions induced by the forces of currents, winds, waves, and earthquakes. Chapters 2 through 5 address single degree of freedom structural models that, together with plane wave loading theories, lead to time history predictions of structural responses. Chapters 6 and 7 extend these analyses to statistical descriptions of both wave loading and structural motion. Chapters 8 and 9 include the analysis and examples of multi-degree of freedom linear structures. Chapter 10 deals with continuous system analysis, including the motion of cables and pipelines. Chapter 11 addresses current practice related to submerged pile design for structures offshore.

I sincerely hope that this book will be useful and serve as an inspiration to engineers and researchers who design and analyze structures for the offshore environment.

JAMES F. WILSON
Chapel Hill, North Carolina

Contributors

Bruce J. Muga, Professor Emeritus of Civil and Environmental Engineering at Duke University, received his B.S. in Civil Engineering from the University of Texas, and his M.S. and Ph.D. degrees in Civil Engineering (Hydrodynamics) from the University of Illinois.

From 1961 to 1967 he was employed as a Project Engineer in the Port and Harbor Division of the U.S. Naval Civil Engineering Laboratory, Port Hueneme, California. In 1964, he was assigned as Consultant to the U.S. Military Assistance Command, Vietnam, to advise on coastal and harbor engineering projects.

In 1967, Dr. Muga accepted a position in teaching and research at Duke University and was Chairman of the Department of Civil Engineering in 1974. He has served as a consultant to many international corporations engaged in offshore and deep ocean engineering activities. He has written numerous technical papers and for seventeen years served on the North Carolina Marine Sciences Council. Prior to retirement, Dr. Muga was a Registered Professional Engineer in California, a member of the American Society of Civil Engineers and the Marine Technology Society. He is a life member of the Permanent International Association of Navigational Congresses.

Lymon C. Reese is the Nasser I. Al-Rashid Chair Emeritus and Professor, Department of Civil Engineering, The University of Texas, Austin, Texas, and is principal at Ensoft, Inc., a distributor of engineering software. Some of his consulting activities are carried out through Lymon C. Reese & Associates, a subsidiary of Ensoft.

He received his Bachelor's and Master's degrees in Civil Engineering from The University of Texas at Austin and his Ph.D. from The University of California at Berkeley. Dr. Reese has had several years of industrial experience and has been a consultant to a number of companies and governmental agencies. He was formerly Assistant Professor of Civil Engineering at Mississippi State University.

Dr. Reese has done extensive research in the field of geotechnical engineering, principally concerning the behavior of deep foundations. He has pioneered in performing field studies of instrumented piles and has developed analytical methods now widely used in the design of major structures. He has authored over 400 technical papers and reports and presented a number of invited lectures and talks in North and South America, Australia, Africa, Asia, and Europe.

Dr. Reese is an Honorary Member of the American Society of Civil Engineers and was selected as Terzaghi Lecturer in 1976; he received the Terzaghi Award in 1983. He received the Distinguished Achievement Award for Individuals from the Offshore Technology Conference in 1985 and was elected to membership in the National Academy of Engineering in 1975. He is a registered professional engineer in Texas and Louisiana.

James F. Wilson earned an A.B. degree from the College of Wooster, a B.S. degree in Mechanical Engineering from MIT in 1956, and a Ph.D. degree in applied mechanics from The Ohio State University, where he was a Ford Foundation Fellow and a Freeman Scholar. He worked in research and development for several companies and government agencies before joining the faculty at Duke University in 1967.

During his academic career, Dr. Wilson was a NASA-ASEE Faculty Fellow, a lecturer at three NATO Advanced Study Institutes, and a Visiting Scholar at Colorado State University and the University of Melbourne, Australia. He has been active in national committees for the American Society of Mechanical Engineers (ASME) and the American Society of Civil Engineers (ASCE), and received national awards for innovative experimental research (ASME, 1977), and the year's best state-of-the-art civil engineering journal publication (ASCE, 1987). He is a Life Fellow in ASME and a retired Fellow of the National Academy of Forensic Engineers. As a registered professional engineer, he regularly serves as an expert witness, testifying on structural failures, product performance, and vehicle accident reconstruction.

He is author or coauthor of over 200 works, which include technical reports on forensic engineering, refereed symposium papers and journal articles, two books on structural dynamics, a three-volume work on experiments in engineering, and two U.S. patents. His experimental research on robotics was highlighted in the 1989 BBC documentary, *Nature's Technology.*

During Dr. Wilson's career at Duke University, he has taught courses in applied mechanics, structural dynamics, and experimental systems, and was the major research advisor for over 35 graduate students, including postdoctoral fellows. He also served as the Director of Graduate Studies for the Department of Civil and Environmental Engineering. As Professor Emeritus since 1998, Dr. Wilson continues to pursue his research and writing interests and consulting practice.

Acknowledgments

I am grateful to those who have made this edition possible: to Robert E. Sandstrom for his contributions to Chapter 1; to Academic Press, Inc., for permission to use parts of Chapter 1, first published in the *Encyclopedia of Physical Science and Technology, III, 2001;* and to Gary Orgill and David M. Wilson who performed many of the calculations for the sample problems.

I especially acknowledge the patience and careful scholarship of the two guest contributors: Bruce J. Muga for Chapters 3 and 6 on fluid mechanics, and Lymon C. Reese for Chapter 11 on soil mechanics. Only by including the mechanics of fluids and soils can realistic analyses of offshore structures be made.

JAMES F. WILSON
Chapel Hill, North Carolina

Structures in the Offshore Environment

James F. Wilson

Offshore structures, constructed on or above the continental shelves and on the adjacent continental slopes, take many forms and serve a multitude of purposes: towers for microwave transmission, installations for power generation, portable pipeline systems for mining the ocean floor, and a few platforms and floating islands that serve as resort hotels. Most structures offshore, however, have been built to support the activities of petroleum industries—activities that include the exploration, drilling, production, storage, and transportation of oil. Exploratory drilling is done from mobile platforms or carefully positioned ships; production and storage operations involve more permanent structures; and pipelines, buoys, and mooring systems for floating structures and ships support all oil acquisition activities.

The design of marine structures compatible with the extreme offshore environmental conditions is a most challenging and creative task for the contemporary ocean engineer. The engineer involved in designing these marine structures must rely on the knowledge and experience of meteorologists, oceanographers, naval architects, geologists, and material scientists. The marine engineer's goal is to conceive and design a lasting structure that can withstand the adverse conditions of high winds and waves, earthquakes, and ice, remaining in harmony with its environment. Mulcahy (1979) expressed this design philosophy as follows:

> Offshore platforms are a bit like space capsules—for each pound of unnecessary deck space that can be trimmed from the structure, the magnitude of the structure needed to support it can be reduced. This is true for a guyed tower, a fixed platform, or a tension leg structure. Decreasing the wave load leads to lower overturning moments, a lesser requirement for pilings, and a smaller number of strength members in the structure. When this is accomplished, smaller launch barges can transport the structure to the work site.

In perspective, offshore structures include a great deal more than the towers and platforms. They include moored or mobile ships whose positions may be precisely controlled. They include the guy lines for compliant towers, the cables for buoys and for tension-leg platforms, and the associated pipelines without which the platforms and submerged oil production systems would be useless. Detailed descriptions of such installations may be found in the references at the end of this chapter. Of particular note is the review article on compliant offshore structures by Adrezin et al.(1996), with its 130 citations to the world literature on the subject up to the mid-1990s. For descriptions of current practice in all types of offshore installations, the reader is referred to the yearly conference proceedings such as found in the References at the end of this chapter.

This chapter begins with a short history of offshore structures, describes typical state-of-the-art installations, and concludes with a discussion of engineering challenges for future designs. Subsequent chapters address in some detail both the mathematical modeling and the environmental loading of offshore structures, together with ways to predict their dynamic responses and structural integrity, from both the deterministic and the statistical viewpoints.

1.1 HISTORICAL PERSPECTIVE

The earliest offshore structure for oil drilling was built about 1887 off the coast of southern California near Santa Barbara. This was simply a wooden wharf outfitted with a rig for drilling vertical wells into the sea floor. More elaborate platforms supported by timber piers were then built for oil drilling, including installations for the mile-deep well in Caddo Lake, Louisiana (1911) and the platform in Lake Maracaibo, Venezuela (1927). Soon after these early pier systems were built, it became apparent that the lifetime of timber structures erected in lakes or oceans is severely limited because of attacks by marine organisms. For this reason, reinforced concrete replaced timber as the supporting structure for many offshore platforms up to the late 1940s. Over the next 50 years about 12,000 platform structures were built offshore, usually of steel but more recently of precast concrete. The chief features of these structures, together with their supporting components such as mooring systems and pipelines, are discussed in this chapter. See also Gerwick (1999) and Will (1982).

Offshore mooring systems have a variety of configurations. All have anchors or groups of piles in the seabed with flexible lines (cables, ropes, chains) leading from them to buoys, ships, or platform structures. The function of a mooring system is to keep the buoy, ship, or platform structure at a relatively fixed location during engineering operations. Engineering efforts in mooring systems have focused in recent years on the development of new anchor configurations with higher pullout loads, larger capacity and lower cost of installation for deeper water applications.

When pipelines were first laid offshore, no extraordinary analyses or deployment techniques were needed since they were in shallow water and were of small diameter, somewhat flexible, and made of relatively ductile steel. As platforms were built in deeper and deeper water with multiple well slots, larger diameter pipelines of higher strength were required. During the 1960s, engineers met this challenge with new designs and with refined methods of analysis and deployment. Pipeline systems evolved into two main types: sea floor and vertical configurations. Both are used to transport gas and oil, but the vertical systems also include risers to carry drilling tools, electric power lines, dredge pipes for deep sea mining, and cold water pipes (CWP) for ocean thermal energy conversion (OTEC).

Throughout the world there are at present about 80,000 km of marine pipelines. Since 1986, the rate of building new marine pipelines has been about 1000 km per year. Individual pipelines on the sea floor vary in length from 1 to 1000 km and in diameter from 7 to 152 cm. For instance, a Norwegian project features a 1000 km line extending from the Troll field to Belgium, which was completed in 1992. At present, Kuwait has the loading line of largest diameter, 152 cm. The pipelines of smaller diameter are used to transport oil and gas from wellheads, and those of larger diameter are used to load and unload oil from tankers moored at offshore terminals. The deepest sea floor pipelines at present are the 46 cm diameter gas lines in the Gulf of Mexico, for which the

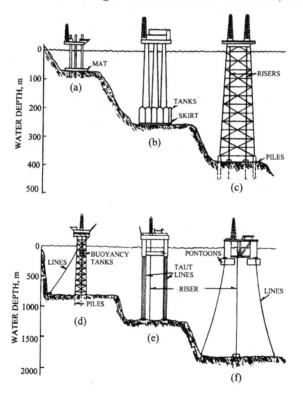

Figure 1.1   Six offshore platforms at their maximum depths: (a) jackup rig; (b) gravity platform; (c) jacket structure; (d) compliant tower; (e) tension leg platform; (f) semisubmersible.

maximum depth is 1400 m. Sea floor pipelines are often anchored to the seabed or buried in trenches for protection from erosion and the undermining effects of currents. Some seabed pipelines have a coating of concrete to add protection and to reduce buoyancy.

Figure 1.2   A mat-type jackup rig at a Louisiana dock.

1.2   PLATFORMS

Six general types of offshore platforms are depicted in Figure 1.1. The first three are designed for depths up to about 500 m, and the last three are for depths to 2000 m. Not shown are subsea production platforms, which are presently rated for 3000 m depths.

Fixed-Bottom Platforms

A mobile structure often used for exploratory oil-drilling operations is the self-elevating platform commonly called a jackup or mat-supported rig. A constructed version of this platform, depicted schematically in Figure 1.1a, is shown in Figure 1.2. Typically, such a platform is supported by three to six legs that are attached to a steel mat resting on the sea floor. In soft soils, the legs pass through the mat and may penetrate the soil to depths of up to 70 m. To the bottom of each leg is attached a steel saucer or *spud can* to help stabilize the structure and to minimize leg penetration into the soil. The height of the platform above the seafloor, up to 100 m, may be adjusted by using motor drives attached to each leg.

Figure 1.3   A jacket-template platform (courtesy of IHI Co. Ltd., Japan).

A platform designed to be used in a fixed location as a production unit is shown in Figure 1.1b. Such a unit, called a gravity platform, consists of a cluster of concrete oil-storage tanks surrounding hollow, tapered concrete legs that extend above the water line to support a steel deck. See Graff and Chen (1981). A typical unit, of which there were 28 operating in the North Sea in 1999, has one to four legs and rests directly on a concrete mat on the sea floor.

With ballast consisting of sand in the bottom of the tanks and seawater in the legs, these structures depend on self-weight alone to maintain an upright position when subjected to the highest waves that are expected to occur in a 100 year time period. A realistic 100 year wave that may occur in the northern North Sea is 27.8 m. At present, the largest concrete gravity platform is the Troll structure, and one of moderate size is the Statfjord-A Condeep structure, both located in the North Sea. The latter structure is 250 m high and has three legs. Located off the coast of Norway, the Statfjord-A Condeep unit has slots for 42 oil wells that reach to depths of 2800 m. When in operation, it accommodates a crew of 200 people who live and work on this structure.

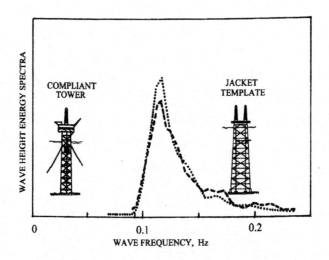

Figure 1.4   Two storm wave height spectra in the Gulf of Mexico, showing two offshore structures with natural frequencies beyond the frequencies of the highest energy waves.

Found more frequently among the permanent, fixed-bottom structures, however, is the steel truss or jacket template structure shown schematically in Figure 1.1c, where an installed structure is depicted in Figure 1.3. As for the gravity platform, each steel jacket unit is designed for a fixed location and a fixed water depth. The first such structure was operational in 1955 in water 30 m deep. By 1999 there were approximately 6500 jacket structures, the tallest of which was the Bullwinkle unit located in the Gulf of Mexico. The common characteristics of these jacket structures are their tubular legs, somewhat inclined to the vertical, and reinforced with tubular braces in K or X patterns. Piles driven through these legs into the sea floor and clusters of piles around some of the legs maintain structural stability in adverse weather. One of the largest jacket structures is the 380 m high Cognac unit, which has 10 legs with 24 piles extending 140 m into the soft clay of the Gulf of Mexico. As with all jacket template structures, its natural or fundamental bending frequency of 0.17 Hz is *above* the 0.11 Hz frequency of the highest energy sea waves in the Gulf of Mexico during storm conditions, as depicted in Figure 1.4.

Compliant Platforms

An alternative class of offshore structures meant for depths from 300 to 800 m is the compliant tower such as that shown in Figure 1.1d. Such a tower may or may not have mooring lines. It is a pile-supported steel truss structure designed to comply or flex with the waves and has considerably less structural material per unit height when compared with a common jacket template tower.

The first compliant tower was the Lena, which was installed in the early 1980s in the Gulf of Mexico. Including its three-level drilling and production deck and its drilling rigs, this tower reaches a total height of 400 m. Each of the 20 stabilizing cables, attached 25 m below the water line and arranged symmetrically about the structure, extends a horizontal distance of about 1000 m to a line of clumped weights that rest on the sea floor, to an anchor cable and an anchor pile. Under normal weather or small storm conditions, the cables act as hard springs, but with severe storms or hurricanes, the cable restraints become softer or compliant. That is, the amplitude of tower rotation increases at a rate greater than that of the loading, since the clumped weights lift off the sea floor to accommodate the increased storm loads on the tower. When storms or hurricane conditions are anticipated, operations on compliant towers cease and the crew is evacuated.

Installation of the Lena cables was more difficult and costly than anticipated. Subsequently, compliant towers without cables have been designed by Exxon, and two such designs were installed in 1999 in the Gulf of Mexico. Unlike the jacket-template structures, the compliant towers have natural frequencies in bending or sway near 0.03 Hz, or well *below* the 0.05 Hz frequency of the highest energy sea waves in the Gulf of Mexico during storm conditions. Thus, an important feature of such structures is that they are designed to have natural sway frequencies well removed from the frequency range of the highest energy waves for normal seas (0.1 to 0.15 Hz) and for storm seas (0.05 to 0.1 Hz). This frequency spread is necessary to avoid platform resonance, which can lead to failure. The sway frequencies of two platforms in comparison to the frequency range for the spectrum of the highest energy storm waves in the Gulf of Mexico are depicted graphically in Figure 1.4. The measurement and meaning of this wave height spectra, which is highly site-dependent, will be discussed in detail in subsequent chapters.

Buoyant Platforms

The tension leg platform (TLP) can be economically competitive with compliant towers for water depths between 300 m and 1200 m. The schematic design of the TLP is depicted in Figure 1.1e. In such designs, the total buoyant force of the submerged pontoons exceeds the structure's total gravity or deadweight loading. Taut, vertical tethers extending from the columns and moored to the foundation templates on the ocean floor keep the structure in position during all weather conditions. The heave, pitch, and roll motion are well restrained by the tethers; but the motions in the horizontal plane, or surge, sway, and yaw, are quite compliant with the motion of the waves. The first production TLP was built 150 km off the coast of Scotland in the mid-1980s. Conoco installed the Julliet in 1989, and Saga Petroleum installed the Snorre near Norway in 1991. The tethers for the Snorre are 137 cm in diameter. By the late 1990s, a total of eleven TLPs were installed, three in the North Sea and eight in the Gulf of Mexico.

For water depths of about 1500 m, a subsea production system provides an excellent alternative to a fixed surface facility. Much of a subsea system rests on the ocean floor, and its production of oil and gas is controlled by computer from a ship or other buoyant structure above the subsea unit. The buoyant structure and the subsea unit are often connected by a marine riser, which will be discussed presently.

Figure 1.5   A semisubmersible platform (courtesy of the builder, Mitsubishi, Ltd., Tokyo and the owner, Japan Drilling Co.).

A popular buoyant structure is the floating production system. Such a structure is practical for water depths up to 3000 m, and also at lesser depths where the field life of the structure is to be relatively short. An example of a buoyant structure is the semisubmersible with fully submerged hulls, shown schematically in Figure 1.1f , with an installed design shown in Figure 1.5. Other examples include ships converted to floating production systems. In the late 1960s, companies initiated research and design for these semisubmersible, multi-hull tubular structures and ships that would remain relatively stable in rough seas. In the late 1990s, the first three draft caisson vessels, or *spars*, were installed for use in 180 m water depths. Spars are floating vertical cylinders that support production decks above storm waves. These structures are controlled to remain essentially still in stormy seas. Some need to be towed from place to place; others are self-propelled. During drilling and production operations, these structures are kept in place with mooring lines and thrusters. The computer-controlled thrusters monitor the mooring line forces and accurately position the structure over the wellhead. One of the first semisubmersible structures was the Sedco 709 with a water depth rating of 1800 m. By the year 2000, semisubmersibles using dynamic positioning were designed for 3000 m water depths.

1.3 MOORINGS

Temporary Anchor Moorings

A classical example of temporary offshore mooring is the spread mooring configuration for a ship in relatively shallow water. Six to eight cables of wire rope or chain are unreeled from onboard winches symmetrically placed around the perimeter of the ship. Tug boats aid in spread mooring installations. In place, each cable hangs as a catenary curve and is attached either directly to a drag embedment anchor in the seabed or to a buoy that is anchored. An example of a spread mooring configuration is shown in Figure 1.6. Particular mooring configurations were reported by Baar et al. (2000) and O'Brian and Muga (1964).

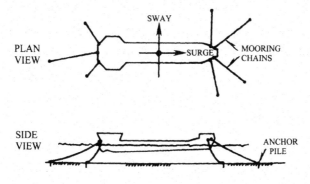

Figure 1.6   A spread mooring configuration.

In a typical spread mooring operation for a semisubmersible in deep water, a work vessel transports each anchor while pulling out its cable attached to the semisubmersible. The vessel lowers the anchor and installs a locating surface buoy just above it. At present, temporary systems for semisubmersible drilling rigs and construction barges are used in water depths of up to 2000 m. An example of an early and successful drilling rig is the Ocean Victory installed at a water depth of 450 m. This rig employed 12 anchors, each with a holding power of 200,000 newton (N) or about 45,000 lb. Each catenary line was about 2500 m long and consisted of two equal segments: one of 8.9 cm diameter steel wire cable and the other of 8.3 cm diameter chain. Newer generation rigs designed for deep water have ten times the anchor-holding capacity of the Ocean Victory. An anchor design based on suction is shown schematically in Figure 1.7.

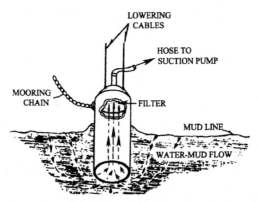

Figure 1.7   A suction anchor showing water flow during embedment.

Platform Pile and Single-Point Moorings

For installing piles for platform moorings, specially fitted derrick barges may use hydraulic hammers, drilling equipment, or possibly a jetting system. In jetting, seawater is forced around the base of a pile, blasting away the soil to make way for pile embedment. When installing mooring piles for tension leg platforms, template structures are carefully positioned at the site, and piles are hammered through the template that serves as a pile guide. Suction piles are employed in deep water where the use of hydraulic hammers is impractical or too costly. Suction piles employ hydrostatic pressure to push the piles to full penetration.

Single-point mooring (SPM) systems are designed to accommodate deep-draft tankers while they transfer crude oil and fuel oil to and from shore. Two typical designs are shown in Figure 1.8: the single anchor leg mooring (SALM), and the catenary anchor leg mooring (CALM). By the year 2000, there were about 50 SALM systems and 150 CALM systems in operation throughout the world. Their common features are the rotating head on the buoy and the vertical chain that anchors the buoy to the sea floor. While a few SALM systems may have taut mooring lines for added buoy stability, all CALM systems have multiple catenary lines (anchor legs). A third type of SPM system, the articulated column, has been designed and laboratory-tested but has yet to be installed offshore.

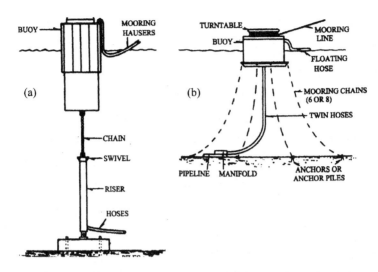

Figure 1.8   Two single-point mooring systems: (a) a single anchor leg mooring (SALM); (b) a catenary anchor leg mooring (CALM).

Successful SPMs are found at the world's largest terminals. One of these is Saudi Arabia's Ju'aymah exporting terminal, which has two SALM buoys and four CALM systems. For this SPM system, crude oil and fuel oil are loaded simultaneously to a moored tanker through the swivel assembly on the seabed. A second example is the CALM system for service vessels associated with the Cognac platform in the Gulf of Mexico. This SPM buoy has 12 catenary lines, each anchored to 0.76 m diameter piles. The water depth here is 275 m.

1.4   PIPELINES

Sea Floor Pipelines

Most of the 80,000 km of offshore pipelines have been installed by one of the following three methods. In these methods, the deployed pipeline forms an S-shape between the vessel and the sea floor.

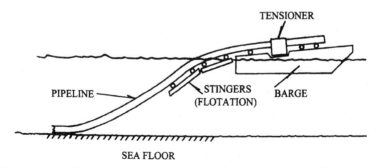

Figure 1.9   The laybarge method of pipeline construction offshore.

1. *Laybarge method.* Pipe sections, which sometimes have been coated previously with concrete for protection, are welded together on the deck of a barge and deployed on rollers over the stern. Near the stern, the pipeline passes over pontoons called stingers that relieve excess bending in the pipeline as it is deployed to the seabed. See Figure 1.9.

2. *Reel barge method.* Small to medium diameter pipe sections (up to 41 cm in diameter) are prewelded and coiled onto a reel mounted to the deployment vessel. As the pipeline is unreeled at sea, it passes through straightening rollers and then deployed as in the laybarge method.

3. *Bottom pull method.* Pipe sections are assembled on shore, and the pipe string is towed into the sea by a barge. During the launch to its place on the seabed, pontoons are often used under the string to avoid excess pipeline bending.

The following two methods of pipeline deployment have been studied extensively but have yet to be used.

1. *J-lay method.* A dynamically positioned vessel such as a drill ship, a converted pipelaying vessel, or a semisubmersible may be used in this operation. On board the vessel a derrick is used to hold the pipeline vertically as it is lowered, and the pipeline forms a J-shape between the vessel and the seafloor. An efficient single station pipe welding procedure has to be developed before this method can be adapted to common practice.

2. *Floating string method.* According to this concept, the pipeline is floated on pontoons at the water surface. Then the pontoons are released successively so that the pipe string gradually sinks to the sea floor.

Vertical Pipelines

One type of vertical offshore pipeline is the marine riser, which is shown for several of the structures in Figure 1.1. Although marine risers make up a fraction of the 80,000 km network, they are nonetheless key components of offshore structures and serve a variety of functions. For example, a riser may contain a bundle of smaller pipelines connecting a wellhead to its platform, or it may transport oil directly from its platform to a sea floor pipeline. A riser may act as a drilling sleeve, or it may contain electric power lines for operating seafloor mining vehicles or other subsea facilities. A typical drilling riser, with a ball joint at the bottom and a telescoping joint at the top, is maintained under tension to ensure stability and is kept to within 8 degrees of the vertical by computer-controlled positioning of its parent drilling vessel. In the mid-1980s, the longest riser was operating off the east coast of the United States, where depths of 2100 m were reached. At present, risers for depths approaching 3000 m are becoming a reality.

Vertical pipelines are used in deep-sea dredging operations. A most challenging engineering problem, which began to receive serious attention in the late 1960s, was that of designing dredge pipes to suck up and transport manganese nodules from depths of 3000 to 5500 m to the ship. See Lecourt and Williams (1971).These nodules, about the size of a man's fist and found in a monolayer on the sea floor in many parts of the world, contain other minerals as well, including cobalt, copper, and nickel with traces of molybdenum, vanadium, and titanium. By the mid-1980s, remote-controlled mining of the nodules was achieved with a 5500 m long dredge pipe. In this design, air bubbles are pumped into the string at various points along its length to aid in pumping the nodules to the surface. The dredge pipe's position is computer-controlled, synchronized to the movement of both the self-propelled mining scoop at one end and the ship at the other end. A schematic diagram of this system is shown in Figure 1.10a.

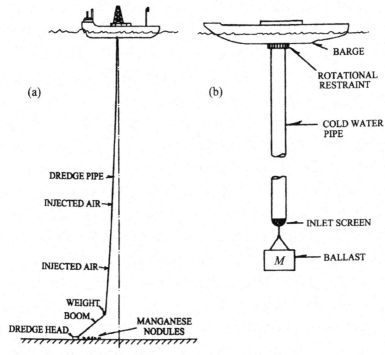

Figure 1.10   Two vertical pipeline systems: (a) a dredge pipe for mining manganese nodules; (b) a cold water pipe for ocean thermal energy conversion.

Vertical cold water pipe systems for ocean thermal energy conversion (OTEC) have been studied extensively. See Wilson et al. (1982). Calculations show that practical full-scale pipe designs are 10-20 m in diameter, are about 1000 m long, and may be used to raise the cooler water in the deep ocean to the warmer surface water. With the differential temperature between the ends of the pipe (10-20°C), it is possible to produce net power through heat exchange. Such a system is depicted in Figure 1.10b. In 1979, a small-scale OTEC power plant operated successfully on a barge 2.3 km offshore of Kona, Hawaii. This pipe was 61 cm in diameter, 660 m long, and made of polyurethane. The CWP was supported by a surface buoy at the barge stern and was tension-moored with a smaller diameter polyurethane pipeline, a seafloor wire rope to shore, and concrete blocks on the sloping sea floor. This was the world's first and so far the only OTEC system at sea to generate net useful power. In the year 2001, further development work on OTEC systems was underway in India.

1.5   CHALLENGES

To meet the need for new sources of energy and minerals, marine engineers must work at the frontiers of known technology. Their main challenges are to design, deploy, and operate facilities and equipment in environments where none have before existed—in deeper and deeper water, on the slopes of the continental shelves, and in the hostile Arctic seas. To meet the challenges, marine engineers continue their research and development efforts in the following four broad and necessarily overlapping topics of concern.

Environmental Forces

Every offshore facility is subjected to several types of environmental loads during its lifetime. Thus, site-dependent databases are being developed to characterize the time-varying fluid-induced loads of winds, currents, and waves. Such loads occur both on a short time scale of seconds and minutes, induced by periodic vortex shedding, wind gusts, and wave slamming; and over longer periods of hours, days, and even years, where the loads are induced by steady waves, tides, and hurricanes. At some sites, data are also needed on subsea earthquake intensity, on the scouring of sea floor foundations by currents, and on the total and differential settlements of the sea floor due to the withdrawal of hydrocarbons during the lifetime of the structure. At present, there is the particular challenge of minimizing differential settlements for gravity platforms. For the Ekofisk structural complex in the North Sea, however, this challenge appears to have been met by employing water injection. In Arctic regions, data are needed on the rates of ice accretion and on the velocity, yield strength, and mass of floating ice. Reliable deterministic and statistical methods are being developed to measure and interpret time-dependent field data suitable for predicting structural loadings.

Structural Materials

An offshore structure should have a high ratio of strength to self-weight. For instance, for each added unit of deck weight for a tension leg platform, an additional 1.3 units of hull weight are required for buoyancy support, and an additional 0.65 unit of mooring pretension force is needed. In this case, high-strength steels with high fracture toughness are being investigated for the purpose of reducing hull weight. Hollow cylindrical steel link chains or synthetic mooring line materials, such as Kevlar with an abrasion-resistant polyethylene cover, are being developed to increase mooring capacity. To determine the suitability of new, high-strength steels and composite materials in the offshore environment, test data are being generated. These data involve measures of corrosion fatigue, fracture toughness, stress corrosion cracking, and weldability for steels and the reliability of several types of high-strength, light-weight synthetic rope mooring lines.

Modeling and Analysis

Once the site, the environmental load conditions, and the preliminary structural design are determined, a mathematical model of the structure is formulated, and computer-aided analyses are performed to determine the overall motion, the critical stresses, and the reliability of the design. The fundamental mechanics of fatigue and fracture were reviewed by Petroski (1984), who discussed fatigue failures that have occurred at the welded tubular joints, which can be subjected to millions of loading cycles during a structure's lifetime. It is particularly important to include fatigue in the structural reliability analysis. Novel and improved methodology for reliability assessment of offshore structural welded joints, with applications to jackup rigs, was discussed by Etube (2001). Computer graphics and finite element methods of structural analysis that aid in this design process are continually being improved. Structural optimization, least weight criteria, nonlinear dynamics, fluid-structure-soil dynamic interactions, and statistical methods are being added and refined in the mathematical models. The aims are to achieve more accurate representations for structural dynamic behavior and improve the predictions of structural lifetimes. See Buchholt (1997) and Jin and Bea (2000).

Experimental Evaluations

Before a structure is installed offshore, extensive tests are made on its component materials and also on its overall dynamic behavior using scaled-down laboratory models with simulated environmental load conditions. Present laboratory-scale efforts are focused on two issues: fluid-structural load interactions, especially the effects of fluid vortices on structural motion; and the mechanics of structural-soil foundation interactions.

Once installed in the field, the structure may be instrumented for a period of time to determine whether its behavior under a variety of natural forces has been accurately predicted by the mathematical models and analyses used in its design. Such data are useful for future similar designs. In addition, for platforms, buoys, and pipeline systems nearing the end of their lifetimes, frequent inspection and maintenance are required to assure continuing performance. Improvements are being made in the computer systems used for the retrieval, storage, and analysis of large amounts of data from both laboratory scale and field tests of offshore structural systems.

REFERENCES

Adrezin, R., Bar-Avi, P. and Benaroya, H., Dynamic Response of Compliant Offshore Structures—Review, *Journal of Aerospace Engineering*, October, 114-131, 1996.

Baar, J. J. M., Heyl, C. N., and Rodenbusch, G., Extreme Responses of Turret Moored Tankers, *Proceedings of the Offshore Technology Conference*, OTC-12147, 2000.

Buchholt, H. A., *Structural Dynamics for Engineers*, Thomas Telford Publications, London, 1997.

Etube, L. S., *Fatigue and Fracture Mechanics of Offshore Structures*, Professional Engineering Publishing Limited, London and Bury St. Edmunds, Suffolk, UK, 2001.

Gerwick, B. C., *Construction of Marine and Offshore Structures*, second ed., CRC Press, Boca Raton, FL, 1999.

Graff, W. J., and Chen, W. F., Bottom-Supported Concrete Platforms: Overview, *Journal of the Structural Division, ASCE* **107**(ST6), 1059-1081, June 1981.

Jin, Z., and Bea, R., Enhancements of TOPCAT-3 Dimensional Loadings, Reliability, and Deck Structure Capabilities, *Proceedings of the Offshore Technology Conference*, OTC-11939, 2000.

Lecourt, E. J., and Williams, D. W., Deep Ocean Mining—New Applications for Oil and Field Marine Equipment, *Proceedings of the Offshore Technology Conference*, OTC-1412, 1971.

Mulcahy, M., Fixed Offshore Structure Design Responds to Tougher Operating Requirements, *Sea Technology,* 10-12, June 1979.

O'Brian, J. T., and Muga, B. J., Sea Tests on a Spread-Moored Landing Craft, *Proceedings of the Eighth Conference on Coastal Engineering,* Lisbon, Portugal, 1964.

Petroski, H.J., Fracture Mechanics and Fatigue in Offshore Structures, chapter 13 of *Dynamics of Offshore Structures,* J. F. Wilson, editor, Wiley, New York 1984.

Will, S. A., Conventional and Deep Water Offshore Platforms, *Civil Engineering* **58**, February 1982.

Wilson, J. F., Muga, B. J., and Pandey, P., Dynamics During Deployment of Pipes for Ocean Thermal Energy Conversion, *Proceedings of the First Offshore Mechanics/Arctic Engineering/ Deepsea Systems Symposium, ASME,* New Orleans, LA, 53-58, March 1982.

2

Structure-Environmental Force Interactions

James F. Wilson

This chapter has three main objectives: (1) to discuss basic ideas, limitations, and physical laws involved in modeling complex offshore structures as single degree of freedom systems; (2) to quantify the important environmental forces, except for wave action, that occur offshore and show how these forces interact with structures (discussions of wave action are in Chapters 3 and 4); and (3) to illustrate the modeling of structural mass, stiffness, and the structural restraint forces of guy lines and soil foundations. Emphasized in this chapter is the first step in dynamic engineering analysis: the formulation of the structural equations of motion. Solutions for structural motion, both deterministic and stochastic, are presented in later chapters.

2.1 SINGLE DEGREE OF FREEDOM STRUCTURES

Before the motion of an offshore structure can be calculated, an analytical representation is needed for the structure (or part of the structure), together with the loadings and restraints. This representation, called the *mathematical model*, has two parts: a simplified schematic diagram or *free body sketch* of the structure, and the associated equations of motion. The free body sketch shows a typical dynamic position or mode shape of the structure, relative to its static equilibrium position; it describes the necessary and sufficient independent coordinates, equal in number to the *degrees of freedom* needed to describe motion uniquely; and it shows as arrows four classes of generalized forces, which include moments. These forces are: (1) self-weight; (2) externally applied environmental forces mentioned briefly in Chapter 1; (3) reaction forces (due to system restraints) that tend to restore the structure to its static equilibrium position; and (4) damping forces that mitigate motion. Based on the free body sketch, the equations of motion are formulated in a straightforward manner, using either Newton's second law, as is done in this chapter, or Lagrange's energy methods, as is done in Chapter 8. Although these two methods may be employed for multi-degree of freedom problems in three-dimensional physical space, the mathematical models in this text are limited to motion in the vertical plane, a representation that is generally adequate for a preliminary dynamic design of offshore structures.

The simplest mathematical model has just one equation of motion, written as a differential equation in terms of only one time-dependent scalar coordinate which uniquely describes the structure's position. Such a model defines a *single degree of freedom* system. Assume that a structure or portion of a structure is rigid, or nearly so, and has a *virtual* mass m. Because of structure-fluid interactions, discussed later in this chapter, virtual mass includes all or part of the structural mass, together with some water that the structure drags with it during motion. Assume further that m is sufficiently higher than the mass of the system restraints (guy lines or soil foundation) that limit its motion. Let the motion of m be restricted to a plane on which the absolute displacement coordinate of its mass center G is $v = v(t)$. Let ΣF_v denote the sum of the four types of external force components on m, in line with v and positive in the positive v direction. In these terms, Newton's second law states

$$\sum F_v = m\ddot{v} \qquad (2.1)$$

where $\ddot{v}$ is the absolute acceleration of G. This equation is particularly useful in single degree of freedom models involving rigid body translation only in which the flexible supports restraining the motion of m are in line with v.

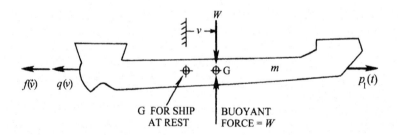

Figure 2.1   Free body sketch for the spread moored ship of Figure 1.6.

Example Problem 2.1.   Consider the surge motion for the spread moored ship shown in Figure 1.6. This ship is asumed to be a rigid structure for which the motion is described solely by its horizontal displacement coordinate $v = v(t)$ of its mass center at G. Sway, pitch, heave, and yaw motions are neglected. The free body sketch of the ship in the vertical plane, shown in Figure 2.1, depicts the four types of externally applied loads: self-weight or ship displacement W, which is balanced by its buoyant force equal in magnitude to W; the equivalent of all v-directed, time-dependent environmental forces $p_1(t)$; the equivalent v-directed mooring line restraint force of the general form $q(v)$; and the velocity-dependent damping force of the general form $f(\dot{v})$. When equation (2.1) is applied in the v direction to this free body sketch, the resulting equation for surge motion becomes

$$m\ddot{v} + f(\dot{v}) + q(v) = p_1(t) \tag{2.2}$$

In the first term of equation (2.2), the virtual mass m for surge motion is about 15 percent higher than the actual ship's mass. Quantitative values for the remaining terms of this equation are discussed later in this chapter.

Other examples of single degree of freedom offshore structures include the pure, plane rotational motion of buoys such as shown in Figures 1.8 and of gravity platforms rocking in the vertical plane, such as depicted in Figure 2.2. With all out of plane motion and translational motion supressed, the angular displacement of these structures, modeled as rigid bodies of virtual mass m, the angular displacement is uniquely described by the single coordinate $\theta = \theta(t)$. Suppose that such a structure rotates about a fixed point 0 in the plane. Let ΣM_0 denote the sum of all external moments in the plane of motion, acting on m, positive in the positive direction of θ. These moments, which are due to the four types of external forces discussed above, are all expressed with respect to the same fixed point 0. In this case, the equation of motion has the general form

$$\sum M_0 = J_0\ddot{\theta} \tag{2.3}$$

in which $\ddot{\theta}$ is the absolute angular acceleration of the rigid body and J_0 is the virtual mass moment of inertia of this body with respect to the reference axis through point 0 and perpendicular to the plane of motion. The value of J_0 is defined as

$$J_0 = \int_m r^2 dm \tag{2.4}$$

where r is the distance from that reference axis to the virtual mass element dm, and the integration is over the whole rigid body. In applications it is often convenient to express J_0 in terms of J_G, or the value of J_0 when the point 0 coincides with the mass center G. The connection is through the parallel axis theorem, or

$$J_0 = J_G + mh_G^2 \tag{2.5}$$

where h_G is the distance between 0 and G. Values of J_G for a variety of solids of uniform density are listed in most elementary texts on rigid body dynamics. For a relatively rigid structure composed of such shapes, the structure's total J_0 value can be estimated by calculating J_0 for each elementary component using equation (2.5) and then superimposing the results.

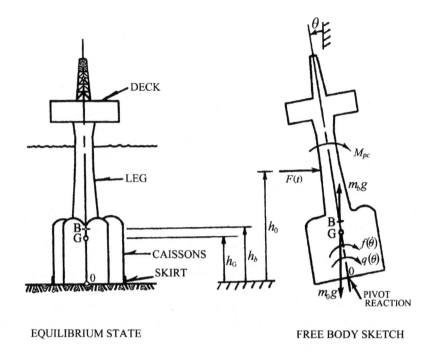

DECK

LEG

M_{pc}

$F(t)$

$m_b g$

B
G

CAISSONS

SKIRT

h_0

h_G h_b

B
G

$f(\dot\theta)$

$q(\theta)$

0

0

$m_o g$

PIVOT
REACTION

EQUILIBRIUM STATE   FREE BODY SKETCH

Figure 2.2   Monopod gravity platform in pure rotation.

Example Problem 2.2.   Consider the rotational motion of a rigid gravity
platform shown in Figure 2.2. This structure has an actual mass of m_0, a
buoyant mass of m_b, and is supported at the sea floor by a soil foundation. Let
the motion be limited to rotations θ about the base pivot point 0. Define M_{pc}
as the net moment about point 0 due to the pressure differences across the top
of the caissons, and let $F(t)$ represent the net horizontal load due to currents,
winds, and waves, located at height h_0 above 0. Let $f(\dot\theta)$ and $q(\theta)$ be the
respective foundation reaction moments for damping and rotational restraint.
In these terms, the application of equation (2.3) to the free body sketch of the
gravity platform of Figure 2.2 leads to the following equation of motion:

$$J_0\ddot\theta + f(\dot\theta) + q(\theta) - (m_0 g h_G - m_b g h_b)\sin\theta = -F(t)\,h_0 - M_{pc} \qquad (2.6)$$

In equation (2.6), J_0 is based on the *virtual* mass of the submerged portion of the
structure. For instance, for a submerged cylinder, virtual mass is approximately
the cylinder mass plus the mass of the water displaced by the cylinder. Quantitative calculations for J_0 and the other terms of equation (2.6) are illustrated
in *Example Problem 5.2*.

In the last example problem, the rigid body assumption may not always be
a realistic one. The rigid body model would be entirely useless, for instance,
if the dynamic flexural stress were needed in the legs of the gravity platform.

Nonetheless, the rigid body assumption may be warranted if only an estimate of the overall dynamic stability of the platform-soil foundation system is needed. Thus the choice of the mathematical model is strongly tempered by the particular goals of the analysis.

2.2 FLUID-INDUCED STRUCTURAL FORCES

There is a wealth of literature on the theory and measurement of forces on solid bodies moving or at rest in dynamic fluid fields. An excellent critique of this literature that extends back to the early nineteenth century and up to 1981 is given by Sarpkaya and Isaacson (1981). The essential physical ideas and governing nondimensional loading parameters presently perceived as characterizing these dynamic forces are now highlighted. Most of the following discussion is limited to an isolated, fully submerged, right circular cylindrical solid for which the incident fluid velocity is perpendicular to its longitudinal axis. In the plane flow cases considered, shown in Figures 2.3 through 2.6, the fluid or cylinder motion is in line with its net force per unit length, $\bar{q}$. The fluid is assumed to be incompressible, and for the present, effects of nearby objects and solid boundaries are not included. As restrictive as these assumptions may seem, the results none the less demonstrate the basic ideas of fluid loading for a major portion of offshore structures and their components, including pipelines, cables, tubular structural members, and many types of submerged tanks and caissons.

Classical Inviscid Fluid Flow

One classical loading parameter is the *inertia coefficient*, C_M, alternatively denoted as C_I or C_i, a parameter that originated with hydrodynamics or the theory of ideal, inviscid flow, formulated during the nineteenth and early twentieth centuries (Batchelor, 2000; Lamb, 1945). This coefficient relates the force per unit length $\bar{q}_I$ that is required to hold a rigid cylinder stationary in a fluid of uniform, constant free stream acceleration of magnitude $\dot{u}$. That is,

$$\bar{q}_I = C_M \, \rho\pi \frac{D^2}{4} \dot{u} \tag{2.7}$$

where ρ is the fluid density and D is the cylinder diameter. Shown in Figure 2.3 is this case of unseparated, unsteady, ideal flow, together with values of C_M for several ratios of cylinder length to diameter, ℓ/D. These results, reported by Wendel (1956), are based on theoretical values of another nondimensional parameter, the added mass coefficient C_A, defined by

$$C_A = C_M - 1 \tag{2.8}$$

It is observed from the theoretical data in Figure 2.3 that as the cylinder length becomes much larger than its diameter, the value of C_M approaches the limit of 2, for which C_A approaches unity by equation (2.8).

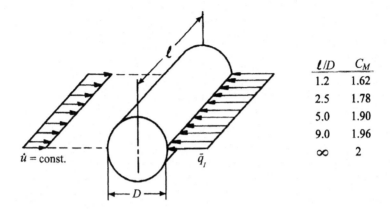

ℓ/D	C_M
1.2	1.62
2.5	1.78
5.0	1.90
9.0	1.96
∞	2

Figure 2.3   A rigid, stationary cylinder in an ideal, accelerating fluid.

Consider now the case of Figure 2.4, a rigid cylinder with a mass per unit length $\bar{m}_0$, immersed in a fluid. This cylinder has an absolute translational displacement $v = v(t)$ in the fluid medium that would normally be at rest, except for the presence of the cylinder. Based on Newton's second law of motion and classical hydrodynamic theory, the force per unit length required to achieve an acceleration $\ddot{v}$ for the cylinder is

$$\bar{q}_I = \left(\bar{m}_0 + C_A \, \rho\pi\frac{D^2}{4} \right) \ddot{v} = \bar{m}\ddot{v} \tag{2.9}$$

Equation (2.9) defines the cylinder's *virtual mass* per unit length, $\bar{m}$. Thus, $\bar{m}$ is the sum of $\bar{m}_0$ *in vacuo* and the added or apparent mass per unit length, $C_A\rho\pi D^2/4$, resulting from those fluid particles that are pulled along by the intruding cylinder.

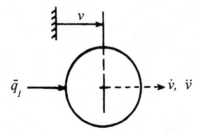

Figure 2.4   A rigid cylinder accelerating in an ideal fluid.

Real Viscous Fluid Flow

Measurements show that C_M and C_A are time-dependent, which is due to fluid viscosity. There is flow separation behind the cylinder, accompanied by differential pressure forces opposing cylinder motion. Such forces are referred to as *form drag*. In applications, a root-mean-square (rms) measured average of each coefficient C_M and C_A is used. If such data are lacking, it is appropriate to choose $C_A = 1$ for design purposes, provided that the geometric ratio ℓ/D is much greater than one.

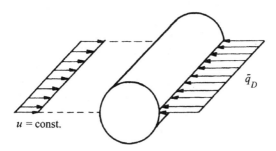

Figure 2.5   Viscous drag on a rigid, stationary cylinder.

Another classical loading parameter is the viscous, frictional drag coefficient, C_D. Define $\bar{q}_D$ as the force per unit length necessary to hold a fully immersed cylinder stationary as it is subjected to a constant free stream fluid velocity, u. In these terms, measurements show that

$$\bar{q}_D = C_D \rho \frac{D}{2} |u| u \tag{2.10}$$

The use of the absolute value sign on one of the velocity terms guarantees that q_D will always oppose the direction of u, as shown in Figure 2.5. For this flow case, the experimental relationships of C_D to two nondimensional parameters, cylinder roughness, and the Reynolds number, are well known (Schlichting, 1968). Here the Reynolds number is defined by

$$\text{Re} = \frac{\rho u D}{\mu} \tag{2.11}$$

where μ is the absolute viscosity of the fluid. For a smooth cylinder subjected to this constant, uniform, free stream flow, the value of C_D is approximately unity for Re in the range of about 1000 to 200,000.

If the cylinder were rotating about its longitudinal axis or if it were not circular, or if other solid elements or rigid boundaries were nearby, one would need an additional loading parameter, the lift coefficient C_L. In such cases, the lift force per unit length, which is perpendicular to u and $\bar{q}_D$, has the same form as equation (2.10), where C_L depends on cylinder roughness, Reynolds number, and the proximity of objects nearby.

In a uniform flow field, both separation and periodic wakes or vortices may form behind the stationary cylinder. This phenomenon, discussed extensively by Blevins (1977), is depicted in Figure 2.6. The vortices behind the cylinder detach alternately. Accompanying this is a periodic pressure fluctuation, top to bottom, at a characteristic frequency of f_s, typically expressed in units of Hz (cycles per second). Periodic vortices or vortex sheets occur for Reynolds numbers in the range of 60 to 10,000 and sometimes even higher. (Swimmers can observe this phenomenon, for instance, by moving a hand downward through the water with fingers spread and feeling a tendency for the fingers to vibrate horizontally or side to side).The nondimensional parameter that correlates vortex-shedding data for the flow of Figure 2.6 is the Strouhal number, defined by

$$S = \frac{f_s D}{u} \qquad (2.12)$$

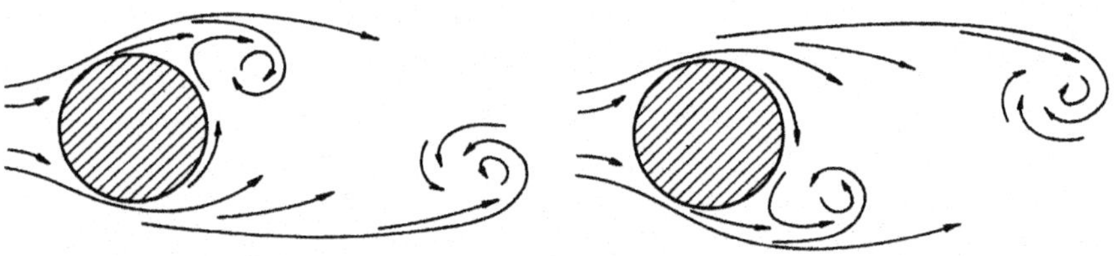

pattern times: $t = 1/f_s, 2/f_s, ...n/f_s$ pattern times: $t = 1.5/f_s, 2.5/f_s, ...(n+0.5)/f_s$

$(n = 1, 2, ...)$ $(n = 1, 2, ...)$

Figure 2.6   Periodic vortices trailing behind a rigid, stationary cylinder.

Generally, S correlates well with the Reynolds number. For instance, correlations of S with Re showing the apparent effects on vortex-shedding frequency of a cylinder in proximity to a twin cylinder and a ground plane are reported by Wilson and Caldwell (1971). In some instances, vortex-induced pressure forces on a cylindrical structure could be large enough to destroy the structure. For instance, periodic vortices behind the cylindrical piles supporting an offshore pier lead to a complete destruction of that pier in a tidal current of two knots. Since vortex-induced loading is unpredictable, one generally makes provisions to avoid periodicity of the vortices. Although many methods have been proposed to suppress vortices (Hafen et al., 1976), a particularly practical one is the addition of helical strakes around the cylinder, as shown in Figure 2.7. Optimal strake geometries for the least vortex-induced loads on circular cylinders were studied by Wilson and Tinsley (1989).

Figure 2.7   A cylinder with a helical strake to negate periodic vortices.

The last nondimensional fluid-loading parameter highlighted herein is the Keulegan-Carpenter number, Kc, which arises when a free stream, plane, periodic flow is imposed on a stationary cylinder. If neither u or $\dot{u}$ is constant, but both are described by a single, simple, plane wave of time period T, then Kc correlates well with the force data on the cylinder. This flow parameter is

$$\mathrm{Kc} = \frac{u_0 T}{D} \qquad (2.13)$$

where u_0 is the amplitude of the wave velocity (Keulegan and Carpenter, 1958).

Conservation of Linear Momentum and Flow Superposition

For a stationary cylinder in a plane flow field with the free stream velocity $u = u(t)$, the total time-varying load per unit length on the cylinder may be expressed by superimposing the two flow models described by Figures 2.3 and 2.5. Thus, by adding their respective drag and inertial loadings expressed by equations (2.10) and (2.7), the result becomes

$$\bar{q} = C_D \rho \frac{D}{2} |u| u + C_M \rho \pi \frac{D^2}{4} \dot{u} \qquad (2.14)$$

Equation (2.14) was first proposed by Morison and his colleagues (1950) as an empirical result, and that equation now bears his name.

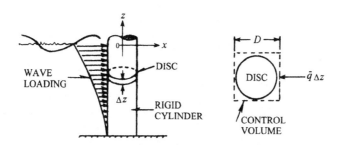

Figure 2.8   A rigid, circular cylinder in a flow field.

There is a theoretical basis for Morison's equation, the general form for which can be deduced by applying the principle of conservation of linear fluid momentum. This conservation principle is based on Newton's second law as applied to the fluid occupying a fixed control volume $\mathcal{V}$ at any instant of time. In this case, $\mathcal{V}$ is chosen as the imaginary rectangular box surrounding a disc element of a submerged circular cylinder, as shown in Figure 2.8. Here, $\mathcal{V} = D^2 \Delta z$, where D is the disc diameter and Δz is the disc height. For water of density ρ and with a horizontal velocity u along x, the net horizontal shear load ΣF_x on the disc is given by

$$\sum F_x = \frac{\partial}{\partial t} \int_{\mathcal{V}} \rho \, u \, d\mathcal{V} + \int_{A_0} \rho \, u \cdot u \, dA_0 \qquad (2.15)$$

where $d\mathcal{V}$ is an element of the control volume and dA_0 is the element of area on the surface of the control volume perpendicular to the flow. The reader is referred to a standard text such as Munson et al. (1998) for a general derivation of equation (2.15).

The terms of equation (2.15) are now evaluated. First, $\Sigma F_x = -\bar{q} \Delta z$ where $\bar{q}$ is the loading per unit length. The negative sign is chosen since the net shear reaction load must oppose the direction of flow. The first integral on the right is approximated by

$$\frac{\partial}{\partial t} \int_{\mathcal{V}} \rho u \, d\mathcal{V} \simeq -\rho D^2 \, \Delta z \, \dot{u} \qquad (2.16a)$$

The negative sign is chosen since the fluid *decelerates* inside the control volume surrounding the rigid disc. The remaining integral represents the net momentum flux along x, or the difference between the out-flowing momentum and the in-flowing momentum through area $A_0 = D \, \Delta z$. Thus

$$\int_{net} \rho u \cdot u \, dA_0 = \int_{out} \rho u \cdot u \, dA_0 - \int_{in} \rho u \cdot u \, dA_0 = 0 - \rho u \cdot |u| \, D \, \Delta z \qquad (2.16b)$$

where $|u|$ is the absolute value of u and is used to preserve the sign of $\bar{q}$. When u reverses direction, so does $\bar{q}$. With equations (2.16), equation (2.15) becomes

$$\bar{q} \simeq \rho D \, u \, |u| + \rho D^2 \dot{u} \qquad (2.17)$$

To correlate equation (2.17) with experimental results, the coefficients $C_D/2$ and $C_M/4$ are inserted as multiples of the two respective terms on the right side of equation (2.17), a result that then agrees with Morison's form, equation (2.14).

According to the database summarized in Chapter 4, C_D and C_M both have a range from 0.4 to 2.0. However, based on a multitude of experiments compiled by the British Ship Research Association (1976) and by Sarpkaya and Isaacson (1981), it is quite apparent that C_D and C_M are not simple constants. In the case of an imposed *periodic* plane flow such as offshore waves of period T, if the root-mean-square (rms) value is chosen for each over a time that is sufficiently large compared to T, then data show that C_D and C_M are functions of three parameters, expressed as

$$C_D = C_D(\text{Re, Kc, cylinder roughness}) \qquad (2.18a)$$

$$C_M = C_M(\text{Re, Kc, cylinder roughness}) \qquad (2.18b)$$

Further, Sarpkaya (1976) proposed that a frequency parameter β replace Re in those relationships, where

$$\beta = \frac{\text{Re}}{\text{Kc}} = \frac{\rho D^2}{\mu T} \qquad (2.19)$$

$$C_D = C_D(\beta, \text{ Kc, cylinder roughness}) \qquad (2.20a)$$

$$C_M = C_M(\beta, \text{ Kc, cylinder roughness}) \qquad (2.20b)$$

The main advantage of Sarpkaya's forms is that u_0, the free stream amplitude of the periodic velocity, then appears only once in each function of equations (2.20), instead of twice in each function of equations (2.18). For periodic flows, this alternative gives efficient correlations of measured cylinder forces with C_D and C_M. Among the phenomena neglected in both of these functional relationships are cavitation, fluid compressibility, three-dimensional flow, proximity effects, and all movement of the cylinder. In the following example problem, Morison's equation is modified to account for one of these neglected factors: the lateral vibrations of the cylinder.

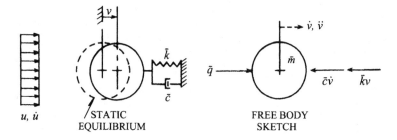

Figure 2.9   Cylinder model used to characterize fluid-structure interactions.

Example Problem 2.3.   Consider fluid-solid interactions for the vibrating cylinder based on the single degree of freedom model shown in Figure 2.9. The cylinder is modeled as a rigid body with an elastic restraint of stiffness $\bar{k}$ per unit length and with a linear viscous structural damping constant of $\bar{c}$ per unit length. This cylinder is an approximate model of a flexible, tubular cross-member of an offshore platform whose legs provide the cylinder's end restraint.

The free stream horizontal flow velocity $u = u(t)$ is in line with the cylinder's translational motion $v = v(t)$. As shown on its free body sketch, the restraint and damping forces per unit length that oppose the cylinder motion are $\bar{k}v$ and $\bar{c}\dot{v}$, respectively. The virtual mass per unit length, $\bar{m}$, is deduced from equation (2.9). The fluid loading per unit length is given by equation (2.14), modified so that the drag force is based on the *relative* velocity $(u - \dot{v})$ between the fluid and the cylinder. When Newton's second law, equation (2.1), is applied to this cylinder, the equation of motion becomes

$$\left(\bar{m}_0 + C_A \, \rho\pi \frac{D^2}{4} \right) \ddot{v} + \bar{c}\dot{v} + \bar{k}v = C_D \rho \frac{D}{2} |u - \dot{v}|(u - \dot{v}) + C_M \, \rho\pi \frac{D^2}{4} \dot{u} \quad (2.21)$$

The term involving C_M results from fluid motion only, such as wave action, where $\dot{u}$ is the absolute acceleration of the fluid.

Equation (2.21) is nonlinear as a consequence of the drag force term. Berge and Penzien (1974) linearized this equation for *small* motion, or

$$C_D' = C_D |u - \dot{v}| \simeq \text{ constant} \quad (2.22)$$

With this assumption, equation (2.21) becomes

$$\left(\bar{m}_0 + C_A \, \rho\pi \frac{D^2}{4} \right) \ddot{v} + \left(\bar{c} + C_D' \rho \frac{D}{2} \right) \dot{v} + \bar{k}v = C_D' \, \rho \frac{D}{2} u + C_M \, \rho\pi \frac{D^2}{4} \ddot{u} \quad (2.23)$$

Equation (2.23) clearly shows that the damping of a moving cylinder is increased due to the fluid drag force, a force that in general overwhelms the internal structural damping $\bar{c}$. The conclusion holds true for its nonlinear counterpart also, equation (2.21).

Thus, to solve equation (2.21) or (2.23) for the structural displacement $v = v(t)$, one needs to know four structural parameters: m_0, D, $\bar{c}$, and $\bar{k}$; the fluid density ρ; the free stream flow field $u = u(t)$; and the three empirical constants C_A, C_D and C_M. In subsequent examples, $\bar{k}$ and $\bar{c}$ will be estimated for particular cases, and the dependency of the latter three empirical constants on offshore waves will be discussed in greater detail. Other environmental factors affecting the motion of offshore structures are now quantified.

Buoyancy and Gravity

It is well known that a solid object can be lifted much more easily when in water than in air. This is because the water pressure exerts an upward or buoyant force on the submerged solid. The Greek mathematician Archimedes (287-212 B.C.) stated this principle in precise terms:

A solid body partially submerged in a fluid is buoyed up by a force $m_b g$ equal to the weight of the fluid displaced. (m_b is the mass of the fluid displaced.)

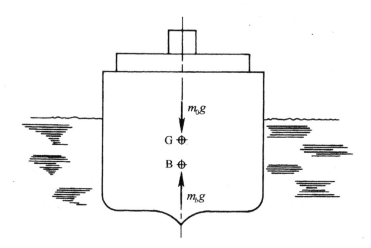

Figure 2.10   Cross section of a ship showing buoyant and gravitational forces.

For example, consider the ship shown in Figure 2.10 which has an actual weight (in air at sea level) of $m_0 g$. The volume of water that the ship displaces is $\mathcal{V}$ and the weight density of the water is denoted by γ_w. The result of applying Archimedes principle to this system is

$$m_b g = \gamma_w \mathcal{V} = m_0 g \qquad (2.24)$$

This buoyant force $\gamma_w \mathcal{V}$ acts at point B on the ship, in a vertical direction opposite to that of the resultant gravitational force $m_0 g$. The latter force acts at the ship's center of gravity. Thus, a ship's actual weight is generally referred to as its *displacement*.

In general, point B is not coincident with point G of the solid body. In precise terms:

> *If the displaced fluid is of constant density, B is located at the centroid of the displaced fluid volume $\mathcal{V}$.*

This statement can be proved in general for solid body of arbitrary shape (Batchelor, 2000). Its validity can be simply illustrated for the following special case of a solid, prismatic block partially submerged in water. Refer now to the sequence of illustrations (a) through (f) in Figure 2.11. Suppose that the solid block (a) is removed from the water and then replaced by pressure forces around the void of volume $\mathcal{V}$ so that the water remains undisturbed (b). The pressure distribution, shown as "gage" pressure, is constant on any horizontal plane and varies linearly with depth. The pressure distribution on the solid block (c) is identical to that on the void. In fact this identical pressure distribution would also occur on a water block (d) of volume $\mathcal{V}$ which would just fill the same void.

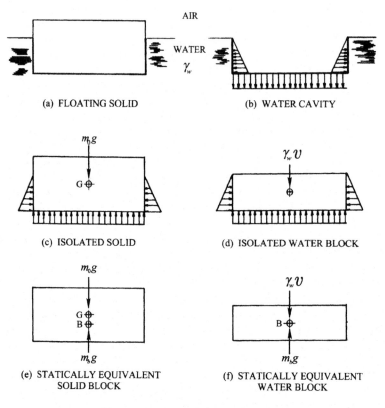

Figure 2.11   Illustration of Archimedes's principle.

To maintain static equilibrium of this water block, two conditions must be met. First, the force of the water block in (d), or $m_b g$ which acts at B, the centroid of $\mathcal{V}$, must be balanced by its net pressure forces along its horizontal boundaries. Since the pressure forces on the vertical boundaries balance, they are of no consequence in this problem. Second, the resultant of the boundary pressure forces must pass through B to avoid rotation of the water block. Since the solid block (e) and the water block (f) have identical boundary pressure distributions, the buoyant force $m_b g$ acts at B on the solid block also.

Example Problem 2.4.   Consider a gravity platform partially submerged in soft mud, as shown in Figure 2.12. Here the buoyant force of air may be neglected because the gravitational force $m_0 g$ is defined as the structure's weight in air. Assume that the water-saturated mud layer behaves as a liquid of approximately constant density γ_m and thus contributes to the structure's buoyancy. Liquefaction of the mud foundations of gravity platforms can occur during a storm due to caisson vibrations and repeated shear stress reversals at the soil-caisson interface (Graff and Chen, 1981). This leads to a gradual increase in pore water pressure which reduces the shear strength of the mud foundation, causing the foundation to behave as a liquid.

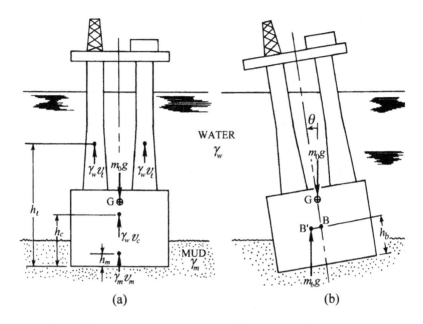

Figure 2.12   A gravity platform partially submerged in a mud layer or in liquefied soil.

The three buoyant forces on this platform are: (1) a buoyant force due to the mud of $\gamma_m \mathcal{V}_m$, which is located at a distance of h_m from the base and acts upward at the centroid of the displaced mud volume $\mathcal{V}_m$; (2) a buoyant force for the portion of the caisson in water of $\gamma_w \mathcal{V}_c$, which is located at h_c from the base and acts upward at the centroid of its displaced water volume $\mathcal{V}_c$; and (3) a buoyant force on the submerged portion of each leg of $\gamma_w \mathcal{V}_\ell$, which is located at h_ℓ from the base and acts at the centroid of its displaced water volume of $\mathcal{V}_\ell$. If the structure has N identical legs, the total buoyant force $m_b g$ and its statically equivalent location h_b on the structure are, respectively

$$m_b g = \gamma_m \mathcal{V}_m + \gamma_w \mathcal{V}_c + N \gamma_w \mathcal{V}_\ell \qquad (2.25)$$

$$h_b = \frac{1}{m_b g} \left(h_m \gamma_m \mathcal{V}_m + h_c \gamma_w \mathcal{V}_c + N h_\ell \gamma_w \mathcal{V}_\ell \right) \qquad (2.26)$$

Here, h_b is located on the vertical centerline of the upright structure, where the identical legs are symmetrically placed with respect to this axis. However, that if the structure is tipped at angle θ, the resultant buoyant force retains its same magnitude for all practical purposes, but its location shifts to an off-center position, from B to B'. The location of B' relative to the mass center G determines the static stability of the whole structure, as the following example problem demonstrates.

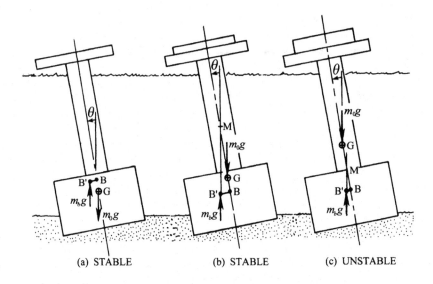

(a) STABLE   (b) STABLE   (c) UNSTABLE

Figure 2.13   Static stability of three gravity platforms supported by a liquefied soil foundation.

Example Problem 2.5.   Consider now the stability of the monopod gravity platforms shown in Figure 2.13, as they rock in plane motion on the liquified mud foundation. Assume that the rocking motions are very slow so that the structures' inertias can be neglected, for which $m_b g \simeq m_0 g$. If B and also B' remain always *above* G as a monopod tips to a small angle θ, the structure will return to vertical equilibrium ($\theta = 0$). This configuration, shown in Figure 2.13(a), is stable because the resultant buoyant force and gravity force produce a couple opposite to the direction of rotation. The restoring moment imposed by the liquified mud foundation would be relatively insignificant in this case. If the design is such that B and B' remain below G, the structure may still be stable, but only if the metacenter M is above G. The metacenter is the point of intersection of the vertical line through B' with the original centerline, as shown in Figure 2.13(b). When M is above G, the *floating* structure is dynamically stable because the restoring, gravity-produced couple is in a direction that will reduce θ. If, however, M is below G, as shown in Figure 2.13(c), then the couple due to the buoyancy and gravity forces will increase θ, and the structure will topple in the absence of the restoring forces of the foundation.

The qualitative analysis presented in the last example problem is analogous to the elementary static stability analysis of ships and other floating objects as presented, for instance, by Munson et al. (1998). In practice, a more thorough analysis of structural stability is needed, which is accomplished by studying the differential equations describing structural motion. Such dynamic analyses will be illustrated in the forthcoming chapters.

Winds and Currents

For a typical fixed-bottom offshore platform, the static drag force due to wind on the superstructure amounts to about 15 percent of the total force on the structure (Muga and Wilson, 1970), and often accounts for about 25 percent of the total overturning moment (Graff, 1981). The wind-induced overturning moment increases linearly with the height of the structure, and thus, as these structures are built in deeper and deeper water, the effects of wind drag then become increasingly significant in design.

A measure of wind velocity is needed to predict both wind loading on the superstructure and to predict the magnitude of the wind-generated wave forces on the submerged portion of the structure. A windstorm is often described as airflow with a mean or steady velocity $\bar{u}(z)$, with a superimposed fluctuating velocity. Here z is the height above the still water level. Gould and Abu-Sitta (1980) point out that the averaging period of one hour has been used in Europe and Canada in presenting data for $\bar{u}(z) = \bar{u}(h)$, for a reference height of either $z = h = 30$ ft or 10 m. Gaythwaite (1981) states that in Great Britain values of $\bar{u}(h)$ chosen for structural design have traditionally been averaged over only one minute. In the United States, however, the concept of the fastest mile of wind speed is used to define $\bar{u}(h)$. That is, measures are made of wind velocity during the time it takes for a mile of air to pass a fixed point, and the annual extreme condition is used as the reference value. Sachs (1972) and Simiu (1976) discuss methods of converting such data to a mean velocity. Gaythwaite (1981) suggests that all but temporary marine structures should be designed using the mean, fastest mile of wind speed associated with return periods of 50 to 100 years.

Once $\bar{u}(h)$ is established for a particular offshore site, the mean horizontal wind velocity at height z above the sea surface is

$$\bar{u}(z) = \left(\frac{z}{h}\right)^{1/n} \bar{u}(h) \qquad (2.27)$$

The exponent n depends on many factors. For instance, $n = 3$ fits data for rough coastal areas; $n = 7$ to 8 for sustained winds over an unobstructed sea; and $n = 12$ to 13 for gusts. At heights of 100 ft or more above the surfaces, the vertical gust-velocity becomes about the same as its horizontal value. Wind data $\bar{u}(h)$ and further discussions of equation (2.27) are given by Muga and Wilson (1970), Sherlock (1953), Simiu and Scanlan (1978), and Vellozzi and Cohen (1968).

On a member of the superstructure which is nine or more diameters removed from neighboring structural elements, $\bar{u} = \bar{u}(z)$ calculated from equation (2.27) can be used with equation (2.10) to estimate wind drag force, where C_D is based both on Reynolds number and on the cross-sectional shape of the member. Values of C_D for common shapes are readily available (Hoerner, 1965, and Pattison et al., 1977).

For a truss structure in wind, the sum of the drag forces on each individual member may give a low estimate of the total drag force. This effect, called solidification, modifies the drag coefficient as follows:

$$C_D = 1.8\phi, \quad \text{for } 0 < \phi < 0.6 \tag{2.28a}$$

$$C_D = 2, \quad \text{for } \phi \geq 0.6 \tag{2.28b}$$

where the solidity ratio ϕ is given by

$$\phi = \frac{\text{projected area of truss members only}}{\text{projected area of the enclosed solid}} \tag{2.29}$$

Here the projected area is in a plane normal to the prevailing direction of the wind.

For members of the superstructure closer than nine diameters, shielding effects (sometimes referred to as sheltering) are apparent. For instance, a cylinder just behind a lead cylinder facing the wind experiences a considerable drop in drag force, and sometimes a negative drag force at close spacings. Wind tunnel data for the shielding of cylinders are reported by Pagon (1934) and Wilson and Caldwell (1971). Numerous references to calculated results based on classical hydrodynamics are discussed by Muga and Wilson (1970). Values of shielding factors varying from 0 to 1.0 and useful for design purposes are recommended by Graff (1981). To obtain the drag force for a shielded component, the shielding factor is multiplied by the static drag force of its unshielded counterpart.

In addition to these static loads, the dynamic effects of wind on offshore structures should be considered also. For instance, for a moored structure whose fundamental period in free oscillation is close to the period of wind gusts, the dynamic deflection of the structure could become significant. However, for a truss structure with fixed legs, the overall dynamic response due to gusts alone are generally insignificant. This is because the time lag between gust arrival at the leading edges and its arrival on the downstream portions of the open structure tends to minimize the overall loading. Static loading of individual members to gusts could be significant since wind gust speeds may be quite high. For instance, a value of 120 knots is often used for the design of structures in the North Sea, a value about 1.4 times the steady or sustained wind speed $\bar{u}(z)$ at a height of 30 m (Graff, 1981).

On the other hand, adverse vibrations of a structural component may arise in steady winds due to vortex shedding, if these vortex frequencies are *in tune* or in resonance with a free vibration frequency of a structural member. This may lead to large displacements or flutter of platelike members and to galloping beam and cable components. As discussed earlier in this chapter, helical strakes or other spoilers can be used to eliminate periodic vortices on tubular members. Vortex-induced resonance is further discussed by Blevins (1977), Gould and Abu-Sitta (1980), and Simiu and Scanlan (1978). Extended discussions concerning the physical basis for both wind and ocean currents are given in the classical treatise of Neumann and Pierson (1966).

In his summary of ocean currents, Gaythwaite (1981) indicates that *tidal currents* and *wind-stress currents* are the two most relevant ones in the structural design of floating and fixed structures. Offshore tidal currents, or the horizontal water flow due to the vertical rise and fall of tides, often attain a maximum velocity of 1 to 2 knots and may even reach 10 knots in some locations. Tidal currents are higher in the spring than in any other time of the year. Data on tidal currents for many offshore locations on the coasts of the United States and Asia are published annually by the National Oceanic and Atmospheric Administration, U.S. Department of Commerce. Data for the tidal current $u_t(z)$ as a function of water depth z generally follows a power law similar to that for wind, equation (2.27). That is

$$u_t(z) = \left(1 + \frac{z}{d}\right)^{1/7} u_t(0) \qquad (2.30)$$

where d is the total water depth, z is the coordinate of the tidal velocity (a negative number measured downward from the sea surface), $u_t(0)$ is the tidal velocity at the surface $z = 0$, and $u_t(-d) = 0$ at the seafloor.

The other important current is $u_w(z)$, the current generated by a sustained wind blowing over the sea surface. The velocity profile of this wind-stress current is approximated as linear with depth, with a maximum value $u_w(0)$ at the sea surface, where $u_w(-d) = 0$ at the seafloor. That is

$$u_w(z) = \left(1 + \frac{z}{d}\right) u_w(0) \qquad (2.31)$$

The magnitude of $u_w(z)$ is generally about 1 to 5 percent of the sustained wind speed.

In the absence of vortex shedding, the steady-state drag force per unit length at the depth location z of a stationary, submerged, tubular member can be calculated from equation (2.10) in which the effective fluid velocity u is the sum of the two current velocities given by the last two equations, and the horizontal wave particle velocity, u_{wave}, is discussed in Chapter 3. That is,

$$u = \left(1 + \frac{z}{d}\right)^{1/7} u_t(0) + \left(1 + \frac{z}{d}\right) u_w(0) + u_{wave} \qquad (2.32)$$

With the design values for $u_t(0)$, $u_w(0)$, and u_{wave}, and with a knowledge of the geometry for all the structure's tubular members whose longitudinal axes are perpendicular to u, the total horizontal drag force from those tubular members of the structure is obtained by using equation (2.32) with equation (2.10) and integrating the result over those tubular members.

Currents do affect structural integrity in other ways. Current-induced scouring, for instance, can undermine pile-supported jacket template platforms and gravity platforms by eroding surrounding sand and soil. Currents carry ice that can impact and damage structures. High currents accelerate the corrosion rate of submerged metal structures. Currents also modify waves and wave loading of structures (Tung, 1974). Except for the impact of ice, which is discussed briefly in the next section, these environmental hazards to offshore structures are subjects that are beyond the scope of this book.

2.3   EARTHQUAKES, ICE IMPACT, AND WAVE SLAMMING

Earthquake Forces

The engineer who is designing jacket-template or gravity platforms which will resist marine seismic disturbances finds that records of strong ground earthquake motion on the sea floor are generally nonexistent in locations where such structures are contemplated (Page, 1975). The needed data are the time histories of velocity or acceleration of the sea floor, in both the vertical and horizontal directions. With such data, the engineer can calculate the structural dynamic responses, estimate seismic damage, and evaluate the possibilities of structural survival (Bea et al., 1979).

Lacking the needed data and faced with the need to design offshore platforms in the earthquake-prone Gulf of Alaska, Wiggins et al. (1976) rationalized the use of California seismic records as a first approximation to the phenomena expected in the Alaskan Gulf. Wiggins et al. compared various measured values for focal depths, which are locations below the earth's surface where strong earthquakes have originated. This average depth below ground level is 16 km for California earthquakes and 26.5 km for the severe 1964 shocks in Alaska. Since ground motion attenuates rapidly with the distance from the focus, ground level motion should be less in Alaska than in California, given earthquakes of equal intensity and assuming identical soil characteristics. Based on this argument, the engineer can employ California earthquake data in the design of Alaskan offshore structures.

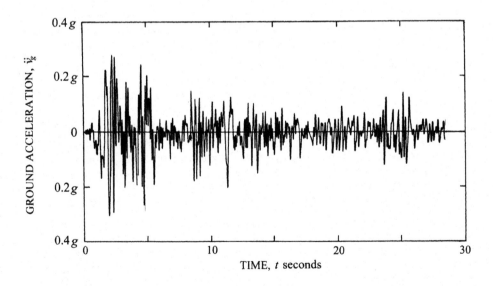

Figure 2.14   Accelerogram for the El Centro earthquake, south east component, May 18, 1940.

A typical earthquake-induced ground acceleration time history, which is sometimes used as input to the base of a structure to evaluate its earthquake resistance, is shown in Figure 2.14. Here the horizontal ground acceleration $\ddot{v}_g$ is expressed as a multiple of g, the acceleration due to gravity. The instantaneous upper bound on this acceleration is approximately $0.3g$. For the largest California earthquakes on firm, deep alluvium, the total time for destructive shaking is about 45 seconds. The following example illustrates how such data can be used as a driving force for a simple model of an offshore structure.

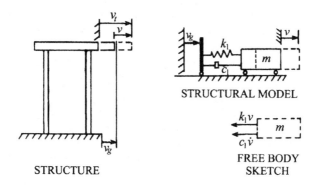

STRUCTURAL MODEL

FREE BODY
SKETCH

STRUCTURE

Figure 2.15   Offshore structural model with horizontal sea floor motion (an earthquake).

Example Problem 2.6.   A single degree of freedom model of a jacket-template structure is shown in Figure 2.15. The motion of the deck relative to the sea floor is $v = v(t)$ and this motion is assumed to be in the horizontal direction. The time history of motion for an earthquake at the sea floor is given as $v_g = v_g(t)$, and this motion is assumed to be in the horizontal direction also. The equivalent virtual mass, leg stiffness, and fluid-structural damping constants are m, k_1, and c_1, respectively, and are depicted on the simple damped, spring-mass model of this structure and on its free body sketch in Figure 2.15. (Methods for calculating m, k_1, and c_1 for particular structures are considered in the next section and in Chapter 5). The total or absolute values for the deck displacement and its absolute acceleration are, respectively:

$$v_t = v + v_g \tag{2.33}$$

$$\ddot{v}_t = \ddot{v} + \ddot{v}_g \tag{2.34}$$

The restoring forces due to structural stiffness and damping depend only on the *relative* displacement v and the *relative* velocity $\dot{v}$. Thus, when Newton's second law of motion is applied to the spring-mass model, the equation of motion becomes

$$m\ddot{v}_t + c_1\dot{v} + k_1v = 0 \tag{2.35}$$

Substituting $\ddot{v}_t$ from equation (2.34) into (2.35), the result is

$$m\ddot{v} + c_1\dot{v} + k_1 v = -m\ddot{v}_g \qquad (2.36)$$

In this structural model, the excitation force $p_1(t)$ is identified as the horizontal ground acceleration of magnitude $-m\ddot{v}_g$. In using this model to compute the structural responses $v = v(t)$ to typical earthquake motion, which will be done in Chapter 5, the negative sign on the right side of equation (2.36) is of no consequence and is generally ignored.

Base shaking of an offshore structure is just one of several types of seismic-induced ground motions (Hudson, 1970). For instance, piles may shift or the structure's foundation may be undermined and fail due to fault displacements or large-scale mud slides. Such large-scale earth motions occur over a period of time which is relatively long compared to the natural period of an offshore structure. Thus these earth motions are essentially static as far as the structure is concerned and are not a part of a structural dynamic analysis.

There is a vast amount of literature describing the geological nature of earthquakes and the effects of soil and rock strata on structural response, including linear and nonlinear effects. For such information the reader may consult the following classical works: the book edited by Wiegel (1970); the concise presentation of the seismic phenomenon, including the deterministic and statistical analyses of structural responses, by Gould and Abu-Sitta (1980); and the compilations of references pertinent to earthquake engineering of offshore structures by Bea et al. (1979) and Marshall (1981).

Ice Impact Forces

Ice is a hazard to fixed offshore structures which are located in polar seas such as the Gulf of Alaska. Gaythwaite (1981) summarized ice hazards and ways to minimize associated structural damage. Current field studies about ice are published yearly in the Proceedings of the Offshore Technology Conference, Houston, Texas.

To analyze for the impact hazard, it is necessary to know ice speed, size, and material properties. Drifting ice travels at speeds from 1 to 7 percent of the wind speed. A typical ice island in Cook Inlet, Alaska, for instance, may be 1 km in diameter, 1 m thick, and travel with a speed of 3 knots. In general, the ratio of the height of a drifting ice block above water to that below is about 1:2; but may vary from 1:1 up to 1:7. With the usual concentrations of Na_2SO_4, sea ice has a compressive or rupture strength from 200 to 400 psi, but this does vary with salt concentration and the rate of impact loading (Peyton, 1968). The American Petroleum Institute (1979, 1997) has recommended the following formula for calculating the horizontal force, F_h, on structures subjected to the impact of ice:

$$F_h = C_i \sigma_{ci} A_0 \qquad (2.37)$$

Here C_i is a coefficient in the range of 0.3 to 0.7 which accounts for loading rate; σ_{ci} is the compressive strength or rupture stress for the ice; and A_0 is the area of the structure exposed to the impacting ice. An example of a structure that tolerates impacting ice is a concrete gravity platform with one wide-based, cone-shaped leg (Bercha and Stenning, 1979). This leg geometry leads to effective stress rupturing of the impacting ice with minimal damage to the structure.

Besides the impact of floating ice, there are other ice hazards to a structure in the polar seas. Most of these hazards are well understood and can be minimized by careful structural design. For instance, one of these hazards is the uplift force on the deck due to the buoyancy of accumulated ice attached to the structure around the water line. (Note that the specific gravity of sea ice ranges from 0.89 to 0.92). Although this uplift force may be offset by added gravity loads on the superstructure due to ice accumulation there, this accretion also increases wind loads because of the increased exposed area. Further, ice accumulation on the legs increases wave and current loads for the same reason. Ice also causes abrasion in various forms. Cyclic freezing and thawing leads to cracking and spalling of offshore concrete structures, phenomena common in our highways. This is usually due to the expansion of freezing water in cracks, pores, or capillary cavities. Some of this entrained water is the excess required for hydration of the cement and may be minimized by careful choice of the mix ingredients.

Using a fracture mechanics approach, the quasi-static penetration and fracture of floating ice plates was investigated by Bazant and Kim (1998). Related studies were reported by DeFranco and Dempsey (1994). These ideas are beginning to be applied to the design of offshore structures in arctic regions, to mitigate their vulnerability to the multiple hazards of ice.

Wave Slamming Forces

Although general descriptions of offshore waves and their associated loadings of structures are considered in Chapters 3 and 4, wave slamming is an important enough hazard that it is now considered separately. Unlike the steady wave train models addressed in Chapters 3 and 6, slamming refers to the impact of a single, occasional wave with a particularly high amplitude of energy. Sarpkaya and Isaacson (1981) reviewed the classical research on the slamming of water against circular cylinders, of which the work of Miller (1977, 1980) seems particularly applicable. Based on water-tank experiments, Miller found that the peak wave slamming force on a rigidly held, horizontal, circular cylinder is correlated by the following equation:

$$F_s = \frac{1}{2} C_s \rho D \, \ell \, u^2 \tag{2.38}$$

Here, the coefficient C_s is in the range of 3.5 to 3.6; D and ℓ are the cylinder diameter and length, and ρ and u are the water mass density and the peak horizontal water particle velocity, respectively. If the cylinder is not rigid but a flexible, elastic body (a tubular brace of a jacket-template platform, for instance), then Sarpkaya and Isaacson (1981) recommend the following procedure for computing the cylinder load: let $C_s = 3.2$ and then multiply the resulting force calculated from equation (2.38) by the force-impact magnification factor calculated through a dynamic response analysis. Magnification factors will be discussed in chapter 5. The alternative is to use $C_s = 5.5$ if no dynamic response analysis is made for the flexible cylinder.

Wave slamming on an offshore structure in which the waves are underneath the deck and in the vertical direction, can be a further design consideration. One way to calculate the resulting sudden uplift force on the deck is to use equation (2.38) in which u is the vertical wave particle velocity and the product $D\ell$ is replaced by the deck's area of impact. However, further research is still needed to determine the range of the fluid coefficient C_s for this particular type of wave impact. For a detailed analysis of wave forces on decks of offshore platforms, see Bea, et al. (1999).

2.4 STRUCTURAL MASS, DAMPING, AND RESTRAINT

Structural Mass and Stiffness

The following two example problems illustrate the modeling of single and multiple beams as a point mass located by the single coordinate $v = v(t)$. Classical beam theory gives the bending stiffness as $k_1 = CEI/\ell^3$. For a single beam, EI is the flexural stiffness, ℓ is the length, and C is a constant that depends on the end fixity of the beam ($C = 3$ for a cantilevered beam with full fixity at the base and no moment at its tip). For a tubular beam with an outside diameter D and an inside diameter D_i, then $I = \pi(D^4 - D_i^4)/64$.

Example Problem 2.7. Consider the horizontal motion of a tubular cross brace welded to the relatively rigid and stationary legs of a jacket platform. This structural element, defined in Figure 2.16a, has full fixity at its ends and is subjected to the uniform horizontal load per unit length $\bar{q}$, given by the right side of equation (2.23). The total horizontal load, modeled as a single point load at midspan, is $\bar{q}\ell = p_1(t)$, or

$$p_1(t) = C_D' \rho\ell\frac{D}{2}u + C_M \rho\ell\pi\frac{D^2}{4}\ddot{u} \qquad (2.39)$$

The dominant mode of motion $\psi(x)$ is shown by the broken lines of Figure 2.16b. The midspan coordinate is $v = v(t)$, which locates the lumped, virtual mass m of the massless cross brace of bending stiffness EI, as shown in Figure 2.16c. The virtual mass is deduced from equation (2.23), or

$$m = \left(\bar{m}_0 + C_A \rho\pi\frac{D^2}{4}\right) f_1\ell \qquad (2.40)$$

in which $\bar{m}_0$ is the actual mass per unit length. Note that $f_1 = 1$ gives an upper bound for m, but this is a bad choice since it is obvious that all of the mass along the length does not have the same displacement as that of the lumped mass. In Chapter 5, *Example Problem 5.3*, it is shown that $f_1 = 0.370$ for this structural element. The bending stiffness of the brace is derived from elementary beam theory and is

$$k_1 = 192\frac{EI}{\ell^3} \tag{2.41}$$

Here, k_1 can be interpreted as the lateral force that, when applied to the midspan, will produce a static deflection of $v = 1$ at midspan. The structure-fluid damping is assumed to be linear-viscous which, referring to equation (2.23), has the following form:

$$c_1 = \bar{c}\ell + C'_D\rho\ell\frac{D}{2} \tag{2.42}$$

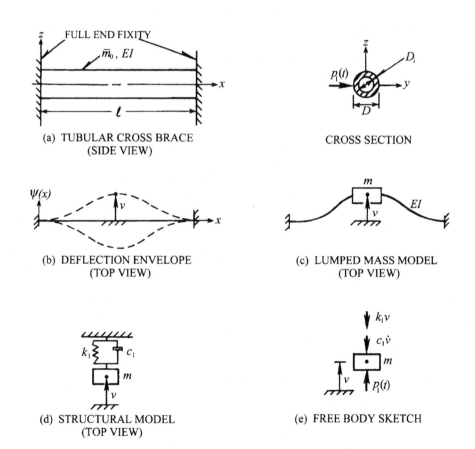

(a) TUBULAR CROSS BRACE
(SIDE VIEW)

CROSS SECTION

(b) DEFLECTION ENVELOPE
(TOP VIEW)

(c) LUMPED MASS MODEL
(TOP VIEW)

(d) STRUCTURAL MODEL
(TOP VIEW)

(e) FREE BODY SKETCH

Figure 2.16   Model of a tubular cross brace of an offshore structure.

In summary, the tubular brace of Figure 2.16a is modeled as the damped spring-mass, single degree of freedom system depicted in Figure 2.16d. The forces on m are shown on the free body sketch in Figure 2.16e, from which the equation of motion is deduced as

$$m\ddot{v} + c_1\dot{v} + k_1 v = p_1(t) \tag{2.43}$$

Here, since $p_1(t)$ is all lumped at midspan instead of being uniformly distributed, the solution $v = v(t)$ to equation (2.43) will be on the high side. This particular mathematical model is thus a conservative one. Needed for $p_1(t)$ are the explicit forms for the flow field $u = u(t)$ and its associated constants C_D and C_M, topics that are deferred to Chapters 3 and 4.

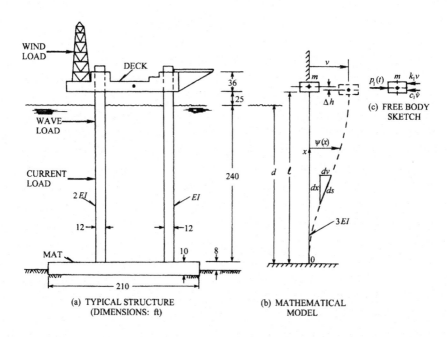

Figure 2.17   Model of a jackup drilling rig.

Example Problem 2.8.   Consider the horizontal motion of the jackup drilling rig for which a simplified diagram is shown in Figure 2.17a. This structure has three identical tubular legs (only two are shown). These legs have full end fixity in that they are clamped at the mat or mudline and also at the deck level. Of the three types of environmental loading, wind, wave, and current, assume that the wave loading dominates. Apply the total wave load as a horizontal load $p_1(t)$ acting at the deck level. Since cross braces are absent, the overall leg bending stiffness is three times that for a single leg, or $3EI$. Assume that the amplitude of the dominant dynamic deflection mode $\psi(x)$ of the legs follows the broken lines shown in Figure 2.17b, a shape that is consistent with the deck loading and the structural restraints. For simplicity, approximate the deck motion $v = v(t)$ as translational only, in which the deck's vertical drop, Δh, is always negligibly small by comparison to v. Now lump a fraction f_1 of the virtual mass of all three flexible legs with the rigid deck mass, m_d. From the coefficient of $\ddot{v}$ in equation (2.23), the equivalent virtual mass for this structural system is deduced as

$$m = 3\left(\bar{m}_0 + C_A\rho\frac{\pi}{4}D^2\right)f_1 d + 3\bar{m}_0(\ell - d)f_1 + m_d \qquad (2.44)$$

in which $\bar{m}_0$ is the actual mass per unit length of a single leg, d is the water depth, and $(\ell - d)$ is the length of a leg between the still water line and the bottom of the deck. (*Example Problem 5.4* will show that $f_1 = 0.375$.) Assume that structural damping is mainly produced by the submerged portions of the legs. From the coefficient of $\dot{v}$ in equation (2.23), the damping coefficient is deduced as

$$c_1 = 3\bar{c}d + 3C'_D\rho\, d\frac{D}{2} \qquad (2.45)$$

From classical beam theory, the restoring force constant is calculated as

$$k_1 = 36\frac{EI}{\ell^3} \qquad (2.46)$$

which is the magnitude of the horizontal force at deck level that produces a unit deflection ($v = 1$) at that point. (For a single leg, $k_1 = 12EI/\ell^3$). The value of k_1 given by equation (2.46) is an upper bound value for two reasons. First, this stiffness is decreased as the full fixity conditions on the legs are relaxed at the mud line or at the deck. Second, this leg stiffness is also reduced as the magnitude of the deck load approaches the Euler buckling load for this structure, an effect that will be explored in a subsequent example problem.

In summary, the jackup drilling rig of Figure 2.17a and Figure 2.17b is modeled as a single degree of freedom system whose free body sketch is shown in Figure 2.17c. When Newton's second law is applied to the equivalent virtual mass in this latter sketch, the governing equation (2.43) is obtained. With the respective values of m, c_1, and k_1 given by equations (2.44), (2.45), and (2.46), the explicit form of equation (2.43) becomes

$$\left[m_d + 3\left(\bar{m}_0 + C_A\rho\frac{\pi}{4}D^2\right)f_1 d + 3\bar{m}_0(\ell - d)f_1\right]\ddot{v}$$

$$+ \left(3\bar{c}d + \frac{3}{2}C'_D\rho\, dD\right)\dot{v} + 36\frac{EI}{\ell^3}v = p_1(t) \qquad (2.47)$$

The modeling of $p_1(t)$ its deferred to later chapters.

Cable Restraints

Cables or guy lines are employed to restrain the motion for several types of offshore structures. The simplest cable configuration is the vertical one used to restrain floating, tension leg platforms such as in Figure 1.1e. In addition, there are two classes of flexible mooring systems in which stationary guy lines hang as catenary curves from the structures to the sea floor. The first is the single line that constrains a vessel or buoy. The second is the multi-line system that constrains vessels, semisubmersible platforms, and compliant towers. The lines in present use are ropes of metallic wire or of synthetic fiber such as nylon, Dacron, or Kevlar; and steel chains with solid or hollow links. Practical aspects of guy line design are discussed in the U.S. Navy publication NAVFAC DM-26 (1968), in the four papers by Childers (1973-1975), and in the work of Niedzwecki and Casarella (1975). These references include analyses of multi-line systems with inextensible or nonstretching lines. Refined analyses which include clumped weights and additional anchors along the cables, as well as the effects of elastic cable stretching, are presented by Adrezin et al. (1996), Ansari (1980), and Wilson and Orgill (1984).

For a taut cable with negligible sag, the longitudinal extension δ depends on both the applied longitudinal force F_e and the material properties of the cable. For instance, Wilson (1959) used the following power law to correlate the load-extension behavior of both steel wire and synthetic fiber line employed in mooring ships:

$$F_e = C_0 \delta^n \qquad (2.48)$$

Here, C_0 and n are constants depending on the material, its length, and its cross-sectional area. If the deflections are sufficiently small, $n = 1$ and the force-deflection relationship based on elementary theory is given by

$$F_e = \frac{A_0 E_e}{\ell} \delta = k_1 \delta \qquad (2.49)$$

where A_0 is the cross-sectional area, ℓ is the length, and E_e is the equivalent Young's modulus for longitudinal extension of the line. In this case, equation (2.49) defines the longitudinal stiffness constant $k_1 = A_0 E_e / \ell$.

A more convenient form of equation (2.48), which also includes equation (2.49) and approximately represents the behavior of an assembly of taut cables tied to a common point whose deflection is δ under load F_e, is

$$F_e = k_1 \delta + k_2 |\delta| \delta + k_3 \delta^3 + \cdots \qquad (2.50)$$

where $k_1, k_2, \ldots$ define the stiffness. The absolute value sign on each even-order term in δ forces F_e to be antisymmetrical about $\delta = 0$, assuring that the restraint stiffness is the same for loading and unloading. Example problems will show that the approximation of equation (2.50) facilitates the dynamic analysis of offshore structures with both extensible and inextensible supporting cables.

One should keep in mind that equations (2.48)-(2.50) apply only when cable dynamics can be neglected; that is, in cases where the fundamental cable frequency in both longitudinal and transverse vibration is much higher than the free vibration frequency of the structure that it restrains. In applications, this frequency criterion should always be checked. Methods to calculate the structural frequency for single degree of freedom systems are given in Chapter 5; and methods to calculate cable frequencies are discussed in Chapter 10.

The catenary is the curve formed by suspending a uniform cable of zero bending stiffness between two points. Classical theory for the static catenary shape forms the basis for an upper bound calculation on the restraint stiffness for cablestayed offshore structures. In this theory, longitudinal cable extension is neglected, as are the effects of cable dynamics. Consider the cable segment of length ℓ and of weight (in the water) of w per unit length, as shown in Figure 2.18. Since the bending stiffness EI is zero, such a cable achieves its stiffness only through a change in shape as the tension forces F_0 and F are changed at each end. Classical theory leads to the equation of the catenary curve and the relationships amoung the system variables $(\ell, w, F_0, F, \theta_0, \theta)$, which in turn are used to compute the structural cable restraints. This analysis is summarized.

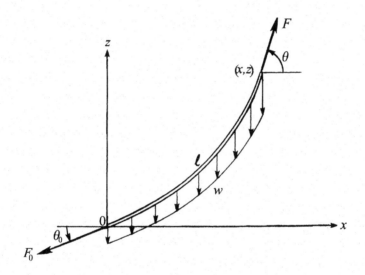

Figure 2.18 Freely hanging cable segment in static equilibrium.

The governing differential equation for the catenary segment, expressed in terms of the (x, z) coordinates defined in Figure 2.18, is

$$\frac{d^2 z}{dx^2} = \frac{w}{F_x} \left[1 + \left(\frac{dz}{dx} \right)^2 \right]^{1/2} \tag{2.51}$$

where F_x is the horizontal component of the tension force. Since the cable's bending stiffness is neglected, the resultant end tensions F_0 and F are in a direction tangent to the catenary curve. For static equilibrium, then, the horizontal component of tension remains unchanged, or

$$F_x = F_0 \cos \theta_0 = F \cos \theta \tag{2.52}$$

For vertical equilibrium of this segment, it follows that

$$F \sin \theta - F_0 \sin \theta_0 = w\ell \tag{2.53}$$

in which ℓ is the length of the segment, given by

$$\ell = \int_x \left[1 + \left(\frac{dz}{dx} \right)^2 \right]^{1/2} \tag{2.54}$$

A closed form solution to equation (2.51), or the (x, y) coordinates of the catenary curve, is

$$x = \frac{F_x}{w} \sinh^{-1} \left(\frac{w\ell}{F_x} + \tan \theta_0 \right) - \frac{F_x}{w} \sinh^{-1}(\tan \theta_0) \tag{2.55a}$$

$$z = \frac{F_x}{w} \cosh \left[\frac{wx}{F_x} + \sinh^{-1}(\tan \theta_0) \right] - \frac{F_x}{w} \cosh \left[\sinh^{-1}(\tan \theta_0) \right] \tag{2.55b}$$

Equations (2.55) can be used to describe the static shapes of guy lines formed of multiple uniform segments placed end-to-end in which the segments have different geometric and material properties. In such configurations, compatibility of both slope and tension force at the junction of each adjacent segment needs to be maintained (Ansari, 1980). The following example problem illustrates the use of the catenary solutions to solve for the restraining stiffness $q(v)$ of a floating platform stayed with identical guy lines.

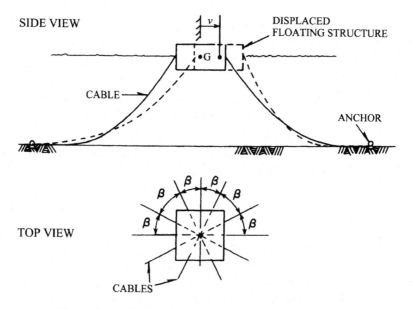

Figure 2.19   Model of a cable stayed floating platform.

Example Problem 2.9. The floating structure of Figure 2.19, representing a moored ship or a semisubmersible platform, is restrained by symmetrically placed, uniform cables separated by equal angles β. Assume that the platform motion $v(t)$ is not excessive so that a portion at the lower end of each cable always remains flat. Thus, vertical pull forces on the anchors do not occur. The problem is to calculate the stiffness $q(v)$, first for one of the pair of opposing cables in line with the deflection coordinate $v(t)$, and then for the other cable of the pair. The calculation of the stiffness due to the full array of cables is left to the reader.

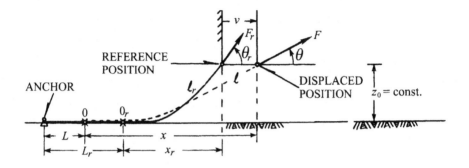

Figure 2.20   Single cable of a floating platform.

Consider the stiffness of the single cable defined in Figure 2.20. For the static equilibrium state $(v = 0)$, the origin of the cable coordinates is at 0, for which the flat length, the suspended length, and the horizontal projected lengths are $L_r, \ell_r,$ and x_r, respectively; and the tension force is F_r at angle θ_r at the top suspension point. At the bottom of the cable, the slope is zero at both 0_r and at the shifted origin 0 for $v \neq 0$, at which points the condition $F_0 = F_x$ is always true. The vertical projected length remains constant, or $z = z_0$. Since the cable is assumed to be inextensible, the reference lengths subscripted r can be expressed in terms of their corresponding unsubscripted values for $v \neq 0$ as

$$L_r + \ell_r = L + \ell = \text{ constant} \tag{2.56}$$

Also, the distance between the anchor and the origin of the platform displacement coordinate v remains constant, or

$$L_r + x_r = L + x - v \tag{2.57}$$

When $(L - L_r)$ is eliminated between equations (2.56) and (2.57), the platform displacement is

$$v = \ell_r - x_r + x - \ell \tag{2.58}$$

Two additional equations, deduced from equations (2.55) for $\theta_0 = 0$ and $z = z_0$, are respectively

$$\ell = \frac{F_x}{w} \sinh \left(\frac{wx}{F_x} \right) \tag{2.59}$$

$$z_0 = \frac{F_x}{w} \cosh \left(\frac{wx}{F_x} \right) - \frac{F_x}{w} \tag{2.60}$$

From the overall equilibrium conditions, equations (2.51) and (2.52), it follows that $F_x = F \cos \theta$ and $w\ell = F \sin \theta$. The ratio of these latter equations leads to

$$\tan \theta = \sinh \left(\frac{wx}{F_x} \right) \tag{2.61}$$

For a fixed value of z_0 and w and an initially fixed value of $\theta = \theta_r$, then the three initial values $F_x = F_{xr}$, $x = x_r$, and $\ell = \ell_r$ can be calculated from equations (2.59), (2.60), and (2.61). For values of $\theta \neq \theta_r$, the corresponding values of F_x, x, ℓ, and v are calculated from equations (2.58)-(2.61). The restoring force of a single cable is simply

$$q(v) = F_x = F_x(v) \tag{2.62}$$

One can now deduce that the restoring force for the pair of identical cables (the single cable of Figure 2.20 and its mirror image across the plane $v = 0$) is given by the superposition of the results just derived for the single cable, or

$$q(v) = F_x(v) - F_x(-v) \tag{2.63}$$

To facilitate the calculations and interpretations of the restoring forces, equations (2.58) through (2.61) can be cast in nondimensional form with the aid of the following definitions:

$$\overline{F}_x = \frac{F_x}{wz_0}; \qquad \overline{F}_{xr} = \frac{F_{xr}}{wz_0} \tag{2.64a}$$

$$\overline{\ell} = \frac{\ell}{z_0}; \qquad \overline{\ell}_r = \frac{\ell_r}{z_0} \tag{2.64b}$$

$$\overline{x} = \frac{x}{z_0}; \qquad \overline{x}_r = \frac{x_r}{z_0}; \qquad \overline{v} = \frac{v}{z_0} \tag{2.64c}$$

With equations (2.64), equations (2.58)-(2.61) become, after some rearrangement,

$$\frac{\overline{x}}{\overline{F}_x} = \sinh^{-1}(\tan \theta) \tag{2.65}$$

$$\overline{F}_x = \left[\cosh\left(\frac{\overline{x}}{\overline{F}_x}\right) - 1\right]^{-1} \tag{2.66}$$

$$\overline{\ell} = \overline{F}_x \tan\theta \tag{2.67}$$

$$\overline{v} = \overline{\ell}_r - \overline{x}_r + \overline{x} - \overline{\ell} \tag{2.68}$$

For any fixed value of θ (including θ_r), the values on the left sides of equations (2.65)-(2.68) can be calculated explicitly, in sequence.

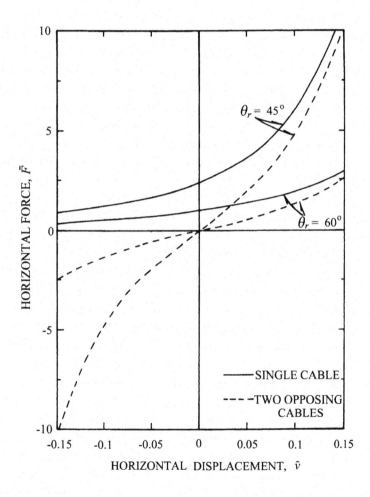

Figure 2.21   Static behavior for a single and two opposing cables of a cable stayed floating platform.

Numerical results for $\theta_r = 45$ and 60 deg are shown in Figure 2.21. For the single cable, the nondimensional horizontal restoring force $\bar{F}_x$ is small for negative nondimensional displacements $\bar{v}$ and increases at a growing rate as $\bar{v}$ becomes positive. For the broken-line curves which correspond to opposing pairs of cables 180 deg apart, $\bar{F}_x$ is antisymmetric about $\bar{v} = 0$ and behaves as a "hardening" spring for both positive and negative displacements $\bar{v}$. For these opposing cables, if the actual displacement v is less than about 5 percent of the attachment height z_0, the horizontal force-displacement relationship is linear, for practical purposes. In general for such diametrically opposed cable systems, the horizontal restoring force $q(v)$ can be approximated by an odd order polynomial in v, or

$$q(v) = k_1 + k_3 v^3 \tag{2.69}$$

Calculations of $q(v)$ for other θ_r values and for muliple pairs of identical, symmetric arrays of cables, with their ultimate approximation in the form of equation (2.69), are left for the reader. In making such calculations, it is recalled that this analysis is valid only for cables that are relatively inextensible. Calculations by Irvine (1981) imply that for steel cables over 13 cm in diameter, where $45 \leq \theta_r \leq 85$ deg, the cable extension strain is less than 0.1 percent. In such cases the assumption of cable inextensibility is a reasonable one and the results obtained herein are valid.

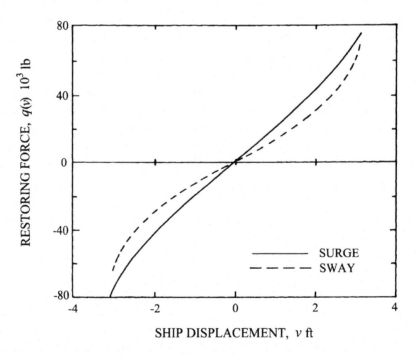

Figure 2.22   Static behavior of a moored LST (O'Brien and Muga, 1964).

Example Problem 2.10. Consider the explicit forms for the cable restraint functions $q(v)$ of the spread moored ship depicted in Figure 1.6 and previously introduced in *Example Problem 2.1.* As in the original experimental study and analysis by O'Brien and Muga (1964), traditional English units are used to express the numerical results. The subject ship was an LST (Landing Ship, Tank) moored in 45 ft of water in the Gulf of Mexico. The calculations for the separate restraint functions $q(v)$ for surge and sway were based on the catenary theory just discussed and the actual mooring geometry of all seven chains for this experimental study. The results are shown in Figure 2.22. These results typify a *hard* spring nonlinear restraint system and are similar to the two-cable example shown in Figure 2.21. When each curve of Figure 2.22 is fit to the odd order cubic polynomial of equation (2.69), the restraining force for surge (longitudinal) displacement and for sway (lateral) displacement become, respectively

$$q(v) = 20,300v + 400v^3 \, \text{lb} \tag{2.70}$$

$$q(v) = 12,700v + 950v^3 \, \text{lb} \tag{2.71}$$

The coefficient of v in each case is k_1, or the slope of the curve at $v = 0$.

Example Problem 2.11. The purposes of this example are to compute the virtual masses for the moored ship described in *Example Problem 2.10*, and to set up the uncoupled equations of motion in surge and sway. Traditional English units are employed for the purpose of clarifying the unit of mass. As previously, the ship is assumed to be a rigid body with an actual mass m_0. Based on its given displacement (weight) of 4400 long tons, the ship's actual mass is

$$m_0 = \frac{1}{32.1 \, \text{ft/sec}^2} \times 4400 \, \text{long tons} \times 2240 \frac{\text{lb}}{\text{long ton}} = 3.06 \times 10^5 \, \text{slug}$$

in which the mass unit of lb-sec^2/ft is designated as *slug*. Experimental evidence shows that for surge motion only, the virtual mass is approximately 15 percent greater than m_0, or $m = 1.15m_0 = 3.52 \times 10^5$ slug; and for sway motion only, the virtual mass for this ship (unstreamlined for sway) is about twice m_0, or $m = 2m_0 = 6.12 \times 10^5$ slug. For either motion, the governing equations are of the form of equation (2.2). Assume negligible damping, $f(\dot{v}) = 0$, and calm seas, $p(t) = 0$. Under these conditions, with $q(v)$ given by equations (2.70) and (2.71), and with the virtual masses just calculated, the respective equations of motion for surge and sway are as follows:

$$3.52 \times 10^5 \ddot{v} + 20,300v + 400v^3 = 0 \tag{2.72}$$

$$6.12 \times 10^5 \ddot{v} + 12,700v + 950v^3 = 0 \tag{2.73}$$

In these equations of motion, $\ddot{v}$ has units of ft/sec^2 and v has units of ft.

Soil Foundation Restraints

The properties of the soils on the sea floor are needed to predict the dynamics of fixed offshore structures. In some cases, it may be appropriate to assume that the soil foundation for such structures behaves elastically and that the soil properties are unaffected by the motion of the contacting structure. Consider, for example, the single cantilevered pile that penetrates the flexible foundation of the sea floor to a depth ℓ_u, as shown in Figure 2.23a. The pile's bending stiffness is EI and its height above the sea floor is ℓ. Under the horizontal tip load F_x, the horizontal tip displacement is δ. Using a static analysis, Kocsis (1976) computed the equivalent length $\ell_e > \ell$ for a uniform pile partly submerged in sandy or in clayey soil of constant, elastic properties. This equivalent length, depicted in Figure 2.23b, is that for a hypothetical pile with *full* fixity at the base, which gives the same horizontal deflection δ under the same horizontal load F_x as for the pile with the flexible soil foundation. For a sandy soil, that length is

$$\ell_e = \ell_u \left[0.4 + 1.353 \left(\frac{\ell}{\ell_u} \right) + 1.875 \left(\frac{\ell}{\ell_u} \right)^2 + \left(\frac{\ell}{\ell_u} \right)^3 \right]^{1/3} \qquad (2.74)$$

in which

$$\ell_u = \left(\frac{102.9 EI}{N_0} \right)^{1/5} \qquad (2.75)$$

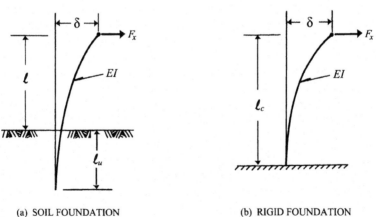

(a) SOIL FOUNDATION   (b) RIGID FOUNDATION

Figure 2.23   Static model for pile-soil stiffness.

For submerged sandy soil, the horizontal subgrade reaction constant N_0 in the last equation has a range of 4 tons/ft^3 to 34 tons/ft^3 for relatively loose to dense sand, respectively (Terzaghi, 1955). With the value of ℓ_e from equations (2.74) and (2.75), k_1 for the pile can calculated from classical beam theory for a cantilevered beam of length ℓ_e as: $k_1 = 3EI/\ell_e^3$. Kocsis (1976) also presented equations similar to (2.74) and (2.75) for the calculation of ℓ_e for a pile in clayey soil foundations.

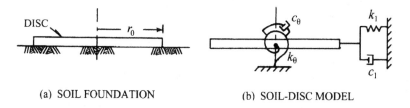

(a) SOIL FOUNDATION   (b) SOIL-DISC MODEL

Figure 2.24 Dynamic model for disc-soil interactions.

Consider now the measured effects soil behavior to the motion of a contacting structure. Veletsos and Wei (1971) performed extensive laboratory experiments on soils in contact with a disc of radius r_0, as depicted in Figure 2.24a. In a typical experiment, this disc was subjected to a harmonic frequency ω, first in the direction of sliding, v, and then in pure rotation θ. Nataraja and Kirk (1977) correlated these data for applications to gravity platform dynamics. For disc sliding motion only, the respective constants for soil stiffness and damping are k_1 and c_1; and for rotational motion only, these respective constants are k_θ and c_θ. These constants, shown in the disc model of Figure 2.24b, are as follows:

$$k_1 = \frac{8G_s}{2-\nu}\left(1 - 0.05\omega r_0\sqrt{\frac{\rho_s}{G_s}}\right) r_0 = a_1 - b_1\omega \qquad (2.76)$$

$$c_1 = \frac{8}{2-\nu}\sqrt{\rho_s G_s}\left(0.67 + 0.02\omega r_0\sqrt{\frac{\rho_s}{G_s}}\right) r_0^2 \qquad (2.77)$$

$$k_\theta = \frac{8G_s}{3(1-\nu)}\left(1 - 0.215\omega r_0\sqrt{\frac{\rho_s}{G_s}}\right) r_0^3 = a_0 - b_0\omega \qquad (2.78)$$

$$c_\theta = \frac{0.375}{1-\nu}\omega\rho_s r_0^5 \qquad (2.79)$$

Here G_s, ν, and ρ_s are the shear modulus, Poisson's ratio, and mass density, respectively, of the soil. The parameter ω (rad/sec) is the frequency of the disc. The use of this dynamic soil model is illustrated in *Example Problem 5.2*.

The validity of employing equations (2.76)-(2.79) for full-scale design of gravity platforms still needs to be shown through full-scale testing. Efforts to employ the generalizations of continuum mechanics to characterize the dynamic, mechanical properties of soil, including the formulation of constitutive equations from carefully designed experiments, are discussed by Zienkiewicz et al. (1978, Chapters 10-16) and Prevost et al.(1981).

PROBLEMS

2.1   The nondimensional parameter *cylinder roughness* affects the values of the coefficients C_D and C_M as indicated by equations (2.18) and (2.20). Define precisely this measure of cylinder roughness by consulting a standard reference book in fluid mechanics. Then based on reported experimental results, discuss briefly the effect of this parameter on the coefficients C_D and C_M for both the steady flow and the periodic flow of water which flows normal to a submerged stationary, circular cylinder.

2.2   To linearize equation (2.21), Berge and Penzien (1974) made certain statistical assumptions concerning the nature of the constant C_D of equation (2.22). Review this paper, especially their equations (10)-(15), and describe the method and assumptions leading to this linearized equation of motion.

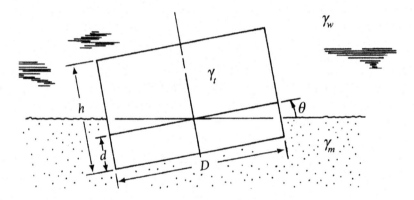

Figure 2.25   Cross section of the cylindrical tank of Problem 2.3.

2.3   The cylindrical tank of Figure 2.25 has diameter D, height h, and an average weight density in air of γ_t when filled with solid, radioactive wastes. The tank penetrates to a depth d into the mud of weight density γ_m at the bottom of the sea. The mud layer behaves as a liquid.

(a) Derive an expression for the tank dimension ratio d/h in terms of the system densities. If $\gamma_w = 64$ lb/ft^3, $\gamma_m = 95$ lb/ft^3, and $\gamma_t = 85$ lb/ft^3, what is this ratio?

(b) Locate the center of buoyancy for the cylinder in its upright position, where $D = 8$ ft and $h = 10$ ft. Discuss the stability of this cylinder for small angles of tilt, θ.

(c) For $D = 8$ ft and the densities given in part (a), calculate the maximum height h for which the cylinder will remain stable for small θ.

2.4   The concrete monotower pictured in Figure 2.26 is in 160 m of water. The caisson is filled with oil, and the leg is filled with seawater up to the still waterline. The pertinent data are:

Deck mass $= 1.5 \times 10^7$ kg
Caisson mass (filled) $= 2 \times 10^8$ kg
Concrete density $= 2500$ kg/m^3
Sea water density $= 1025$ kg/m^3

(a) Does this structure float if it is raised just slightly from the bottom? If so, what is the minimum deck mass needed to keep it from floating?

(b) Locate the center of mass G and the center buoyance B for this structure when in the upright position. If the answer to part (a) is positive, use your recommended deck weight to locate G.

(c) Discuss the stability of this structure, or the structure as modified in part (a), if a thin layer of the soil foundation behaves as a liquid.

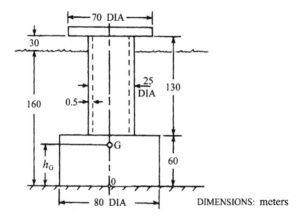

Figure 2.26   Cross section of an idealized concrete monotower, Problem 2.4.

2.5   The mean wind velocity is used to calculate the wind-induced load on an offshore structure. Discuss briefly a method for converting experimental measures of the fastest mile of wind speed to a mean wind velocity. State your assumptions and cite pertinent references.

2.6   Using equation (2.27), plot the normalized horizontal wind velocity $\bar{u}(z)/\bar{u}(h)$ as a function of the height parameter z/h. Show three typical curves: one for a rough coastal area, one for sustained wind over an unobstructed sea, and one for a gusty wind. Summarize your observations about these particular wind velocity profiles.

2.7   The addition of helical strakes to a cylindrical structure such as shown in Figure 2.7 is not the only way to negate trailing periodic vortices. Describe other practical, alternative devices that may be added to the structure to eliminate such vortices. Use the references cited in the text as a starting point.

2.8   On the same graph, plot two flow velocity profiles as a function of water depth: a tidal current and a wind-stress current. For this graph, assume that both flows are in the same direction. If the wind velocity is offset by an angle θ to the tidal current, what is the equation for the *total* velocity in terms of θ.

2.9 A relatively isolated, vertical cylindrical pile is fully submerged and subjected to a steady tidal current of magnitude 4 knots. The pile has diameter $D = 0.3$ m and height $h = 5$ m. Compute the total current-induced drag force on this pile. State all assumptions you make in arriving at your answer.

2.10 Based on studies reported in the most recent literature (for instance, *The Proceedings of the Offshore Technology Conference*), discuss briefly both the advantages and dangers of using onshore seismic data for ground motion in the design of fixed-bottom offshore structures for earthquake resistance.

2.11 The coefficient C_i of equation (2.37), which is used to predict the horizontal impact of ice on structures, varies by more than a factor of two. Investigate the physical reasons for this variability.

2.12 After consulting the recent literature on the subject, write a summary report comparing the particular design features that allow the offshore platforms near Alaska to resist ice hazards. How do these platforms differ from the jacket-template structures in the Gulf of Mexico?

2.13 For the case of two opposing cables shown in Figure 2.19, where the equilibrium angle θ_r is 45 deg, calculate k_1 and k_3 for a best fit to equation (2.69). If $w = 39$ lb/ft and $z_0 = 1000$ ft, what is $q(v)$?

2.14 Assume that the floating structure of Figure 2.19 is supported by four identical, symmetrically arranged cables, 90 deg apart. Assume that the only motion is horizontal and along the line of two opposing cables. Show that the restoring force as a function of deflection as given by the curves of Figure 2.21 is still quite accurate because the cables perpendicular to the direction of motion contribute very little resistance to this motion. For what range of initial angles θ_r is this approximation *not* a good one?

2.15 Assume that N pairs of identical cables are arranged at equal angles β, as shown in Figure 2.19. Derive a general expression for the restoring force along the line of one of these pairs. State all your assumptions. Then plot the antisymmetric curve in the nondimensional form of Figure 2.21 for $N = 6$, $\theta_r = 45$ deg, and $\beta = 30$ deg.

2.16 Based on the analysis of Kocsis (1976), calculate ℓ_e, the equivalent length of a solid concrete pile submerged in medium stiff sand. Use the following data: $\ell = 30$ ft, $D = 4$ ft, $E = 3000$ kips/in^2, and $N_0 = 14$ tons/ft^3. (Hint: First convert to a consistent set of units.)

REFERENCES

Adrezin, R., Bar-Avi, P., and Benaroya, H., Dynamic Response of Compliant Offshore Sructures–Review, *Journal of Aerospace Engineering*, 1996.

American Petroleum Institute, *Recommended Practice for Planning, Designing, and Constructing Offshore Platforms,* Dallas, TX. API, RP-2A, 1979; API twentieth ed., 1997.

Ansari, K. A., "Mooring with Multi-Component Cable Systems," *Journal of Energy Resources Technology, ASME* **102**, 1980.

Batchelor, G. K., *An Introduction to Fluid Mechanics*, Cambridge University Press, Cambridge, UK, 2000.

Bazant, Z. P., and Kim, J. J. H., Size Effect in Penetration of Sea Ice Plate with Part-Through Cracks. I: Theory, II: Results, *ASCE Journal of Engineering Mechanics* **124** (12), December 1998.

Bea, R. G., Audibert, J. M. E., and Akky, M. R., Earthquake Response of Offshore Platforms, *Journal of the Structural Division, ASCE* **105** (ST2), 1979.

Bea, R. G., Xu, T., Stear, J., and Ramos, R., Wave Forces on Decks of Offshore Platforms, *ASCE Journal of Waterway, Port, Coastal, and Ocean Engineering* **125** (3), 1999.

Bercha, F. G., and Stenning, D. G., Arctic Offshore Deepwater Ice-Structure Interactions, *Proceedings of the Offshore Technology Conference*, OTC-3632, Houston, TX, 1979.

Berge, B., and Penzien, J., Three-Dimensional Stochastic Response of Offshore Towers to Wave Forces, *Proceedings of the Offshore Technology Conference*, OTC-2050, Houston, TX, 1974.

Blevins, R. D., *Flow-Induced Vibrations*, Van Nostrand Reinhold, New York, 1977.

British Ship Research Association, *A Critical Evaluation of Data on Wave Force Coefficients*, Report No. 278.12, August, 1976.

Childers, M. A., Mooring Systems for Hostile Waters, *Petroleum Engineer*, 1973.

Childers, M. A., Deep Water Mooring—Part I, Environmental Factors Control Station Keeping Methods, *Petroleum Engineer* 1, 1974.

Childers, M. A., Deep Water Mooring—Part II, The Ultradeep Water Spread Mooring System, *Petroleum Engineer* 2, 1974.

Childers, M. A., Deep Water Mooring—Part III, Equipment for Handling the Ultradeep Water Spread Mooring System, *Petroleum Engineer* 5, 1975.

DeFranco, S. J., and Dempsey, J. P., Crack Propagation and Fracture Resistance in Saline Ice, *Journal of Glaciology* **40**, 1994.

Gaythwaite, J., *The Marine Environment and Structural Design,* Van Nostrand Reinhold, New York, 1981.

Gould, P. L., and Abu-Sitta, S. H., *Dynamic Response of Structures to Wind and Earthquake Loading*, Wiley, New York, 1980.

Graff, W. J., *Introduction to Offshore Structures,* Gulf Publishing Co., Houston, TX, 1981.

Graff, W.J., and Chen, W.F., Bottom-Supported Concrete Platforms: Overview, *Journal of the Structural Division, ASCE* **107** (ST6), 1981.

Hafen, B.E., Meggitt, D. J., and Liu, F. C., *Strumming Suppression—An Annotated Bibliography*, N-1456, U.S. Civil Engineering Laboratory, Port Hueneme, CA, 1976.

Hoerner, S. F., *Fluid Dynamic Drag*; Published by author, New York, 1965.

Hudson, D. E., Ground Motion Measurements, *Earthquake Engineering*, R. L. Wiegel, editor, Prentice-Hall, Englewood Cliffs, NJ, 1970.

Irvine, H. M., *Cable Structures*, The MIT Press, Cambridge, MA, 1981.

Keulegan, G. H., and Carpenter, L. H., Forces on Cylinders and Plates in an Oscillating Fluid, *Journal of Research of the National Bureau of Standards* **60** (5), 1958.

Kocsis, P., The Equivalent Length of a Pile or Caisson in Soil, *Civil Engineering* **63**, 1976.

Lamb, H., *Hydrodynamics*, sixth ed., Dover, New York, 1945.

Marshall, P. W., Fixed Pile-Supported Steel Offshore Platforms, *Journal of the Structural Division, ASCE* **107** (ST6), 1981.

Miller, B. L., Wave Slamming Loads on Horizontal Circular Elements of Offshore Structures, RINA-5, *Journal of the Royal Institute of Naval Architecture*, 1977.

Miller, B. L., *Wave Slamming on Offshore Structures*, NMI-R81, National Maritime Institute, 1980.

Morison, J. R., O'Brien, M. P., Johnson, J. W., and Schaaf, S. A., The Forces Exerted by Surface Waves on Piles, *Petroleum Transactions, AIME* **189**, 1950.

Muga, B. J. and Wilson, J. F., *Dynamic Analysis of Ocean Structures*, Plenum, New York, 1970.

Munson, B. R., Young, D. F., and Okushi, T. H., *Fundamentals of Fluid Mechanics*, Wiley, New York, 1998.

Nataraja, R. and Kirk, C. L., Dynamic Response of a Gravity Platform under Random Wave Forces, OTC-2904, *Proceedings of the Offshore Technology Conference*, 1977.

NAVFAC DM-26, *Design Manual-Harbor and Coastal Facilities*, U.S. Navy Publications and Forms Center, Philadelphia, PA 19120, 1968.

Newman, J. N., *Marine Hydrodynamics*, The MIT Press, Cambridge, MA, 1977.

Neumann, G., and Pierson, W. J., Jr., *Principles of Physical Oceanography*, Prentice-Hall, Englewood Cliffs, NJ, 1966.

Niedzwecki, J. M., and Casarella, M. J., On the Design of Mooring Lines for Deep Water Applications, ASME paper No. 75-WA/OCE-1, 1975.

O'Brien, J. T., and Muga, B. J., Sea Tests on a Spread-Moored Landing Craft, *Proceedings of the Eighth Conference on Coastal Engineering*, Lisbon, Portugal, 1964.

Page, R. A., Evaluation of Seismicity and Earthquake Shaking at Offshore Sites, OTC-2354, *Proceedings of the Offshore Technology Conference*, Houston, TX, 1975.

Pagon, W. W., Drag Coefficients for Structures Studied in Wind Tunnel Model Studies, *Engineering News-Record* **113** (15), McGraw-Hill, New York, 1934.

Pattison, J. H., Rispin, P. P., and Tsai, N., *Handbook of Hydrodynamic Characteristics of Moored Array Components*, NSRDC Report No. SPD-745-01, 1977.

Peyton, H. R., Ice and Marine Structures, *Ocean Industry*, Houston, TX, 1968.

Prevost, J. H., Cuny, B., Hughes, T. J. R. and Scott, R. F., Offshore Gravity Structures: Analysis, *Journal of the Geotechnical Engineering Division, ASCE* **107** (GT2), 1981.

Sachs, P., *Wind Forces in Engineering*, Pergamon, Oxford, UK, 1972.

Sarpkaya, T., Vortex Shedding and Resistance in Harmonic Flow About Smooth and Rough Cylinders at High Reynolds Numbers, NPS-59SL76021, Naval Postgraduate School, Monterey, CA, 1976.

Sarpkaya, T., and Isaacson, M., *Mechanics of Wave Forces on Offshore Structures*, Van Nostrand Reinhold, New York, 1981.

Schlichting, H., *Boundary-Layer Theory*, sixth ed., McGraw-Hill, New York, 1968.

Sherlock, R. H., Variations of Wind Velocity and Gusts with Height, *Transactions ASCE* **118**, 1953.

Simiu, E., Equivalent Static Wind Loads for Tall Building Design, *Proceedings of the Structural Division, ASCE* **102** (ST4), 1976.

Simiu, E., and Scanlan, R. H., *Wind Effects on Structures*, Wiley, New York, 1978.

Terzaghi, K., Evaluation of Coefficients of Subgrade Reactions, *Geotechnique* **5**, 1955.

Tung, C. C., Peak Distribution of Random Wave-Current-Force, *Journal of the Engineering Mechanics Division, ASCE* **100** (EMS), 1974.

Veletsos, A. S. and Wei, Y. T., Lateral and Rocking Vibration of Footings, *Journal of the Soil Mechanics and Foundations Division, ASCE* **95** (SM9), 1971.

Vellozzi, J., and Cohen, E., Gust Response Factors, *Journal of the Structural Division, ASCE* **94** (ST6), 1968.

Wendel, K., Hydrodynamische Masses, Tragheits-Momente, *Jahrbuch der schiffbautechnischen Gesellschaft* **44**, 1950. Translation No. 260, DTNSRDC, 1956.

Wiegel, R. L., ed., *Earthquake Engineering*, Prentice-Hall, Englewood Cliffs, NJ, 1970.

Wiggins, J. H., Hasselman, T. K., and Chrostowski, J. D., Seismic Risk Analysis for Offshore Platforms in the Gulf of Alaska, OTC-2669, *Proceedings of the Offshore Technology Conference*, Houston, TX, 1976.

Wilson, B. W., Case of Critical Surging of a Moored Ship, Paper No. 2318, *ASCE Proceedings*, 1959.

Wilson, J. F. and Caldwell, H. M., Force and Stability Measurements on Models of Submerged Piplines, *ASME Journal of Engineering for Industry* **93** (4), 1971.

Wilson, J. F., and Orgill, G., Optimal Cable Configurations for Passive Dynamic Control of Compliant Towers, *ASME Journal of Dynamic Systems, Measurement and Control* **106**, 1984.

Wilson, J. F., and Tinsley, J. C., Vortex Load Reduction Experiments in Optimal Helical Strake Geometry for Rigid Cylinders, *ASME Journal of Energy Resources Technology* **111**, 1989.

Zienkiewicz, O. C., Lewis, R. W., and Stagg, K. G., eds., *Numerical Methods in Offshore Engineering*, Wiley-Interscience, New York, 1978.

3

Deterministic Descriptions of Offshore Waves

Bruce J. Muga

To evaluate the fluid-induced forces acting on a structure or on any of its components, we need to know its surrounding hydrodynamic flow field. For offshore structures, this flow field arises from time-varying natural processes: winds, currents, and surface gravity waves. Although these processes nearly always occur in various combinations, the scope of this chapter is limited to descriptions of surface gravity waves, which are usually wind-generated. It has been observed that these time-dependent waves occur on two different scales. The shorter time scale, measured in minutes or seconds, is useful for describing detailed features such as wind gusts and surface wave periods. This shorter time scale corresponds most closely to the response time of fixed offshore structures. The longer time scale, measured in terms of hours, days, or even years, is useful for describing variations in the wave intensity and in its statistics. The longer time scale is important too because structures may fail in low cycle fatigue fracture after months or years of service, a topic discussed by Etube (2001). In this present chapter, we give a brief synopsis and critique of the classical wave theories used most often by engineers for preliminary calculations of forces on offshore structures. Here, the focus is on the shorter time scale. For comprehensive bibliographies of wave theories, see Sarpkaya and Isaacson (1981) and Young (1999).

Illustrated in Figure 3.1 are the two fundamentally different descriptions of surface gravity waves: deterministic and probabilistic. Deterministic descriptions, analytical or numerical, are used to characterize the short time scale features of waves. Deterministic analytic descriptions encompass classical wave theories, which in turn are subdivided into linear and nonlinear types. Probabilistic descriptions are used to characterize the long time scale features of offshore waves. In both the deterministic and probabilistic descriptions, linear wave theory is important to engineers for two reasons: it is simple to apply when estimating forces on offshore structures during the preliminary phases of design; and it affords a simple basis for estimating the probability of failure of a given structure. Examples in future chapters illustrate these ideas.

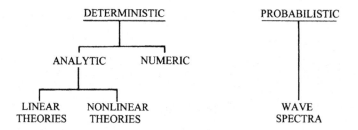

Figure 3.1   Alternative approaches for describing surface gravity waves.

3.1   DESCRIPTION OF PLANE WAVES

Before discussing the most important classical wave theories, we first introduce some fundamental definitions and concepts. The natural occurrence of waves is usually complicated by the simultaneous superposition of waves of many different shapes and energies. Thus we define a single wave or a train of waves of pure form that always behaves in exactly the same way on passing any given point, and that propagates without change of shape when the surrounding conditions remain the same. Further, we consider only plane or long-crested waves of permanent form, waves that are independent of the coordinate normal to the wave propagation direction. Therefore their crests or troughs can be represented by straight lines in a horizontal plane. By permanent form, we mean that the field of motion, pressure distribution, and surface configuration are maintained as one follows the wave at a speed c, the designation of the phase velocity or celerity. (In a strict sense all real waves have finite length crests, a factor that is ignored herein.) Consider a wave form that is simple harmonic so that at any time t the wave has a sinusoidal shape with reference to the still water line (SWL), or the x-axis as shown in Figure 3.2. At $t = 0$, the instantaneous surface elevation η is then

$$\eta_{t=0} = A \cos kx \qquad (3.1)$$

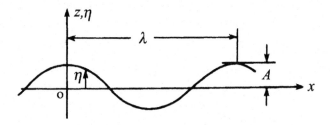

Figure 3.2   Definition of a simple harmonic wave.

In equation (3.1), A defines the amplitude of the wave and k is the wave number. In radian measure, $kx = 2\pi x/\lambda$; or x/λ is the multiple of the angle 2π, where λ is the particular value of x for which $\cos kx = 1$, yielding the next crest of the wave in advance of the first at $x = 0$. The wave length λ is expressed in terms of wave number by

$$k = \frac{2\pi}{\lambda} \tag{3.2}$$

The general expression for η in a progressive sinusoidal wave moving at velocity c in the positive x direction is

$$\eta(x,t) = A \cos k(x - ct) \tag{3.3}$$

Now, if we let $x = 0$ and consider the variation of the instantaneous surface elevation with time as the wave passes the origin 0, we have from equation (3.3)

$$\eta_{x=0} = A \cos(-kct) = A \cos kct \tag{3.4}$$

which is depicted in Figure 3.3. Again, the significance of kc is that of converting kct to radian measure, or

$$kct = 2\pi \frac{t}{T} \tag{3.5}$$

where T is the particular value of t that makes the cosine term unity, giving the next crest of the wave in succession to the first at $t = 0$. The quantity T is the period of the wave and

$$kc = \frac{2\pi}{T} = \omega \text{ rad/sec} \tag{3.6}$$

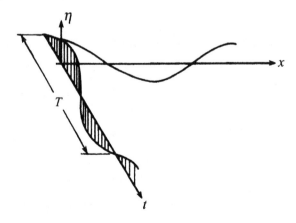

Figure 3.3   Variation of instantaneous surface elevation with time.

where ω is the angular frequency. From equations (3.2) and (3.5) it follows that

$$\lambda = cT \tag{3.7}$$

Equation (3.7), which could have been inferred directly, is a fundamental relationship in wave theory and has general application regardless of wave form.

Equation (3.6) enables us to rewrite equation (3.3) as

$$\eta(x,t) = A\cos(kx - \omega t) \tag{3.8}$$

If we now compare two similar waves of identical form that pass the same place at different times, we have for one at $t = 0$

$$\eta'_{t=0} = A\cos kx \tag{3.9a}$$

and for the other

$$\eta''_{t=0} = A\cos(kx + \varepsilon) \tag{3.9b}$$

where ε is the phase displacement, as illustrated in Figure 3.4. This is positive if η'' lags the first wave, but negative if η'' leads η'. The interpretation of sign is the same as for ωt in $\eta = A\cos(kx \pm \omega t)$: negative for a forward wave, and positive for a rearward wave, where *forward* means motion in the direction of x positive.

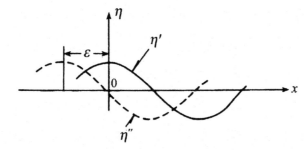

Figure 3.4   Illustration of phase displacement for two waves of identical form.

The general expression for a sinusoidal wave in terms of a phase displacement ε which preceeds the origin is thus

$$\eta(x,t) = A\cos\{(kx - \omega t) - \varepsilon\} \tag{3.10a}$$

If $\varepsilon = \pi/2$, then equation (3.10a) becomes

$$\eta(x,t) = A\sin(kx - \omega t) \tag{3.10b}$$

which describes a progressive harmonic wave moving in the positive x direction. In summary, a three-dimensional space-time representation of the surface elevation for a progressive, plane wave is given by

$$\eta(x,t) = A\cos(kx - \omega t) \tag{3.10c}$$

which is shown in Figure 3.5.

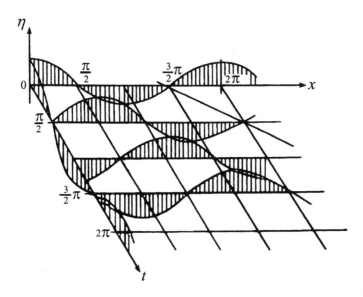

Figure 3.5 Space-time representation for a progressive plane wave.

The foregoing discussion includes the fundamental concepts and most of the definitions needed to describe classical wave theories. It is noted that the symbol H, often used for wave height, has yet to be defined. The reason for this will soon become apparent.

3.2 LINEAR PLANE WAVES

The theory of linear waves is alternatively known as Airy's theory, small amplitude theory, and first-order theory. Developed primarily by Airy (1845) and by Laplace (1816), it is the most important of the classical theories because it is both easy use and it forms the basis for the probabilistic spectral description of waves. For a full development and discussion of the theory, the reader should consult the intellectually entertaining work by Kinsman (1965). We now summarize the assumptions, the governing equations, and the solutions for the wave velocity and pressure profiles useful for predicting wave-induced forces on offshore structures.

Suppose that the simple harmonic plane wave defined in Figure 3.6 is propagating in the positive x direction in water of density ρ. The vertical coordinate is z, directed positive upward; and the origin is located at the still water line (SWL) or the mean surface level. The nine assumptions inherent in linear theory are as follows:

1. The amplitude A of the surface disturbance is very small relative to the wave length λ and the water depth d.

2. The velocity head $(u^2 + w^2)/2g$ is small compared with the hydrostatic pressure head $\rho g z$. Here u and w are the horizontal and vertical water particle velocities, respectively.

3. The water depth d is uniform.
4. The water is nonviscous and irrotational.
5. The water is incompressible and nonstratified (homogeneous).
6. The Coriolis forces due to the earth's rotation are negligible.
7. Surface tension is negligible.
8. The sea floor is smooth and impermeable.
9. The sea level atmospheric pressure p_a is uniform. Here, the hydrostatic pressure is $-\rho g z$, and the dynamic water pressure is denoted by p.

Assumption 1 actually implies 2, although this is not immediately obvious. Assumption 5 excludes acoustic and internal wave phenomena. It is known that Assumption 6 is valid if the very long waves associated with tides and seiches in large seas are excluded. It is also known that surface tension effects are negligible, Assumption 7, for all but very short wave lengths. The other assumptions, needed to make the governing equations tractable for closed form solutions, are reasonable approximations for a large number of applications.

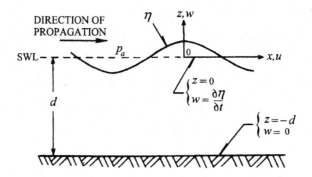

Figure 3.6   Schematic representation of a simple harmonic wave. The vertical scale of the surface profile is exaggerated for clarity.

When the nine assumptions are invoked, the differential equations and boundary restraints for the water particle velocity and pressure reduce to the following forms:

$$\frac{\partial u}{\partial z} - \frac{\partial w}{\partial x} = 0 \qquad (3.11a)$$

$$\frac{\partial u}{\partial x} + \frac{\partial w}{\partial z} = 0 \qquad (3.11b)$$

$$\frac{\partial u}{\partial t} = -\frac{1}{\rho}\frac{\partial p}{\partial x} \qquad (3.11c)$$

$$\frac{\partial w}{\partial t} = -\frac{\partial p}{\partial z} - g \qquad (3.11d)$$

$$w = \frac{\partial \eta}{\partial t} \quad \text{at} \quad z = 0 \tag{3.12}$$

$$w = 0 \quad \text{at} \quad z = -d \tag{3.13}$$

$$p = p_a \quad \text{at} \quad z = 0 \tag{3.14}$$

Equation (3.11a) is the zero vorticity or irrotational condition, which follows from Assumption 4. Equation (3.11b) is the continuity condition. Equations (3.11c) and (3.11d) are the momentum conservation equations, otherwise known as the Eulerian equations of motion. Equations (3.12) and (3.14) are the boundary conditions at the surface, and equation (3.13) is the boundary condition at the sea floor. The free surface is defined by $z = \eta$. However, in view of Assumption 1, the conditions on w and p of equations (3.12) and (3.14) are applied at the mean water level, $z = 0$.

A particular solution that satisfies the linear equations (3.11) is the plane wave form for surface elevation $\eta = \eta(x, t)$, already presented as equation (3.10b). With this result, together with equations (3.11)-(3.14), the water particle velocities u and w, their respective accelerations $\partial u/\partial t$ ($= \dot{u}$) and $\partial w/\partial t$ ($= \dot{w}$), and the dynamic pressure p can be deduced. All of these quantities are summarized in Table 3.1. It is noted that the total or absolute water pressure is the sum of three pressure terms: the atmosphere pressure, p_a; the hydrostatic pressure, $p_s = -\rho g z$; and the dynamic pressure, or

$$p_{total} = p_a + p_s + p \tag{3.15}$$

In calculating the forces on offshore structures, the atmospheric pressure is of no consequence.

The wave number and frequency relation that is compatable with the solutions just presented is

$$\omega^2 = gk \tanh kd \tag{3.16}$$

From equation (3.6), the wave phase velocity or celerity is given by $c = \omega/k$. With this and equation (3.16), the celerity becomes

$$c = \left(\frac{g}{k} \tanh kd\right)^{1/2} \tag{3.17}$$

This equation for celerity reduces to rather simple limiting forms for cases in which k is either very large (short wave lengths) or very small (long wave lengths). Such examples are included as problems at the end of this chapter.

TABLE 3.1   Results for Linear Small Amplitude Wave Theory

Parameter	Formula
Surface wave profile	$\eta = A \cos(kx - \omega t)$
Horizontal particle velocity	$u = \frac{2\pi A}{T} \frac{\cosh k(z+d)}{\sinh kd} \cos(kx - \omega t)$
Vertical particle velocity	$w = \frac{2\pi A}{T} \frac{\sinh k(z+d)}{\sinh kd} \sin(kx - \omega t)$
Horizontal particle acceleration	$\dot{u} = \frac{4\pi^2 A}{T^2} \frac{\cosh k(z+d)}{\sinh kd} \sin(kx - \omega t)$
Vertical particle acceleration	$\dot{w} = -\frac{4\pi^2 A}{T^2} \frac{\sinh k(z+d)}{\sinh kd} \cos(kx - \omega t)$
Hydrostatic pressure	$p_s = -\rho g z$
Dynamic pressure	$p = \rho g A \frac{\cosh k(z+d)}{\cosh kd} \cos(kx - \omega t)$
Wave celerity	$c = \left(\frac{g}{k} \tanh kd\right)^{1/2}$
Wave group velocity	$c_g = \frac{c}{2}\left(1 + \frac{2kd}{\sinh 2kd}\right)$

Consider a few comments about the results for linear wave theory that are summarized in Table 3.1. First, note that the water particle velocities, the wave celerity, and the wave group velocity (derived by Sarpkaya and Isaacson, 1981) are all different in form and have different physical meanings. The water particle velocities and particle accelerations are those used in Morison's equation to compute the drag and inertial forces of these waves on offshore structures. Second, the origin $x = 0$ of the wave is arbitrary, which implies that a constant, arbitrary phase angle can be added to the term $(kx - \omega t)$ in the formulas of Table 3.1. To the casual reader, this may be a source of confusion since many references locate the origin at the trough of the wave rather than at its crest. Third, some classical references define the coordinate z as positive downward, which has the effect of reversing signs for those terms containing z. Further, the vertical coordinate is sometimes labeled y in place of the more common z.

3.3   NONLINEAR WAVES

Two distinguishing features of a small amplitude wave based on linear theory are its sinusoidal surface profile and its circular fluid particle orbit. These two features do not coexist in a finite amplitude wave based on nonlinear theory. Summarized now are the most important features of several nonlinear wave theories: the trochoidal, cnoidal, Stokes, solitary, and numerical theories. For a complete historical background and detailed description of these and other nonlinear wave theories, see Sarpkaya and Isaacson (1981). In the summary that follows, the term *nonlinear wave* implies a wave of finite amplitude.

Trochoidal Theory

Gerstner (1802) and Rankine (1863) developed the trochoidal wave theory independently. The three distinguishing features of the trochoidal theory are: circular particle orbits; a rotational fluid; and a trochoidal wave surface profile. Historically, this nonlinear theory found favor with naval architects, who focused not so much on the fluid kinematics with its rotational characteristic, but on the trochoidal surface profile of these finite amplitude waves. With the advent of offshore structures, more realistic wave kinematics were needed, and thus trochoidal theory is not usually used by engineers in offshore structural design. The importance of trochoidal wave theory is that it serves as a link from linear theory to the finite amplitude oscillatory wave theory as developed by Stokes (1845), Levi-Civita (1925), Struik (1926), and Havelock (1914).

Cnoidal Theory

A finite amplitude wave theory appropriate for shallow water is the cnoidal wave theory, first studied by Korteweg and de Vries (1895) and more recently by Masch and Wiegel (1961). As suggested by Sarpkaya and Isaacson (1981), the cnoidal wave parameters are formulated in terms of elliptic cosine functions, from which the term "cnoidal" arises. Tables and charts published in the latter two citations aid in application of this theory.

The cnoidal wave theory was developed from the governing equations for long waves using the assumption that the square of the slope of the water surface, or wave steepness, is small relative to unity. One important feature is that cnoidal waves are periodic. For small values of H/d, where H is the crest to trough wave dimension and d is the water depth, the cnoidal wave profile is sinusoidal. Another limiting case is for very long wave lengths, which yields a solitary wave profile, as discussed below. The use of the cnoidal wave theory is limited to the following range: $0.01 \leq H/d \leq 0.78$ and $\lambda/d < 8$. Within this range, cnoidal theory describes the progression of the periodic waves more accurately than does the theory for Stokes waves. Cnoidal theory bridges the gap between the periodic and the solitary wave theories.

Stokes Theory

The basic assumption in the development of the finite amplitude wave theory is that the fluid motion is irrotational. This assumption can be justified physically if the fluid viscosity is vanishingly small. The governing equations are then formulated in a manner parallel to that for linear wave theory, equations (3.11). Those equations are as follows:

$$\frac{\partial w}{\partial x} - \frac{\partial u}{\partial z} = 0 \tag{3.18}$$

$$\frac{\partial u}{\partial x} + \frac{\partial w}{\partial z} = 0 \tag{3.19}$$

$$\frac{\partial u}{\partial t} + u\frac{\partial u}{\partial x} + w\frac{\partial u}{\partial z} = -\frac{1}{\rho}\frac{\partial p}{\partial x} \tag{3.20}$$

$$\frac{\partial w}{\partial t} + u\frac{\partial w}{\partial x} + w\frac{\partial w}{\partial z} = -\frac{1}{\rho}\frac{\partial p}{\partial z} - g \tag{3.21}$$

Equations (3.18) and (3.19) express the zero vorticity and continuity conditions, respectively, and equations (3.20) and (3.21) express the conservation of linear momentum. Once the velocity field can be computed, then the pressure field can be determined. The boundary conditions are that the pressure on the free surface of the wave is everywhere a constant, or

$$p(x, \eta, t) = p_a = \text{constant} \tag{3.22}$$

Here, $p = p(x, z, t)$ leads to

$$\frac{\partial p}{\partial t} + u\frac{\partial p}{\partial x} + w\frac{\partial p}{\partial z} = 0 \tag{3.23}$$

Thus the free-surface boundary condition is nonlinear with respect to the unknown variables u, w, and p.

Stokes (1847) and others solved equations (3.18)-(3.23) by a successive approximation procedure in which the solutions were formulated in terms of a series of ascending order terms. Solutions to the second and third order are widely available in the open literature. See, for instance, Kinsman (1965), Ippen (1966), and Sarpkaya and Isaacson (1981).

Some frequently used results of the finite amplitude theory to the second order are presented in Table 3.2. When the solutions for the wave surface profiles, the particle velocities, the particle accelerations, and the pressures given in Table 3.2 are compared respecively to those in Table 3.1, it is noted that each first order term in Table 3.2 corresponds to its counterpart given by the linear theory. The remaining terms are the second-order corrections due to the nonlinear convective inertia terms appearing in the governing equations ($u\,\partial u/\partial x$, etc). Higher-order expressions of the Stokes theory are simply those in which the approximations for corrective effects are carried to the corresponding power term. In principle, if Stokes theory is carried to a sufficiently high order, it would be adequate for describing water waves in any depth of water. In practice, this is only possible for waves in deep water. In shallow water the convective terms become relatively large, the series convergence is slow and erratic, and a large number of terms is required to achieve a uniform degree of accuracy. Other classical formulations such as the solitary and cnoidal theories require fewer terms to achieve the desired accuracy.

TABLE 3.2   Results for Stokes Second Order Wave Theory

Parameter	Formula
Surface wave profile	$\eta = \frac{H}{2}\cos(kx - \omega t) + \frac{H^2\pi}{8\lambda}\frac{\cosh kd}{\sin^3 kd}$ $\times [2 + \cosh(2kd)]\cos[2(kx - \omega t)]$
Horizontal particle velocity	$u = \frac{H\pi}{T}\frac{\cosh k(z+d)}{\sinh kd}\cos(kx - \omega t)$ $+ \frac{3H^2\pi^2}{4T\lambda}\frac{\cosh k(z+d)}{\sinh^4 kd}\cos[2(kx - \omega t)]$
Vertical particle velocity	$w = \frac{H\pi}{T}\frac{\sinh k(z+d)}{\sinh kd}\sin(kx - \omega t)$ $+ \frac{3H^2\pi^2}{4T\lambda}\frac{\sinh 2k(z+d)}{\sinh^4 kd}\sin[2(kx - \omega t)]$
Horizontal particle acceleration	$\dot{u} = \frac{2H\pi^2}{T^2}\frac{\cosh k(z+d)}{\sinh kd}\sin(kx - \omega t)$ $+ \frac{3H^2\pi^3}{T^2\lambda}\frac{\cosh 2k(z+d)}{\sinh^4 kd}\sin[2(kx - \omega t)]$
Vertical particle acceleration	$\dot{w} = -\frac{2H\pi^2}{T^2}\frac{\sinh k(z+d)}{\sinh kd}\cos(kx - \omega t)$ $- \frac{3H^2\pi^3}{T^2\lambda}\frac{\sinh 2k(z+d)}{\sinh^4 kd}\cos[2(kx - \omega t)]$
Hydrostatic pressure	$p_s = -\rho g z - \frac{\rho g\pi H^2}{4\lambda\sinh 2kd}\{\cosh[2k(z+d)] - 1\}$
Dynamic pressure	$p = \frac{\rho g H}{2}\frac{\cosh k(z+d)}{\cosh kd}\cos(kx - \omega t) + \frac{3\pi\rho g H^2}{4\lambda\sinh 2kd}$ $\times\left[\frac{\cosh 2k(z+d)}{\sinh^2 kd} - \frac{1}{3}\right]\cos[2(kx - \omega t)]$
Wave celerity	$c = \left(\frac{g}{k}\tanh kd\right)^{1/2}$

Ursell (1953) investigated the accuracy of Stokes second order theory by comparing the amplitude of the second order term to the amplitude of the first order term. He has generalized this comparison and expressed it in terms of the Ursell parameter, defined by

$$U_R = \frac{\eta_0}{\lambda}\left(\frac{\lambda}{d}\right)^3 \tag{3.24}$$

where η_0 is the maximum elevation above the still water level. When the Ursell parameter is very small, linear wave theory is valid. However, one should note that, although the Ursell parameter is a useful guide, it is not the sole measure of determining the relative importance of the nonlinear terms. In shallow water, for instance, the relative amplitude H/d becomes the more important parameter.

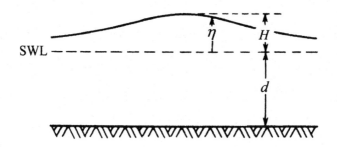

Figure 3.7 Profile of the solitary wave.

Solitary Theory

The solitary wave profile is illustrated in Figure 3.7. This wave has no trough since its profile never extends below the still water level. Solitary wave theory describes such a wave of infinite length that propagates in water of uniform depth. Technically, the solitary wave is the limiting case of a periodic shallow water wave of finite height in which the wave length approaches infinity while the relative height, H/d, is maintained constant. The usual method of generating a solitary wave in the laboratory is by adding a finite volume of water at one end of a closed tank.

Russell (1845) derived the celerity for a solitary wave as $c = \sqrt{g(H + d)}$. Most analyses of the solitary wave superpose the wave celerity $-c$ into the field of fluid motion, thereby reducing the moving solitary wave to a stationary wave for which the origin of the coordinate axes translates with a velocity $-c$. The conditions to be satisfied are those of continuity, equation (3.11b), and zero vorticity, equation (3.11a). The analysis begins with the definition of the velocity potential ϕ, or

$$u = \frac{\partial \phi}{\partial x}; \qquad w = -\frac{\partial \phi}{\partial z} \tag{3.25}$$

It is seen that equations (3.25) satisfy the continuity condition, equation (3.11b). When equations (3.25) are combined with the zero vorticity condition, equation (3.11a), the result is Laplace's equation, or

$$\frac{\partial^2 \phi}{\partial x^2} + \frac{\partial^2 \phi}{\partial z^2} = 0 \tag{3.26}$$

The boundary condition at the free surface is

$$(u - c)\frac{\partial p}{\partial x} + w\frac{\partial p}{\partial z} = 0 \tag{3.27}$$

It is noted that the pressure at the surface is

$$p(x, \eta, t) = p_a \tag{3.28}$$

In general, p must satisfy the following form of Bernoulli's equation:

$$\frac{p}{\rho} - uc + \frac{1}{2}(u^2 + w^2) + gz = \text{constant} \tag{3.29}$$

If the moving horizontal axis is denoted by ξ, where $\xi = x - ct$, then the potential function ϕ, the surface water elevation η, and the vertical water particle velocity w must vanish for large ξ. For these conditions, and provided that H/d is less than approximately 0.7, the solutions for ϕ and η are as follows (Lamb, 1945):

$$\phi = -cd\frac{N}{M}\left[\frac{\sinh M(\xi/d)}{\cos M(1 + z/d) + \cosh M(\xi/d)}\right] \tag{3.30}$$

$$\eta = H\left[\text{sech}\sqrt{\frac{3H}{4(d+H)}}\frac{\xi}{d}\right]^2 \tag{3.31}$$

Here M and N are dimensionless parameters given implicitly by the following two equations:

$$N = \frac{2}{3}\sin^2\left[M\left(1 + \frac{2H}{3d}\right)\right] \tag{3.32}$$

$$\frac{H}{d} = \frac{N}{H}\tan\left[\frac{1}{2}M\left(1 + \frac{H}{d}\right)\right] \tag{3.33}$$

When H and d are specified, M and N are computed by trial and correction from the last two equations, and the particle velocity components are computed directly from equation (3.30) using equation (3.25). It is noted that for sinusoidal waves η is periodic and changes sign, whereas for solitary waves η is always positive. This is consistent with the solitary wave form shown in Figure 3.7.

Further analysis shows that the maximum value of H/d is equal to 0.78, which occurs when $u = c$. Also the expressions for particle velocities and surface elevation are compatible with linear theory when H/d approaches zero. Munk (1949) summarized the work on solitary waves and proposed a modification applicable to periodic waves.

Numerical Theory

With the advent of high speed, high capacity computers, and with the development of efficient programming and numerical techniques, numerical wave theories have become increasingly popular. Such theories, which are more accurately described as procedures, are all based on deterministic solutions of the flow field equations, with statistical features sometimes incorporated in the solution procedures. The distinguishing features are the different treatments of the boundary conditions and the alternative criteria defining accuracy. One popular numerical theory is based on ideal (non-viscous) fluid in plane flow as described by Laplace's equation (3.26). A historical computer code that solves the Navier-Stokes equations for an incompressible fluid, and includes the effects of fluid viscosity, was developed by Hirt et al. (1975).

To illustrate a widely used numerical theory, consider the free stream function approach developed by Dean (1965). The fluid is assumed to be non-viscous, incompressible, and irrotational, with motion limited to the x, z-plane. In this case, the governing differential equation of Laplace can be written in terms of the stream function ψ, or

$$\frac{\partial^2 \psi}{\partial x^2} + \frac{\partial^2 \psi}{\partial z^2} = 0 \qquad (3.34)$$

The velocities in terms of ψ and the velocity potential ϕ are

$$u = -\frac{\partial \psi}{\partial z} = -\frac{\partial \phi}{\partial x}; \qquad w = \frac{\partial \psi}{\partial x} = -\frac{\partial \phi}{\partial z} \qquad (3.35)$$

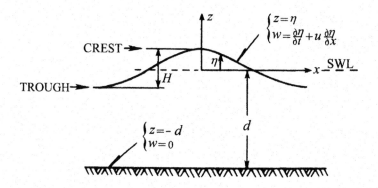

Figure 3.8   Wave boundary conditions used in stream function theory (Dean, 1965).

The boundary conditions to be satisfied are shown in Figure 3.8 and are summarized:

1. At the sea floor where $z = -d$, the boundary is flat, horizontal, and impermeable. Thus

$$w = \frac{\partial \psi}{\partial x} = -\frac{\partial \phi}{\partial z} = 0 \qquad (3.36)$$

2. At the free surface where $z = \eta$, the particles remain on that surface, or

$$\frac{\partial \eta}{\partial t} + u \frac{\partial \eta}{\partial x} = w \qquad (3.37)$$

This is known as the kinematic free surface boundary condition.

3. Also on the free surface at $z = \eta$ the pressure is uniform. This implies that Bernoulli's equation for unsteady flow must be satisfied, or

$$\eta + \frac{1}{2g}(u^2 + w^2) - \frac{1}{g}\frac{\partial \phi}{\partial t} = \text{constant} \qquad (3.38)$$

This is known as the dynamic free surface boundary condition.

If the wave propagates without change in form, then a uniform velocity field of magnitude c can be imposed on the field of motion (Dean, 1967). The boundary conditions of equations (3.37) and (3.38) are thus reduced to

$$\frac{\partial \eta}{\partial t} = \frac{w}{u - c} \qquad (3.39)$$

With this last result, Bernoulli's equation for steady flow becomes

$$\eta + \frac{1}{2g}[(u - c)^2 + w^2] = \text{constant} \qquad (3.40)$$

The algorithm used by Dean (1967) to solve for η, u, and v is outlined as follows. Assume a form of ψ, in terms of undetermined coefficients, which satisfies Laplace's equation (3.34) and also satisfies the boundary conditions of equations (3.36) and (3.37). The form of the solution is such that the coefficients, the wave number, and the free surface value of the stream function are computed based on a least squares fit to the dynamic free surface boundary condition, equation (3.38). This free stream function theory can also be used to compute the characteristics of nonsymmetrical waves for which the surface profiles are specified. Published tables (Dean, 1974) aid in applications. However, such results can be computed directly using numerical methods.

The use of nonlinear wave theory is summarized. Trochoidal wave descriptions, which include fluid rotation, were historically used by naval architects, but are not generally used by offshore structural engineers. Stokes approximate theory is practical for describing short waves of finite height; but this theory becomes cumbersome and impractical for long waves of finite height. Fortunately, alternative theories (cnoidal, solitary, numerical) have been developed for this latter case. The numerical theory based on free stream functions is appropriate for deepwater waves of finite height, for nonsymmetrical waves, and for shallow water waves for which the application of linear theory is often inappropriate.

3.4   DOMAINS OF VALIDITY FOR WAVE THEORIES

A question that frequently arises concerns the selection of a wave theory for a given situation. Unfortunately, there are several bases for evaluating these various theories. No consensus has yet emerged as to a common basis. Dean (1974) and Le Mehaute (1976) have studied the problem and provided Figures 3.9 and 3.10, respectively, to aid in selecting an appropriate theory. Here H is the wave height (twice the amplitude A of Figure 3.2); H_B is the value of H when the wave breaks; and d is the water depth.

Dean's study of 1974 is a quantitative one, based on the closeness of fit of the kinematic dynamic free surface boundary conditions. It is limited to waves lying between the upper breaking limit curve and the lower curve labeled $H = H_B/4$ in Figure 3.9. We see that within this range there are three possibilities: the cnoidal first order, the linear, and the free stream function fifth order theory. If it is desired to use only analytic theories, then the range of application of the cnoidal theory may be extended to cover all of that region lying below the lower dashed curve; the Stokes fifth order may be employed for that region lying below the upper dashed curve; and the linear theory may be extended for application to the region lying between the two dashed curves.

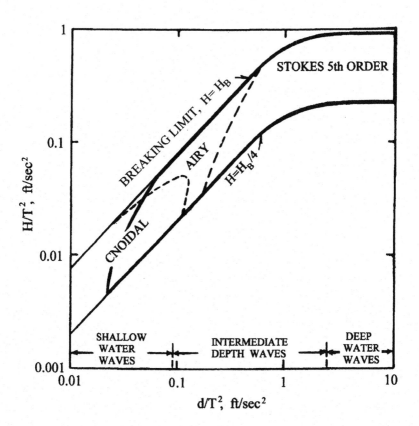

Figure 3.9 Periodic wave theories providing best fit to dynamic free surface boundary conditions. The region of validity for the free stream function fifth order theory is encompassed by the bold boundary lines (Dean, 1974).

Le Mehaute's study of 1976 appears to be based somewhat on subjective considerations, although it does cover the entire range of physically possible waves of permanent form. Only the analytic theories are shown in Figure 3.10. Here, L denotes the wave length λ. When the stream function fifth order is included, there is some overlapping with the higher-order Stokes theory domains and the cnoidal theory domain. Le Mehaute's graph is particularly useful during preliminary engineering calculations since it indicates the possibility of employing simple theories easily solvable on hand-held calculators.

The following numerical examples illustrate the utility of these graphs by Dean and Le Mehaute.

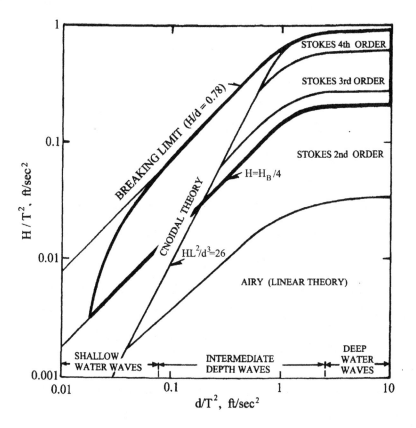

Figure 3.10   Limits of validity for selected wave theories. The region of validity for the free stream function fifth order theory is encompassed by the bold boundary lines (Le Mehaute, 1969).

Example Problem 3.1.   Find the maximum horizontal and vertical water particle velocity at an elevation of 10 ft below the still waterline, in a wave that has a period of 8 sec, a height peak to trough of 2 ft, and is propagating in a constant water depth of 200 ft.

1.   Deduce the appropriate theory by first computing the following flow parameters:

$$d/T^2 = 200/8^2 = 3.125 \text{ ft/sec}^2$$

$$H/T^2 = 2/8^2 = 0.03125 \text{ ft/sec}^2$$

From Figure 3.10, we can see that this wave can be described by linear wave theory, and that it is a deepwater wave.

2. Compute the wave length λ and wave number k by combining equations (3.2), (3.7), and (3.17). In the last of these equations, use the approximation for deep water: $\tanh kd \simeq 1$:

$$\lambda = \frac{g}{2\pi} T^2 = \frac{32.2}{2\pi} 8^2 = 328 \text{ ft}$$

$$k = \frac{2\pi}{\lambda} = \frac{2\pi}{328} = 0.0192 \text{ ft}^{-1}$$

3. Calculate u and w from linear wave theory using Table 3.1:

$$u = \frac{\pi H}{T} \frac{\cosh\left[k(d+z)\right]}{\sinh kd} \cos\left(kx - \omega t\right)$$

This value is a maximum when the cosine term is $+1$. Note that for $z = -10$ ft, $d + z = 190$ ft. Substitute the numerical values to obtain the maximum horizontal water particle velocity, or

$$u_{max} = \frac{\pi(2)}{8} \frac{\cosh\left[0.0192(190)\right]}{\sinh\left[0.0192(200)\right]} = 0.649 \text{ ft/sec}$$

From Table 3.1, we observe that the maximum value of w, the vertical water particle velocity, occurs when the sine term is $+1$. Thus

$$w_{max} = \frac{\pi H}{T} \frac{\sinh\left[k(d+z)\right]}{\sinh kd} = \frac{\pi(2)}{8} \frac{\sinh\left[0.0192(190)\right]}{\sinh\left[0.0192(200)\right]} = 0.648 \text{ ft/sec}$$

We observe that the maximum particle velocities are very nearly identical, where the difference is due to roundoff errors associated with the hyperbolic functions. Further, these maximums do not occur at the same time. Profiles of u_{max} and w_{max} can be generated by varying z from 0 to -200 ft; and the entire history of the wave flow can be mapped by varying x and t in the trigonometric functions for u and w.

Example Problem 3.2.   Determine the surface profile and the pressure variation at an elevation of 30 ft above the seabed in a wave that has a period of 6 sec, a height of 3 ft, and is propagating in a constant water depth of 50 ft.

1. Deduce the appropriate theory by first computing the following two flow parameters:

$$d/T^2 = 50/6^2 = 1.389 \text{ ft/sec}^2$$

$$H/T^2 = 3/6^2 = 0.083 \text{ ft/sec}^2$$

From Figure 3.10, we see that this is an intermediate depth wave and that Stokes second-order theory is appropriate.

2. Compute the wave length λ, the wave number k, and the circular frequency ω. For this intermediate depth wave, use the dispersion relation, equation (3.17), to compute the wave length. Using $k = 2\pi/\lambda$ from equation (3.2) and $c = \lambda/T$ from equation (3.7), the dispersion relation can be expressed implicitly in terms of λ and the given wave parameters, or

$$\lambda = T\sqrt{\frac{g\lambda}{2\pi} \tanh\left(\frac{2\pi d}{\lambda}\right)} = 6\sqrt{\frac{32.2}{2\pi}\lambda \tanh\left(\frac{2\pi(50)}{\lambda}\right)}$$

This equation, when solved for λ by trial and correction, leads to the following results:

$$\lambda = 174.6 \text{ ft}; \qquad k = 2\pi/\lambda = 0.0360 \text{ ft}^{-1}; \qquad \omega = k\lambda/T = 1.048 \text{ rad/sec}$$

3. Use the appropriate expressions from Table 3.2, together the given values: $H = 3$ ft, and $d = 50$ ft. With the values of λ, k, and ω just computed, the surface elevation η is thus:

$$\eta = 1.5 \cos\left(0.036x - 1.048t\right) + 0.0502 \cos\left(0.072x - 2.096t\right)$$

These results for the surface elevation show that, due to the second term on the right, there is approximately a 3 percent correction to the linear theory due to second order effects. The results for the pressure p at the depth $z = 30 - 50 = -20$ ft in water of mass density $\rho = 1.94$ slug/ft^3 are found by substituting the foregoing numerical results into the full expression for p in Table 3.2. In the numerical results below, the four terms on the right side of the equation, from left to right, represent, respectively: the hydrostatic pressure of linear theory, or $-\rho g z = -1.94(32.2)(-20) = 1249.36$ lb/ft^2; the dynamic pressure of linear theory; the second order dynamic correction to the linear theory; and the second order hydrostatic correction to linear theory:

$$p = 1249.36 + 49.52 \cos\left(kx - \omega t\right) + 0.07 \cos 2(kx - \omega t) - 0.47 \text{ lb/ft}^2$$

Again we observe that the second order corrections to linear theory theory are very small, less than 1 percent in this example problem.

Example Problem 3.3. Determine the celerity c for a wave with a period of 15 sec and a wave height of 2 ft, propagating in water 20 ft deep.

1. Deduce the appropriate wave theory by computing the following parameters:

$$d/T^2 = 20/15^2 = 0.0889$$

$$H/T^2 = 2/225 = 0.00889$$

By referring to Figures 3.9 and 3.10, we see that both the cnoidal and free stream function theories are appropriate for describing this wave, for which computer-aided solutions are needed.

2. Rather than generate our own numerical results, refer to results of Masch and Wiegel (1961). That is,

Compute: $T\sqrt{g/d} = 15\sqrt{32.2/20} = 19.03$

Table 2: $c^2/(gd) = 0.9675$

Compute: $c = 24.96$ ft/sec

3. As a check on this result, we find that the wave characteristics place it near Cases 3D and 4D given by Dean (1974), but Dean's tabulated values do not encompass the numerical values of the present problem.

4. From linear theory the wave celerity is $c = \sqrt{gd} = 25.37$ ft/sec.

5. From solitary theory the wave celerity is $c = \sqrt{g(d+H)} = 26.62$ ft/sec.

These last two values of c are reasonably close to the value 24.96 ft/sec obtained from cnoidal theory. This does not imply, however, that the other flow parameters are necessarily in close agreement, and those parameters need to be calculated from cnoidal or free stream function theory.

Linear wave theory will be employed extensively in subsequent chapters to calculate fluid loading on offshore structures.

PROBLEMS

3.1   For simple wave swells with periods $T = 2\pi/\omega = \lambda/c$, $1 < T < 30$ sec, where the depth to wave length ratio satisfies $d/\lambda > 0.5$, the waves are defined as deepwater waves. In this case show that the wave celerity and period have the following approximate forms:

$$c = \left(\frac{g\lambda}{2\pi}\right)^{1/2}; \qquad T = \frac{2\pi c}{g}$$

If $T = 10$ sec and $\lambda = 512$ ft, calculate the wave celerity and the minimum depth for which these relationships are valid.

3.2   A simple, shallow water wave is defined when the ratio of water depth to wavelength satisfies $d/\lambda < 0.05$. In this case, show that the wave celerity is given approximately by $c = \sqrt{gd}$. If $\lambda = 512$ ft, what is the maximum water depth for which this relationship holds?

3.3   For simple linear waves of length λ at intermediate water depths d, then $0.05 < d/\lambda < 0.5$. Deduce that the square of the wave celerity is given by the following formula.

$$c = \frac{g\lambda}{2\pi} \tanh \frac{2\pi d}{\lambda}$$

Plot three curves corresponding to $\lambda = 200$ ft, 400 ft, and 600 ft, which show the wave celerity as a function of water depth in this intermediate range.

3.4   A deepwater wave having a length of 512 ft and a height of 20 ft propagates from deep to shallow water over the sea bottom which has a slope of 1:30. Determine the wave length and celerity as a function of depth.

3.5   In order to collect site-specific oceanographic data, a tower is installed in a water depth of 80 ft. One gage located 50 ft above the bottom senses an average maximum dynamic pressure of 150 lb/ft^2 at a period of 9 sec. Using these data, compute the height and length of the corresponding wave. Is your computed wave height the maximum that is present at the site? Explain.

3.6   We know that surface tension forces become more important as the wave lengths decrease. In particular, as wave lengths approach 1 in., the surface tension forces become important relative to the gravity forces. We wish to make a model study of a semisubmersible platform (200 ft × 200 ft) as it is excited by waves with periods ranging from 4 to 15 sec. What is the minimum scale ratio that is acceptable? What factors other than surface tension effects should be considered in determining this scale ratio?

3.7   Suppose that a deepwater wave having a period of 10 sec and a height of 10 ft is propagating normally toward shore over a gently sloping beach. Determine the depth of water and the wave length when the wave breaks. (Hint: Remember that as the wave length changes, the height must also change.)

REFERENCES

Airy, G. B., On Tides and Waves, *Encyclopoedia Metropolitana* **5**, 1845.

Dean, R. G., Stream Function Representation of Nonlinear Ocean Waves, *Journal of Geophysical Research* **70** (18), 1965.

Dean, R. G., Relative Validities of Water Wave Theories, *ASCE Proceedings: First Conference on Civil Engineering in the Oceans*, San Francisco, CA, 1967.

Dean, R. G., Evaluation and Development of Water Wave Theories for Engineering Application: Presentation of Research Results, Special Report No. 1, U.S. Army Corps of Engineers, Coastal Engineering Research Center, Ft. Belvoir, VA, 1974.

Etube, L. S., *Fatigue and Fracture Mechanics of Offshore Structures*, Professional Engineering Publishing Limited, London and Bury St. Edmunds, Suffolk, UK, 2001.

Gerstner, F., Theorie der Wellen, *Abhandlungen der koniglichen bohmischen Gesellschaft der Wissenschaften*, Prague, 1802.

Havelock, T. H., *The Propagation of Disturbances in Dispersive Media*, Cambridge University Press, London, 1914.

Hirt, C. W., Nichols, B. D., and Romero, N. C., SOLA-A Numerical Solution Algorithm for Transient Fluid Flows, Report LA-5852, Los Alamos Scientific Laboratory, 1975.

Ippen, A.T., ed., *Estuary and Coastline Hydrodynamics*, McGraw-Hill, New York, 1966.

Kinsman, B., *Wind Waves, Their Generation and Propagation on the Ocean Surface*, Prentice-Hall, Englewood Cliffs, NJ, 1965.

Korteweg, D. J., and de Vries, G., On the Change of Form of Long Waves Advancing in a Rectangular Canal, and On a New Type of Long Stationary Waves, *Philolophical Magazine*, fifth series **39**, 1895.

Lamb, H., *Hydrodynamics*, Dover, New York, 1945.

Laplace, P. S., de, Sur la vitesse du son dans fair et dans l'eau, *Annales de chimie et de physique* **2** (3), 1816.

Le Mehaute, B., *An Introduction to Hydrodynamics and Water Waves, Vol. 2: Water Wave Theories*, Essa Technical report ERL 118-POL-3-2, U.S. Dept. of Commerce, Environmental Science Services Administration, Pacific Oceanographic Laboratories, Miami, FL, July, 1969.

Le Mehaute, B., *An Introduction to Hydrodynamics and Water Waves*, Springer-Verlag, Dusseldorf, 1976.

Levi-Civita, T., Determination rigoureuse des ondes permanentes d'ampleur finie mathematica **93**, 264-314, 1925.

Masch, F. D., and Wiegel, R. L., *Cnoidal Waves: Tables of Functions*, Council on Wave Research, The Engineering Foundation, University of California, Richmond, CA, 1961.

Munk, W. H., The Solitary Wave and Its Application to Surf Problems, *Annals of New York Academy of Science* **51** (3), 1949.

Rankine, W. J. M., On the Exact Form of Waves near the Surface of Deep Water, *Philolophical Magazine*, London, 1863.

Russell, J. S., *Report of the Committee on Waves*, Meeting of the British Association for Advancement of Science, 1845.

Sarpkaya, T., and Isaacson, M., *Mechanics of Wave Forces on Offshore Structures*, Van Nostrand Reinhold, New York, 1981.

Stokes, G. G., On the Theories of the Internal Friction of Fluids in Motion, and of Motion of Elastic Solids, *Transactions, Cambridge Philosophical Society* **8** (287), 1845.

Stokes, G. G., On the Theory of Oscillatory Waves, *Transactions, Cambridge Philosophical Society* **8**, 1847.

Struik, D. J., Determination rigoureuse des ondes irrotationnelles periodiques dans un canal a profondeur finie, *Mathematische Annalen* **95**, 595-634, 1926.

Ursell, F., The Long Wave Paradox in the Theory of Gravity Waves, *Proceedings, Cambridge Philosophical Society* **49** (4), 1953.

Young, I. R., *Wind Generated Ocean Waves*, Elsevier Science Ltd., Oxford, UK, 1999.

4

Wave Forces on Structures

James F. Wilson

One approach used to estimate the highest expected wave forces on an offshore structure is based on a single design wave. For a particular wave theory, with a wave height and wave period chosen according to the location of the structure, the corresponding pressure field and horizontal components of wave particle velocity and acceleration are then determined. With this flow information, the distributions of the two governing flow parameters Re and Kc (the Reynolds and Keulegan-Carpenter numbers) are found for the structural components; the flow regime is determined; and the appropriate fluid force coefficients for drag, inertia, and diffraction (wave scattering) are chosen from a database. The structural loading is then computed using these latter coefficients, together with the expressions for wave velocity and acceleration applied to either Morison's loading model, or a modified version thereof, or to a diffraction model. This approach is now illustrated for cases where the fluid and structural motion are limited to the x, z-plane, and where the flows are normal to longitudinal axes of the structural elements, usually right circular cylinders, the basic element of offshore structures. Identified are the flow regimes appropriate for the experimental coefficients C_D, C_M, and the diffraction coefficients, with a brief summary of the uncertainties in these coefficients. Also defined are transfer functions, or functions that relate the wave velocities and accelerations to the structural forces. These transfer functions are employed in later chapters to calculate the responses of linearly-behaving structures to actual sea states.

4.1   WAVE LOADING OF FLEXIBLE CYLINDERS

Suppose that a cylinder in a wave field has sufficient flexibility such that its horizontal velocity v and horizontal acceleration $\dot{v}$ are often significantly higher than the corresponding quantities u and $\dot{u}$ for the water wave. In this case, Berge and Penzien (1974) suggested a modification of Morison's equation (2.14) in which u is replaced by the relative velocity $(u - v)$, and $\dot{u}$ is replaced by the relative acceleration $(\dot{u} - \dot{v})$. Recall *Example Problem 2.3*. A load model that accounts for both cylinder flexibility and the effects of the water wave length λ can be written as follows:

$$\bar{q} = C_{M1}\frac{\pi}{4}\rho D^2 \dot{u} + C_{M2}\frac{\pi}{4}\rho D^2(\dot{u} - \ddot{v}) + C_D\frac{1}{2}\rho D(u - \dot{v})\,|u - \dot{v}| \qquad (4.1)$$

The flow coefficients recommended by Moan et al. (1975) are given by

$$C_M = C_{M1} + C_{M2} \qquad (4.2\text{a})$$

$$C_{M1} = 1 - 0.12\frac{\pi D}{\lambda} \qquad (4.2\text{b})$$

$$C_{M2} = 1.0, \quad \text{for} \quad \frac{\pi D}{\lambda} < 0.5 \qquad (4.2\text{c})$$

$$C_{M2} = 1.54 - 1.08\frac{\pi D}{\lambda}, \quad \text{for} \quad \frac{\pi D}{\lambda} > 0.5 \qquad (4.2\text{d})$$

The following two example problems are applications of this formulation.

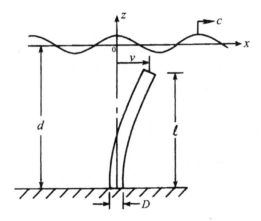

Figure 4.1 A submerged, flexible cantilevered cylinder.

Example Problem 4.1.   A uniform, flexible, cantilevered cylinder of actual mass per unit length $\bar{m}_0$, length ℓ, and diameter D is fully submerged in water of depth d, as shown in Figure 4.1. Calculate the total horizontal force on this structure using equation (4.1). Neglect all drag forces. Then formulate the equation of motion based on the coordinate v, the displacement at the top of the cylinder. Assume linear wave theory. For the horizontal wave particle acceleration $\dot{u} = \dot{u}(x, z, t)$ given in Table 3.1, let $x = 0$. Note that the origin is at the still water line and that z is positive upward.

The total horizontal force $p_1(t)$ is given by integrating equation (4.1) over the length of the cylinder. For $C_D = 0$, this is

$$p_1(t) = \int_{-d}^{(\ell-d)} \bar{q} \, dz$$

$$= C_{M1}\frac{\pi}{4}\rho D^2 \int_{-d}^{(\ell-d)} \dot{u} \, dz + C_{M2}\frac{\pi}{4}\rho D^2 \int_{-d}^{(\ell-d)} (\dot{u} - \dot{v}) dz \qquad (4.3)$$

The stiffness of the cylinder, equal in magnitude to the horizontal static force applied to the top which will produce a unit horizontal deflection, is easily shown to be $k_1 = 3EI/\ell^3$. The equivalent lumped mass, computed in chapter 5, is $m_0 = 0.227\bar{m}_0\ell$. For a conservative (or high) estimate of v, lump $p_1(t)$ at the top of the cylinder, or the coordinate point, $z = \ell - d$. For a material damping constant of c_1, the equation of motion becomes

$$m_0\ddot{v} + c_1\dot{v} + k_1 v = p_1(t) \qquad (4.4)$$

It is further assumed that $\dot{u}$ is much larger than $\ddot{v}$ along the cylinder and that $\ddot{v}$ is independent of the coordinate z, consistent with this single degree of freedom model. Using equation (4.3) with the mass and stiffness results just obtained, equation (4.4) becomes

$$\left(0.227\bar{m}_0\ell + C_{M2}\frac{\pi}{4}\rho D^2\ell\right)\ddot{v} + c_1\dot{v} + \frac{3EI}{\ell^3}v = C_M\frac{\pi}{4}\rho D^2 \int_{-d}^{(\ell-d)} \dot{u} \, dz \qquad (4.5)$$

Using $\dot{u}$ from Table 3.1 (with $x = 0$), the integral in equation (4.5) is easily evaluated. It is seen from this last result that C_{M2} is actually the added mass coefficient C_A. The virtual mass, or the coefficient of $\ddot{v}$ in equation (4.5), thus arises in a natural way in this formulation.

Example Problem 4.2. Solve *Example Problem 4.1* shown in Figure 4.1, but this time include the fluid drag term. Then linearize the resulting equation of motion following the method discussed by Berge and Penzien (1974).

The steps in the solution are summarized as follows. The equation of motion is

$$m_0\ddot{v} + c_1\dot{v} + k_1 v = \int_{-d}^{(\ell-d)} \bar{q} \, dz = p_1(t) \qquad (4.6)$$

The loading term is

$$\bar{q} = C_M\frac{\pi}{4}\rho D^2\dot{u} - C_{M2}\frac{\pi}{4}\rho D^2\ddot{v} + C_D\sqrt{2/\pi}\rho D\left(u - \dot{v}\right)\sigma \qquad (4.7a)$$

in which σ, the standard deviation of the relative velocity $(u - \dot{v})$, is

$$\sigma = \left[\frac{1}{T_0}\int_0^{T_0} (u - \dot{v})^2 dt\right]^{1/2} \qquad (4.7b)$$

Here, T_0 is the time required for several oscillations of the cylinder. When equations (4.6) and (4.7a) are combined, the result is

$$\left(m_0 + C_{M2} \frac{\pi}{4} \rho D^2 \ell \right) \ddot{v} + \left[c_1 + C_D \sqrt{2/\pi} \rho D \ell \, \sigma \right] \dot{v} + k_1 v$$

$$= C_M \frac{\pi}{4} \rho D^2 \int_{-d}^{(\ell-d)} \dot{u} \, dz + C_D \sqrt{2/\pi} \rho D \, \sigma \int_{-d}^{(\ell-d)} u \, dz \qquad (4.8)$$

where m_0 and k_1 are given in *Example Problem 4.1*.

With the values of u and $\dot{u}$ from Table 3.1, together with the chosen system constants, economical numerical solutions to equation (4.8) can be computed by iteration. That is, with an initial guess for the standard deviation σ, solve equation (4.8) for v; recalculate σ; and solve again for v. Continue this procedure until $\sigma \simeq$ const. Suitable convergence can be obtained after four or five iterations (Berge and Penzien, 1974). Numerical solutions to the differential equation (4.8) can be generated using the software package Mathematica[R] (1999).

The results of the two dynamic models expressed by equations (4.5) and (4.8) lead to two important conclusions regarding the fluid-structural interactions. First, the added mass term is the same, whether or not velocity-dependent fluid drag is present. Second, system damping is increased with the addition of fluid drag, as is seen by comparing the coefficients of $\dot{v}$ on the left sides of equations (4.5) and (4.8).

4.2   CLASSIFICATION OF FLUID LOAD REGIMES

The equation or method for calculating the load on a cylindrical structure in a fluid wave flow field depends on the flow regime. Hogben (1976) states:

> Loads on structures in waves may be conveniently classified under three headings: drag, inertia and diffraction. The relative importance of these in a particular case depends on the type and size of the structure and the nature of the wave conditions. Broadly it may be said that drag loads are the result of flow separation induced by the relative velocity of the fluid and are most significant for tubular components of small diameter in waves of large height. Inertial loads are due to the pressure gradient associated with the relative acceleration of the ambient fluid and are most significant for structural components of large sectional dimensions. Diffraction forces are due to scattering of the incident wave by the structure and are only significant when the sectional dimensions are a substantial fraction of the wave length.

The following relationships between cylinder diameter D, the wave height H (peak to trough), and the wave length λ serve to define the flow regimes more precisely.

1. **Drag:** $D/H < 0.1$. In Morison's equation the term involving C_D dominates the term involving C_M. Wave loads on the very small diameter components of offshore structures such as conductor tubes will be drag-dominated. In such cases $\bar{q} \simeq \bar{q}_D$, where $\bar{q}_D$ is given by equation (2.10).

2. **Inertia:** $0.5 < D/H < 1.0$. In Morison's equation the term involving C_M dominates the term involving C_D. The columns supporting the deck of a gravity-type platform are designed to sustain wave forces in this regime. In such cases $\bar{q} \simeq \bar{q}_I$ where $\bar{q}_I$ is given by equation (2.7).

3. **Diffraction:** $D/\lambda > 0.2$. Wave forces on stationary bodies in this flow regime were discussed in detail by Hogben (1976) and Sarpkaya and Isaacson (1981). Generally, diffraction force calculations are based on the Froude-Krylov pressure distributions derived from ideal, hydrodynamic flow and linear wave theory. These forces are then modified by the experimental flow coeffcients C_h, C_v, and C_0.

Consider the application of diffraction theory to the submerged cylinder shown in Figure 4.2.

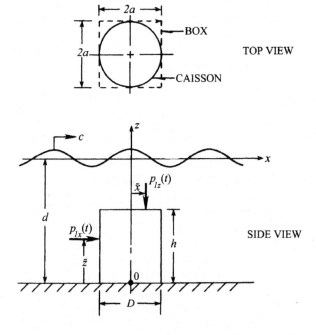

Figure 4.2   Diffraction wave loading on a submerged cylinder or box-type caisson.

For this example, the total horizontal force, vertical force, and overturning moment on the cylinder are

$$p_{1x}(t) = C_h F_{kx}; \qquad p_{1z}(t) = C_v F_{kz} \tag{4.9}$$

$$M_0(t) = C_0 \cdot (\bar{z} F_{kx} + \bar{x} F_{kz}) = C_0 M_k \tag{4.10}$$

The terms F_{kx} and F_{kz}, defined as the Froude-Krylov forces, are the net pressure-induced forces on the vertical sides and on the top horizontal surface, respectively. Those forces, located at the respective centers of pressure $\bar{z}$ and $\bar{x}$, are time-dependent. The basic assumption in the calculation of the Froude-Krylov forces is that the wave pressure field is completely undisturbed by the presence of the structure. Hogben and Standing (1975) recommended the following flow coefficients for a submerged cylinder of diameter D and height h such as shown in Figure 4.2. The water depth is d and the length of the incident wave is λ.

$$C_h = 1 + 0.75 \left(\frac{h}{D} \right)^{1/3} \left[1 - 0.3 \left(\frac{\pi D}{\lambda} \right)^2 \right] \tag{4.11}$$

$$C_v = 1 + 0.74 \left(\frac{\pi D}{\lambda} \right)^2 \frac{h}{D}, \qquad \text{for} \quad \frac{\pi h}{\lambda} < 1 \tag{4.12a}$$

$$C_v = 1 + \frac{\pi D}{2\lambda}, \qquad \text{for} \quad \frac{\pi h}{\lambda} > 1 \tag{4.12b}$$

$$C_0 = 1.9 - 0.35 \frac{\pi D}{\lambda} \tag{4.13}$$

The restrictions on equations (4.11)-(4.13) are

$$\frac{h}{d} < 0.6, \quad \text{for } C_h, C_v, C_0 \tag{4.14}$$

$$0.3 < \frac{h}{D} < 2.3, \quad \text{for } C_h, C_v \text{ only} \tag{4.15}$$

$$0.6 < \frac{h}{D} < 2.3, \quad \text{for } C_0 \text{ only} \tag{4.16}$$

With the complementary results for F_{kx}, F_{kz}, and M_k derived in closed form by Sarpkaya and Isaacson (1981), the total diffraction loading for vertical cylinders can be calculated from equations (4.9) and (4.10).

Economical calculations for diffraction forces can be made on noncylindrical shapes such as the submerged tank cluster of the monotower shown in Figure

2.2. In this procedure, the irregular shape is simply replaced by a rectangular box of the same overall dimensions as the irregular shape, a procedure that greatly simplifies the calculation of the wave pressure loads. The justification for this procedure was given by Hogben and Standing (1975), who found that the total diffraction forces and moments are practically independent of structural planform. The following example, parallel to the analysis of Nataraja and Kirk (1977), illustrates this type of calculation.

Example Problem 4.3.   Calculate F_{kx} and M_k, the Froude-Krylov horizontal force and overturning moment on a rectangular box caisson resting on the ocean floor. The box has cross section dimensions $2a \times 2a$ and height h, as shown in Figure 4.2. Assume that there is a single incident wave and that linear wave theory is appropriate. Relative to the coordinate system of Figure 4.2, the dynamic pressure given in Table 3.1 becomes

$$p(x, z, t) = \rho g \frac{H}{2} \frac{\cosh k(z + d)}{\cosh kd} \cos(kx - \omega t) \tag{4.17}$$

in which $k = 2\pi/\lambda$ and ω is given by equation (3.16). The value of F_{kx} is computed by integrating the pressure difference over the vertical sides of the box normal to a right-traveling wave, or

$$F_{kx} = 2a \int_{-d}^{(h-d)} [p(-a, z, t) - p(a, z, t)]dz \tag{4.18}$$

With equation (4.17), the last equation is integrated and use is made of trigonometric identities, which leads to

$$F_{kx} = -2\rho g a H \frac{\sinh kx}{k \cosh kd} \sin ka \sin \omega t \tag{4.19}$$

The overturing moment is the sum of two integrals. The first is M_{kh}, the moment about 0 due to the side pressure forces. The second is M_{kv}, the moment about 0 due to the pressure forces on the top of the box. Thus

$$M_{kh} = 2a \int_{-d}^{(h-d)} [p(-a, z, t) - p(a, z, t)](z + d)dz \tag{4.20}$$

$$M_{kv} = 2a \int_{-a}^{a} xp[x, (h - d), t]dx \tag{4.21}$$

where

$$M_k = M_{kh} + M_{kv} \tag{4.22}$$

The evaluation of this latter moment in closed form using equations (4.17), (4.20), and (4.21) is a straightforward exercise. With these results, the total horizontal diffraction force $p_{1x}(t)$ and the overturning moment $M_0(t)$ are then given by equations (4.9) and (4.10), using the respective flow coefficients of equations (4.11) and (4.13). The calculation of F_{kz} and its corresponding total vertical load $p_{1z}(t)$ is straightforward. See Problems 4.13 and 4.14.

4.3   FLOW REGIMES FOR OFFSHORE STRUCTURES

A perspective on the size of components (caissons, legs, bracings, etc.) for typical offshore structures is shown in Figure 4.3. The gravity and tethered buoyant platforms, the most massive of the offshore structures, will experience diffraction forces under most wave conditions. At the other extreme, the jackup platforms with their relatively thin legs and small diameter bracing will experience mainly drag forces.

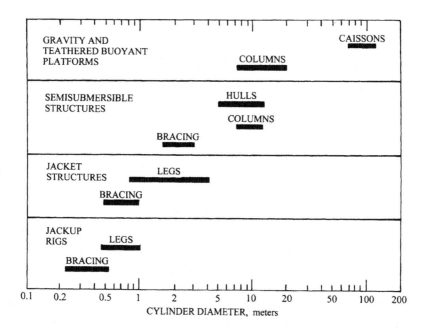

Figure 4.3   Size perspective for cylindrical components of offshore structures (Adapted from Hogben, 1976).

A further insight into the influence of wave height and cylinder diameter on the occurrence of the three flow regimes was presented by Hogben (1976). Such a mapping, shown here only for cylinders at the water surface ($z = 0$), is illustrated in Figure 4.4. The following assumptions were used in deriving these curves.

1. Linear wave theory, Table 3.1.
2. Deep water approximation, where $\omega^2 \simeq gk$.
3. A wave length to wave height ratio of $\lambda/H = 15$.
4. $C_M = 2.0$ for all flow regimes. $C_D = 0.6$ for *postcritical* drag where the drag force changes abruptly at $\mathrm{Re} = 5 \times 10^5$. $C_D = 1.2$ for *subcritical* drag.
5. Re based on the amplitude of the wave's horizontal particle velocity.

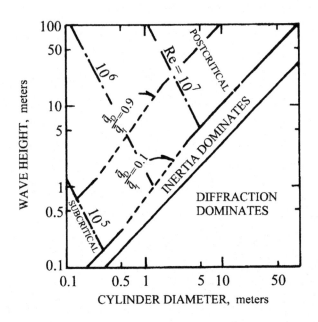

Figure 4.4   Regimes of forces on cylinders located at the sea surface, $z = 0$
(Adapted from Hogben, 1976).

The five main observations of this study are summarized. First, gravity-type platforms experience diffraction at wave heights up to 30 m, the upper extreme for $\lambda = 450$ m. Second, as the depth of the cylinder below the surface is increased, the boundary of the diffraction regime remains fixed, but the boundary of the inertia regime of Figure 4.4 expands upward, a result shown in other studies for which $z > 0$ (Hogben, 1976). Third, towers supporting the deck of a platform generally lie in the inertia flow regime in the extreme wave conditions usually assumed for design purposes. Fourth, loads on very small diameter components such as those of conductor tubes, legs of jackup platforms, and jacket-template structures are generally drag-dominated, but in deeper waters such as in the North Sea (150 m or more in depth) the main structural legs will be large enough to incur both inertial and drag loads. Fifth, the ratio $\bar{q}_D / \bar{q}_I$ in Figure 4.4 shows the relative effect of fluid drag forces to fluid inertia forces. When this ratio is 0.9 (the upper dashed curve), the drag and inertia forces are of comparable magnitude; but when this ratio is 0.1 (the lower dashed curve), the inertia forces dominate.

Because of the inherent assumptions in Hogben's 1976 study, Figure 4.4 for $z = 0$ and his companion results for $z > 0$ should not be used directly to design offshore structures. However, calculated results presented in a similar form, but based on flow coefficients and wave parameters appropriate to the site of the offshore structure, can be quite useful in engineering design.

4.4 SUMMARY OF FLOW COEFFICIENTS C_D AND C_M

A survey of measured values for C_D and C_M made by the British Ship Research Association (1976) is depicted in Figure 4.5. This chart is a composite of historical data for smooth, vertical cylinders. This chart shows the wide variability of these flow coefficients and their dependency on the Reynolds number, the Keulegan-Carpenter number, and the wave height to cylinder diameter ratio, H/D. For the curves showing the detailed behavior of C_D and C_M in the subcritical regime of Reynolds number, the reader is referred to the original report of Keulegan and Carpenter (1958) and the summaries by Sarpkaya and Isaacson (1981). For this lower range of Re, both C_D and C_M depend strongly on Kc.

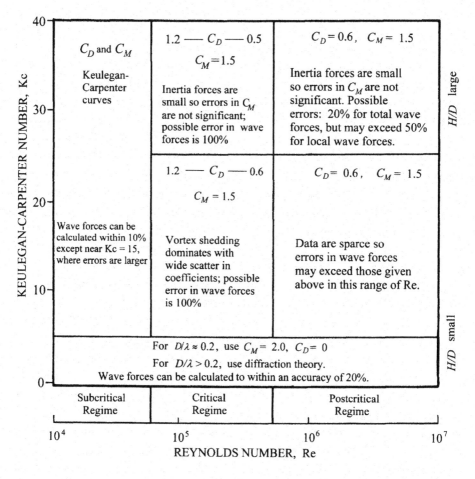

Figure 4.5 Summary of C_D and C_M for smooth, vertical cylinders (British Ship Research Association, 1976).

The most important aspects of the data in Figure 4.5 are summarized.

1. The C_D and C_M values apply only to smooth cylinders in deepwater for which $d/(gT^2) > 0.003$, where d is the water depth, T is the wave period, and g is the acceleration due to gravity.

2. No allowances were made for wave slamming or fluid interactions with other structural members in proximity to the smooth cylinder.

3. No allowances were made for current-wave interactions.

4. Re and Kc were not always well defined in this data survey. This may account somewhat for the wide scatter in the reported results for these coefficients.

5. The C_D and C_M values listed in this table should be used with a wave theory appropriate to environmental conditions, where the wave theory chosen is used as a basis for calculating Re and Kc. Approximate values of these latter parameters are generally adequate for estimating the flow coefficients.

4.5   TRANSFER FUNCTIONS FOR WAVE LOADING

The basis for the formulation of the transfer functions in this section is linear theory for a single water wave, often called a simple wave. Recall that for linear theory, Table 3.1, the wave characteristics are defined by its height $H = 2A$, its period $T = 2\pi/\omega$, and its wave number $k = 2\pi/\lambda$ where λ is its wave length. Recall also that for deep water waves where the water depth $d > 0.5\lambda$, then $k = \omega^2/g$. See Problem 3.1.

The transfer function $G(\omega)$ is defined as the function that relates H of the incident wave to the load it imparts to the structural component. In general, transfer functions are defined in the following harmonic form:

$$G(\omega) = G_0 e^{j\omega t}, \quad \text{for} \quad j = \sqrt{-1} \tag{4.23}$$

Here G_0 is a complex number independent of time. Transfer functions for linear dynamic systems only are defined here, systems for which the drag and restraining force terms are linearized. Since a simple wave is harmonic, the loading functions $\bar{q}$, $p_1(t)$, and M_0 can also be written in the form of equation (4.23). The connection between a loading function and its transfer function is

$$\frac{\text{loading function}}{\text{wave height}} = \text{real}\{G(\omega)\} \tag{4.24}$$

The transfer function for a loading $\bar{q}$ is calculated, for instance, by first finding $\bar{q}/H$, and then casting that result in the form of equation (4.23). The following identities are useful in such a procedure:

$$e^{j\alpha} = \cos \alpha + j \sin \alpha \tag{4.25a}$$

$$\cos(\alpha + \beta) = \cos \alpha \cos \beta - \sin \alpha \sin \beta \tag{4.25b}$$

$$\sin{(\alpha + \beta)} = \sin{\alpha}\cos{\beta} + \cos{\alpha}\sin{\beta} \tag{4.24c}$$

where α and β are real numbers.

The square of the modulus of $G(\omega)$, needed later for statistical response studies of structures, is defined by

$$|G(\omega)|^2 = G(\omega) \cdot G^*(\omega) \tag{4.26}$$

Here $G^*(\omega)$ is the complex conjugate of $G(\omega)$ and is formed by replacing j by $-j$ in the transfer function. The calculation of transfer functions is now illustrated with two example problems.

Example Problem 4.4. Calculate the transfer function for the flexible, cantilevered cylinder of Figure 4.1. Assume that the fluid-induced forces are inertia dominated and that the motion of the cylinder is much smaller than the motion of the water particles. Thus

$$\bar{q} = \frac{\pi}{4}\rho D^2 \dot{u} \tag{4.27}$$

The horizontal wave particle acceleration from Table 3.1, corresponding to $x = 0$ or the average location of the horizontal wave particles on the cantilever, is

$$\dot{u} = -\frac{H}{2}\omega^2 \frac{\cosh{k(z+d)}}{\sinh{kd}}\sin{\omega t} \tag{4.28}$$

Using equation (4.28) with (4.27), it follows that

$$\frac{\bar{q}}{H} = -\frac{\pi}{8}C_M D^2 \rho\omega^2 \frac{\cosh{k(z+d)}}{\sinh{kd}}\sin{\omega t} \tag{4.29}$$

With equation (4.25a), the latter ratio yields the transfer function as

$$G(\omega) = j\frac{\pi}{8}C_M D^2 \rho\omega^2 \frac{\cosh{k(z+d)}}{\sinh{kd}}e^{j\omega t} \tag{4.30}$$

The square of the modulus, calculated from equations (4.26) and (4.30), is

$$|G(\omega)|^2 = \left[\frac{\pi}{8}C_M D^2 \rho\omega^2 \frac{\cosh{k(z+d)}}{\sinh{kd}}\right]^2 \tag{4.31}$$

It is noted that the expression for $G(\omega)$ just derived is for $\bar{q}$ and is a function of the location z on the cantilevered beam. The transfer function required for the lumped mass model, equation (4.4), must be based on the *total* load $p_1(t)$. This latter transfer function can be derived by simply integrating equation (4.30) over the range of $z = -d$ to $z = (\ell - d)$. The resulting transfer function and its corresponding modulus will then be independent of z.

Example Problem 4.5. Calculate the transfer function for the total horizontal load on the box-type caisson shown in Figure 4.2 and described in *Example Problem 4.3*. Assume that the fluid-induced forces are diffraction-dominated. It follows that the total horizontal load $p_{1x}(t)$ is the product of F_{kx} given by equation (4.19) and the diffraction coefficient C_h given by equation (4.11). Thus the total load-wave height ratio is

$$\frac{p_{1x}(t)}{H} = -2\rho g a C_h \frac{\sinh kh}{k \cosh kh} \sin ka \sin \omega t \qquad (4.32)$$

The required transfer function is this ratio expressed in complex notation, or

$$G(\omega) = j2\rho g a C_h \frac{\sinh kh}{k \cosh kh} \sin ka \; e^{j\omega t} \qquad (4.33)$$

These examples illustrate that there is a different transfer function for each flow regime and for each structural component of an offshore structure. In practice, transfer functions for all components are calculated and assembled for the structure, all based on linear wave theory and a single, simple wave. Such results can then superimposed to account for the loading effects of many simple waves selected to simulate the design sea state at the site of the offshore structure. This simulation involving the selection of simple waves over particular ranges of H, ω, and wave phase ε is discussed in Chapter 6. The corresponding structural responses are then discussed in Chapter 7 for single degree of freedom systems and in Chapter 9 for multi-degree of freedom structures.

PROBLEMS

4.1   The total force $p_1(t)$ on a submerged, flexible pile in the inertia flow regime is given by the right side of equation (4.3). For a single, simple wave evaluate $p_1(t)$ by carrying out the necessary integration.

4.2   The diameter of the submerged pile shown in Figure 4.1 is sufficiently small so that the drag forces are comparable in magnitude to the inertia forces. Evaluate $p_1(t)$ in this case by carrying out the integrations of equation (4.8).

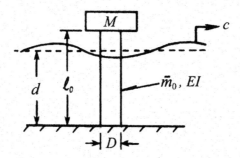

Figure 4.6   Monopod structure for Problems 4.3, 4.4, and 4.5.

4.3   Shown in Figure 4.6 is a simplified model of a cantilevered monopod structure in water of depth d. The deck mass is M, modeled as a point mass at height ℓ_0 above the sea floor. The leg has a diameter D, a uniform mass per unit length of $\bar{m}_0$, and a bending stiffness EI. The leg is subjected to a simple plane wave whose respective horizontal water particle velocity and acceleration are much larger than those of the leg. Neglect drag forces and also neglect any reduction in the leg bending stiffness due to the deck mass. Derive for this structure a single degree of freedom model for the motion of the deck, an equation similar to equation (4.5). Define explicitly the system mass and stiffness. Carry out the integrations needed to describe the wave load, $p_1(t)$.

4.4   Solve Problem 4.3, but now include drag forces on the leg. Where possible, carry out the integrations for wave loading.

4.5   Suppose that the structure of Figure 4.6 is flexible enough so that its horizontal water particle velocity and acceleration are of the same order as those of a simple, incident wave. Deduce the corresponding equation of motion that includes drag forces. Identify the mass, damping, stiffness, and fluid loading terms for the linearized form similar to equation (4.8). Carry out the integrations where possible. Then write down a brief outline of a numerical, computer-aided procedure to calculate the time history of the horizontal displacement for the platform.

4.6   Deduce the form of the wave loading term of *Example Problem 4.1* for which there are N simple waves. The i-th values of frequency, wave height, and wave phase are ω_i, H_i and ε_i respectively, where $i = 1, 2, \ldots, N$. The phase ε_i is added to the argument of the harmonic terms in u and $\dot{u}$ of Table 3.1.

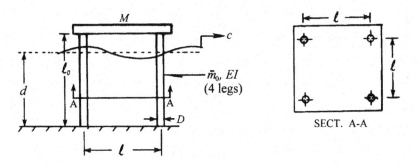

Figure 4.7   Platform structure for Problem 4.7.

4.7   The unbraced platform shown in Figure 4.7 has four legs, each of diameter D and length ℓ_0, separated by distance ℓ. The legs are subjected to a single, simple plane wave for which $\lambda \gg D$. Neglect the fluid drag forces and also neglect the motion $(\dot{v}, \ddot{v})$ of the structure compared to the respective horizontal motion parameters $(u, \dot{u})$ of the wave particles. Derive the expression for wave loading in each leg, accounting for the relative magnitude of the separation length ℓ compared to the wavelength λ. Then discuss the net wave-induced force on the whole structure for the three cases: $\ell/\lambda \ll 1$, $\ell/\lambda = 0.5$, and $\ell/\lambda = 1$. If the *net* structural load is zero for any of these cases, explain why the individual wave loads on the legs can still be significant. Sketch possible leg deformation patterns for each of these three cases.

4.8   Search the recent literature, such as the annual *Proceedings of the Off-shore Technology Conference*, to determine whether diffraction coefficients such as C_h, C_v, and C_0 have been measured for fixed objects other than vertical, right circular cylinders.

4.9   For a deep water wave, show that equation (3.16) reduces to $\omega^2 = gk$. Then, for a vertical cylinder of diameter D and height h, express the diffraction coefficients of equations (4.11) through (4.13) in terms of g, D, h, and ω only.

4.10   Determine an efficient analytical or numerical procedure to calculate wave diffraction loads on a rigid, right circular conical monopod leg of an ice-resistant offshore platform. Consult the classical references of Hogben (1976) and also recent literature on the subject.

4.11   The legs of a jackup platform are cylinders of diameter 0.5 m. Each leg is subjected to a single, simple wave, with each impact occurring separately and at a different time. For each ratio $\lambda/H = 10$, 15, and 20, determine the range of H for which the drag, inertia, and diffraction dominates.

4.12   Based on the data of Sarpkaya and Isaacson (1981), deduce the values C_M and C_D for cylinders in the flow regime of subcritical Reynolds numbers, where $10 < \text{Kc} < 40$. That is, add typical flow coefficients to the left column of Figure 4.5.

4.13   Calculate the explicit expressions for the Froude-Krylov force F_{kz}, and the overturning moment M_k for the solid box caisson of Figure 4.2.

4.14   With the results of *Example Problems 4.3* and *4.5*, together with Problem 4.13, complete the calculations for the three most important wave load transfer functions on the box caisson: those for the total horizontal load, the vertical load, and overturning moment. Use complex notation to describe $G(\omega)$ in each case; then calculate each value of $|G(\omega)|^2$.

REFERENCES

Berge, B., and Penzien, J., Three-Dimensional Stochastic Response of Offshore Towers to Wave Forces, OTC-2050, *Proceedings of the Offshore Technology Conference*, 1974.

British Ship Research Association, *A Critical Evaluation of Data on Wave Force Coefficients*, Report No. 278.12, August 1976.

Hogben, N., Wave Loads on Structures, *Behavior of Offshore Structures (BOSS)*, Norwegian Institute of Technology, Oslo, 1976.

Hogben, N., and Standing, R. G., Experience in Computing Wave Loads on Large Bodies, OTC-2189, *Proceedings of the Offshore Technology Conference*, 1975.

Keulegan, G. H., and Carpenter, L. H., Forces on Cylinders and Plates in an Oscillating Fluid, *Journal of Research of the National Bureau of Standards* **60** (5), 1958.

Mathematica®, version 4, Wolfram Media, Inc., Champaign, IL, 1999.

Moan, T., Haver, S., and Vinje, T., Stochastic Dynamic Response Analysis of Offshore Platforms, with Particular Reference to Gravity Platforms, OTC-2407, *Proceedings of the Offshore Technology Conference,* 1975.

Nataraja, R., and Kirk, C. L., Dynamic Response of a Gravity Platform under Random Waves, OTC-2904, *Proceedings of the Offshore Technology Conference,* 1977.

Sarpkaya, T., and Isaacson, M., *Mechanics of Wave Forces on Offshore Structures,* Van Nostrand Reinhold, New York, 1981.

5

Deterministic Responses for Single Degree of Freedom Structures

James F. Wilson

Once the single degree of freedom dynamic model for an offshore structure is formulated and the loading conditions are identified, the structure's response characteristics are calculated using its equation of motion. The response characteristics are of two types: the natural frequency, and the time history of displacement $v(t)$ or of rotation $\theta(t)$. The frequency calculation is generally necessary to assure the integrity of the structural design. This is because an offshore structure in depths exceeding 70 m generally has a natural frequency falling in the range of the expected wave frequencies, which may lead to a dangerous condition of structural resonance.

The peak displacement or rotation response is sought where several possible extreme environmental loading conditions are applied, generally one at a time. These loads include the effects of high winds, waves, and also earthquake excitation. Typically, the peak dynamic response for each loading is then compared to its static response, or its response if the same loading were applied very slowly. This response ratio, dynamic to static, is called by several names, including the dynamic amplification factor, the dynamic load factor, and the impact factor. This response ratio can be applied to the expected static loads for design purposes.

In this chapter, natural frequencies and dynamic responses of selected linear and nonlinear structural models are evaluated. For linear systems, the classical response functions due to harmonic and impulse loading are derived, followed by the response due to a general, time-dependent loading. The response analysis for nonlinear systems employs first order perturbation theory and numerical methods. Example problems illustrate the calculations of the natural frequency and the dynamic response for several types of offshore structures: a fixed-legged structure such as a jackup rig in response to earthquake excitation at the sea floor; and responses to a single design wave of a spread-moored ship and a SALM buoy, both of which are supported by nonlinear restraints.

5.1   NATURAL FREQUENCIES OF LINEAR SYSTEMS

In this section, two methods for calculating the natural frequency of linear single degree of freedom structural models are presented: the Direct Method, which is based on the equation of motion; and the Rayleigh Method, which is based on energy principles. Several example problems are presented that illustrate the relationships between these two methods.

Direct Method

The direct method of obtaining the natural frequency of a single degree of freedom structure is to use its equation of motion. The structure's natural frequency ω_0 is defined as the frequency compatible with an undamped structure of constant mass, with a restraint force that varies linearly with the displacement coordinate, and with no external excitation force. Under these conditions, the governing equation of motion, written in terms of the displacement coordinate v, is

$$m\ddot{v} + k_1 v = 0 \tag{5.1}$$

which is a special case of equation (2.2). When this linear structure is given an arbitrarily small displacement amplitude v_0 and then released, then v exhibits free harmonic oscillations of the form

$$v = v_0 \sin \omega_0 t \tag{5.2}$$

When equation (5.2) and its second derivative are substituted into equation (5.1), the result is

$$(-m\omega_0^2 + k_1)v_0 \sin \omega_0 t = 0 \tag{5.3}$$

In the latter equation, the term $\sin \omega_0 t$ is not zero for all time t and thus the term in brackets must be zero. This leads the following equation for the natural frequency of the structure:

$$\omega_0 = \sqrt{\frac{k_1}{m}} \tag{5.4}$$

It is noted that the natural frequency is independent of the initial displacement amplitude v_0, which is characteristic of linear systems.

Example Problem 5.1.   Shown in Figure 5.1a is a spherical buoy of mass m, which is half submerged in water in its static equilibrium. The buoy is depressed vertically by a small initial amplitude v_0 and released, after which it undergoes free oscillations along the vertical coordinate v. Derive the equation of motion for the buoy, and from that compute its natural frequency in units of rad/sec and in Hz. The buoy's radius is $R = 3.5$ ft and the water density is $\gamma_w = 64$ lb/ft^3. Neglect system damping and the fluid drag forces. Let $C_M = 0$.

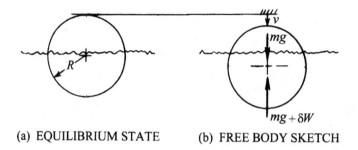

(a) EQUILIBRIUM STATE   (b) FREE BODY SKETCH

Figure 5.1   Spherical buoy in vertical oscillation, *Example Problem 5.1.*

Shown in Figure 5.1b is the free body sketch of the buoy with a vertical displacement v. In the static state, $v = 0$ and according to Archimedes principle, the weight of the buoy floating at its equator is given by the weight of the water displaced by half of the sphere's volume, or $mg = 2\pi R^3 \gamma_w/3$. Also by Achimedes principle, the incremental upward force δW for a downward buoy displacement v corresponding to an approximate incremental water volume displacement of $\pi R^2 v$, is $\delta W = \pi R^2 \gamma_w v$. When Newton's second law is applied to the buoy of Figure 5.1b, the equation of motion for the buoy becomes

$$mg - (mg - \delta W) = m\ddot{v} \tag{5.5}$$

After substituting the loads and rearranging, the above equation becomes

$$\left(\frac{2}{3g}\pi R^3 \gamma_w\right)\ddot{v} + \left(\pi R^2 \gamma_w\right) v = 0 \tag{5.6}$$

When equation (5.2) is used with the above result, the natural frequency of the buoy is computed as

$$\omega_0 = \sqrt{\frac{3g}{2R}} = \sqrt{\frac{3(32.2)\,\text{ft/sec}^2}{2(3.5)\,\text{ft}}} = 3.71\,\text{rad/sec} \tag{5.7}$$

In alternative units, the natural frequency is $f_0 = \omega_0/(2\pi) = 0.591$ Hz.

Example Problem 5.2.   Consider the plane rocking motion of the monopod concrete gravity platform which was first shown in Figure 2.2 and discussed in *Example Problem 2.2.* This structure with typical nominal dimensions is depicted in Figure 5.2. Compute the undamped rocking frequency for this monotower as it interacts with its soil foundation. Assume that the structure is rigid and that the rotations θ in the plane are small, or less than about 10 deg. The equation for its free vibration is a special case of the more general equation (2.6) in which the excitation forces on the right side are zero and on the left side $\sin\theta = \theta$, $f(\dot\theta) = 0$, and $q(\theta) = k_\theta\theta$ (the soil foundation reaction moment). Thus, equation (2.6) reduces to

$$J_0\ddot\theta + (k_\theta - m_0 g h_G + m_b g h_b)\theta = 0 \qquad (5.8)$$

According to equation (2.78), the soil's restoring moment coefficient is $k_\theta = a_0 - \omega b_0$, where a_0 and b_0 are soil constants. Here, $\omega = \omega_0$. Thus, the latter equation of motion becomes

$$J_0\ddot\theta + (a_0 - \omega_0 b_0 - m_0 g h_G + m_b g h_b)\theta = 0 \qquad (5.9)$$

Assume a harmonic solution to equation (5.9), or

$$\theta = \theta_0 \sin\omega_0 t; \quad \ddot\theta = -\omega_0^2 \sin\omega_0 t \qquad (5.10)$$

where θ_0 is a constant. When θ and $\ddot\theta$ from equation (5.10) are substituted into equation (5.9), the result is a quadratic equation in ω_0, which can be solved for the positive root ω_0 using the quadratic formula. The result is

$$\omega_0 = \frac{1}{2J_0}\{[b_0^2 + 4J_0(a_0 - m_0 g h_G + m_b g h_b)]^{1/2} - b_0\} \qquad (5.11)$$

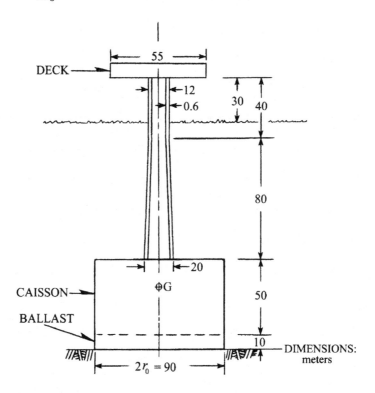

Figure 5.2   Model of a monopod gravity platform with typical dimensions.

Realistic system parameters used to compute numerical values for monopod platform rocking frequencies are listed in Table 5.1. The soil properties are those suggested by Nataraja and Kirk (1977), measured properties that span a range in the North Sea where gravity platforms have been placed. Computations for the structural properties were based on the nominal dimensions given in Figure 5.2, together with the following assumptions concerning the structure's five elementary structural components.

1. The deck and deck equipment are lumped as a thin disc 55 m in diameter.

2. The uniform section of the leg above the still water line is an empty, thin-walled pipe.

3. The uniform submerged section of the leg is a thin-walled pipe filled with water.

4. The tapered, submerged section of the leg is a thin-walled cone filled with water.

5. The caisson is approximated as a cylinder consisting of two parts. Part (a) is a cluster of vertical, cylindrical tanks 50 m high, together displacing 70 percent of the volume of seawater occupied by a cylinder of height $z = 50$ m and radius $r_0 = 45$ m. These tanks are filled with oil, and the average density of the tanks with their contents is 900 kg/m^3. This tank cluster is assumed to be a homogeneous solid of this density. Part (b), the base and ballast, is a homogeneous cylindrical solid 10 m high with a density of 2000 kg/m^3.

The following items were calculated for each of these five elementary structural components: the actual mass and the buoyant mass, with their respective locations from the base point; and the virtual mass moment of inertia about 0, the base centerline point. Also, for the submerged elements, the added mass coefficient was chosen as unity ($C_A = 1$). The properties of the solids in Appendix A, together with the parallel axis theorem, equation (2.5), were used to compute J_0. In these calculations, the mass density of sea water and of concrete were chosen as 1025 kg/m^3 and 2500 kg/m^3, respectively. More detailed calculations for this same problem were presented by Wilson and Orgill (1984).

The composite values of m_0, m_b, h_G, and h_b for the whole structure, together with the rocking frequencies calculated from equation (5.11), are listed in Table 5.1. The important results of these calculations are summarized. First, since the total actual mass (3.56×10^8 kg) is greater than the total buoyant mass (2.59×10^8 kg), the structure will not float. Second, since the center h_G of the actual mass is 1.0 m below the center of buoyancy h_b, this structure is inherently stable under small motion. Note that the moment due to the buoyant force opposes that due to gravity, as shown in Figure 2.12b. Third, a lower bound on the rocking frequency based on $G_s = 10$ MPa is $\omega_0 = 1.41$ rad/s or $f_0 = 0.224$ Hz. Fourth, the upper bound on the rocking frequency based on $G_s = 50$ MPa is $\omega_0 = 3.17$ rad/s or $f_0 = 0.505$ Hz. Finally, since the period T_0 is $1/f_0$, then the bounds on the natural period of the platform are 4.46 s and 1.98 s, which correspond to the reported lower and higher limits for the soil shear modulus in the North Sea.

Table 5.1 System Parameters and Results for *Example Problem 5.2*

Soil properties	$G_s = 10$ MPa to 50 MPa
	$\rho_s = 2000$ kg/m^3; $\nu_s = 0.33$
Soil parameters	$a_0 = 3.63 \times 10^{12}$ N·m for $G_s = 10$ MPa
	$b_0 = 4.97 \times 10^{11}$ N·m·s for $G_s = 10$ MPa
Structural mass, actual	$m_0 = 3.56 \times 10^8$ kg
Structural mass, buoyant	$m_b = 2.59 \times 10^8$ kg
Center of actual mass	$h_G = 30.7$ m
Center of buoyant mass	$h_b = 31.7$ m
Structural inertia, virtual	$J_0 = 1.46 \times 10^{12}$ N·m·s
Minimum rocking frequency	$\omega_0 = 1.41$ rad/s
Maximum rocking frequency	$\omega_0 = 3.17$ rad/s

The Rayleigh Method

In formulating the mathematical models for the cross beam and the jackup drilling rig in *Example Problems 2.7* and *2.8*, only a fraction f_1 of each of the structure's flexible mass was lumped at the coordinate point v. By comparing the expression for natural frequency ω_0 derived by the Direct Method in those examples to that derived by the Rayleigh Method that follows, f_1 can be computed. It will be shown that the accuracy of f_1 depends on a judicious choice of the fundamental mode shape of structural vibration, a choice that is tempered by the beam's end constraints, such as a clamped or free end condition.

Since fundamental mode shapes and end conditions are of basic importance in the Rayleigh Method, consider first some sample shapes for beam-type structures. For an accurate calculation of ω_0, that mode shape should be a simple one with a minimum of reversals in curvature from one end of the beam to the other. Also, the chosen mode shape must match the beam's geometric boundary or end conditions. For instance, a simple form for the mode shape corresponding to the natural frequency in the transverse bending vibration of a cross brace clamped at each end is sketched in Figure 2.16b. One simple approximation to that mode shape is

$$\psi(x) = 1 - \cos \frac{2\pi x}{\ell} \tag{5.12}$$

which satisfies the geometric constraints imposed at the ends: for displacement, $\psi(0) = \psi(\ell) = 0$; and for slope, $\psi'(0) = \psi'(\ell) = 0$. Here (') denotes the operator (d/dx).

For the jackup platform of Figure 2.17, a judicious choice of mode shape is

$$\psi(x) = 1 - \cos \frac{\pi x}{\ell} \tag{5.13}$$

which satisfies the geometric constraints: $\psi(0) = \psi'(0) = \psi'(\ell) = 0$ and $\psi(\ell) = 2$. The last constraint is nonzero, which is consistent with a nonzero amplitude for the deck. If this jackup platform were modified so that the deck was instead hinged instead of clamped to each leg, then an approximate mode shape would be

$$\psi(x) = 1 - \cos\frac{\pi x}{2\ell} \tag{5.14}$$

which satisfies the geometric constraints of full fixity at the sea floor: $\psi(0) = \psi'(0) = 0$; and of nonzero displacement and slope at the deck level: $\psi(\ell) = 1$ and $\psi'(\ell) = \pi/(2\ell)$.

For the Rayleigh Method, the actual magnitudes of the nonzero end slopes and deflections are unimportant, provided that the chosen mode shape $\psi(x)$ satisfies the beam's geometric constraints. Then this method leads to an upper bound value for ω_0. A straightforward proof of this upper bound property was provided by Den Hartog (1947).

The steps of the Rayleigh Method used to calculate ω_0 for beam-type structures are summarized as follows:

1. Choose a simple mode shape $\psi(x)$ that satisfies the geometric boundary conditions of deflection and slope. Express the lateral displacement of the beam at location x along its length as

$$v = v(x, t) = v_0\,\psi(x)\sin\omega_0 t \tag{5.15}$$

where v_0 is an arbitrary constant.

2. From the latter equation, calculate $V = U + V_g$, the total potential energy for the beam. Here U is the elastic strain energy for bending of the beams (legs and/or cross braces) and V_g is the loss in potential energy due to a downward displacement of a mass, as for instance, the decrease in the gravitational energy of a deck mass m_d due to the lateral displacement of its vertical supporting legs.

3. Based on equation (5.15), calculate the total kinetic energy K: that for the beam plus that for its end mass, if any.

4. Calculate ω_0 by equating the expressions for the maximum kinetic energy (when $V = 0$) to the maximum potential energy (when $K = 0$), or

$$K_{\max} = V_{\max} \tag{5.16}$$

Before the Rayleigh method is illustrated, consider some general forms of system energy. From elementary beam theory, the strain energy for bending of a beam element of length dx at position x is

$$dU = \frac{EI}{2}\left(\frac{\partial^2 v}{\partial x^2}\right)^2 dx \tag{5.17}$$

where $v = v(x, t)$ is given by equation (5.15). The total potential energy of N identical beams each of length ℓ, and each with the identical displacement $v(x, t)$ is thus

$$U = \frac{N}{2}\int_0^\ell EI\left(\frac{\partial^2 v}{\partial x^2}\right)^2 dx \tag{5.18}$$

If the structure consists of vertical beams and a deck of weight of $m_d g$ at $x = \ell$, as for the jackup platform of Figure 2.17, then the decrease in the gravitational potential energy of the deck as the horizontal leg displacement increases is

$$V_g = -m_g g \, \Delta h \tag{5.19}$$

where Δh is the vertical drop in height of the deck due to curvature of the elastic beams (legs). The beam element of arc length ds shown in Figure 2.17b can be approximated using a binomial expansion, or

$$ds = (dx^2 + dv^2)^{1/2} \simeq dx \left[1 + \frac{1}{2} \left(\frac{\partial v}{\partial x} \right)^2 \right] \tag{5.20}$$

Thus the drop in height of this arc element is

$$ds - dx = \frac{1}{2} \left(\frac{\partial v}{\partial x} \right)^2 dx \tag{5.21}$$

from which the total drop of m_d is calculated by integrating over the length of the beam, or

$$\Delta h = \int_0^\ell (ds - dx) = \frac{1}{2} \int_0^\ell \left(\frac{\partial v}{\partial x} \right)^2 dx \tag{5.22}$$

With the latter result and equation (5.19), the decrease in the potential energy of the deck mass due to the lateral leg motion is thus

$$V_g = -\frac{1}{2} m_d g \int_0^\ell \left(\frac{\partial v}{\partial x} \right)^2 dx \tag{5.23}$$

Consider the kinetic energy for the beam and deck elements. For a submerged beam element of length dx with a virtual mass per unit length of $\bar{m}$, the kinetic energy is $\bar{m}(\partial v/\partial t)^2 dx/2$. (If the element is not submerged, replace $\bar{m}$ by $\bar{m}_0$, its actual mass per unit length.) The kinetic energy for a fully submerged beam such as in Figure 2.16a is thus

$$K = \frac{1}{2} \int_0^\ell \bar{m} \left(\frac{\partial v}{\partial t} \right)^2 dx \tag{5.24}$$

For N beams or legs only partially submerged, such as in Figure 2.17, the result is

$$K = \frac{N}{2} \int_0^d \bar{m} \left(\frac{\partial v}{\partial t} \right)^2 dx + \frac{N}{2} \int_d^\ell \bar{m}_0 \left(\frac{\partial v}{\partial t} \right)^2 dx \tag{5.25}$$

where d is the water depth. For this jackup platform there is an additional kinetic energy term due to the movement of the deck mass. If the rotational energy of the deck is small in comparison to that in translation along the v direction, then its kinetic energy is

$$K = \frac{1}{2} m_d \left(\frac{\partial v}{\partial t} \right)^2_{x=\ell} \tag{5.26}$$

Example Problem 5.3. Use the Rayleigh Method to determine ω_0 for the fully submerged cross beam clamped at each end, as shown in Figure 2.16a. From this frequency, deduce the lumped mass for the single degree of freedom model equivalent to the first mode of vibration.

1. Since $\psi(x)$ as given by equation (5.12) satisfies the geometric constraints, the harmonic, lateral displacement from equation (5.15) is

$$v(x,t) = v_0 \left(1 - \cos \frac{2\pi x}{\ell} \right) \sin \omega_0 t \tag{5.27}$$

2. Since a single beam ($N = 1$) defines this dynamic system, the total potential energy is due only to strain energy U. From $v(x,t)$ above and equation (5.18) it follows that

$$V_{\max} = U_{\max} = \frac{1}{2} \left(\frac{2\pi}{\ell} \right)^4 v_0^2 EI \int_0^\ell \cos^2 \frac{2\pi x}{\ell} dx = \frac{4\pi^4}{\ell^3} EI \, v_0^2 \tag{5.28}$$

for which $\sin \omega_0 t = 1$ was used to obtain the maximum energy.

3. The maximum kinetic energy for this beam, computed from equations (5.24) and (5.27) for $\sin \omega_0 t = 1$, is

$$K_{\max} = \frac{1}{2} \bar{m} v_0^2 \omega_0^2 \int_0^\ell \left(1 - \cos \frac{2\pi x}{\ell} \right)^2 dx = \frac{3}{4} \bar{m} v_0^2 \omega_0^2 \ell \tag{5.29}$$

4. When the energies of equations (5.28) and (5.29) are equated, then

$$\omega_0 = \sqrt{\frac{519.5 EI}{\ell^4 \bar{m}}} \tag{5.30}$$

Now compare this frequency with that derived for this same problem modeled previously in *Example Problem 2.7* as a lumped mass, single degree of freedom system. There, the equation for undamped motion without external excitation was

$$m\ddot{v} + \frac{192 EI}{\ell^3} v = 0 \tag{5.31}$$

which follows from equations (2.41) and (2.43). The total virtual mass of the *lumped* system is m and is a fraction f_1 of the total virtual mass $\bar{m}\ell$, or

$$m = f_1 \bar{m}\ell \tag{5.32}$$

With the last two equations, the natural frequency from equation (5.5) becomes

$$\omega_0 = \sqrt{\frac{192EI}{\ell^4 f_1 \bar{m}}} \tag{5.33}$$

By equating the frequencies of equations (5.33) and (5.30), then $f_1 = 0.370$. Note that this value of f_1 is tempered by the choice of $\psi(x)$. However, Den Hartog (1947) points out that the exact frequency for this problem is only 1.3 percent lower than that of equation (5.30). Thus, lumping 37 percent of the beam's virtual mass at midspan is a good approximation in this case.

Table 5.2   System Parameters and Results for *Example Problem 5.4*

Height of leg	$\ell = 3180$ in.
Depth of water	$d = 2880$ in.
Leg, Young's modulus (steel)	$E = 30 \times 10^6$ psi
Leg outside diameter	$D = 144$ in.
Leg inside diameter	$D_i = 140.25$ in.
Leg pipe weight density (steel)	$\gamma_p = 0.283$ lb/in.^{3}
Sea water weight density	$\gamma_w = 0.0375$ lb/in.^{3}
Deck weight	$m_d g = 1.02 \times 10^7$ lb
Second area moment, one leg	$I = \pi(D^4 - D_i^4)/4 = 2.114 \times 10^6$ in.^{4}
Added mass coefficient	$C_A = 1$
Virtual weight per unit length, one leg (submerged)	$\bar{m}g = \pi\gamma_p(D^2 - D_i^2)/4$ $+ 1 \cdot \gamma_w \pi D^2/4 = 1417$ lb/in.
Actual weight per unit length, one leg (above water)	$\bar{m}_0 g = \pi\gamma_p(D^2 - D_i^2)/4 = 237$ lb/in.
Acceleration due to gravity	$g = 386$ in./sec^2
Length parameter, equation (5.37)	$\ell' = 897$ in.
Length parameter, equation (5.38)	$\ell'' = 296$ in.

Equation (5.36): $\omega_0 = \left[\dfrac{3.88\times10^{-4}(1.86\times10^8 - 1.02\times10^7)}{9.88\times10^3 + 0.545\times10^3 + 26.4\times10^3}\right]^{1/2} = 1.36$ rad/sec

Example Problem 5.4.   Use the Rayleigh Method to calculate an equation for the natural frequency in sway for the three-legged jackup rig shown in Figure 2.17. Then compare that frequency with ω_0 computed by the Direct Method based on the equation of motion derived in *Example Problem 2.8*. Use the numerical parameters for this structure given in Table 5.2.

The solution using the four-step Rayleigh procedure is outlined below. It is left to the reader to verify the analytical solution and numerical results.

1.   Choose the mode shape $\psi(x)$ given by equation (5.13) for use in the lateral displacement function $v(x,t)$ of equation (5.15).

2.   With $v(x,t)$, calculate the maximum potential energy of the system based on the sum of U and V_g of equations (5.18) and (5.23). Let $N = 3$ and approximate the deck mass as a point mass concentrated at the top of the legs. This

leads to

$$V_{\max} = \frac{3}{2} \int_0^\ell EI \left(\frac{\partial^2 v}{\partial x^2}\right)^2_{\max} dx - \frac{1}{2} m_d g \int_0^\ell \left(\frac{\partial v}{\partial x}\right)^2_{\max} dx \qquad (5.34)$$

3. Using the same function $v(x,t)$, calculate the maximum kinetic energy based on the sum of equations (5.25) and (5.26). The result is

$$K_{\max} = \frac{3}{2} \int_0^d \bar{m} \left(\frac{\partial v}{\partial t}\right)^2_{\max} dx + \frac{3}{2} \int_0^\ell \bar{m}_0 \left(\frac{\partial v}{\partial t}\right)^2_{\max} dx$$

$$+ \frac{1}{2} m_d \left(\frac{\partial v}{\partial t}\right)_{\substack{\max \\ x=\ell}} \qquad (5.35)$$

4. Compute the natural frequency by equating the energies of Steps 2 and 3, and cast the result in the following form:

$$\omega_0 = \left[\frac{\left(3\pi^2 EI/\ell^2 - m_d g\right) \pi^2/(8\ell)}{3\bar{m}\ell' + 3\bar{m}_0 \ell'' + m_d}\right]^{1/2} \qquad (5.36)$$

Here, the length parameters are defined by

$$\ell' = \frac{3d}{8} - \frac{\ell}{2\pi} \sin\frac{\pi d}{\ell} + \frac{\ell}{16\pi} \sin\frac{2\pi d}{\ell} \qquad (5.37)$$

$$\ell'' = \frac{3}{8}(\ell - d) + \frac{\ell}{2\pi} \sin\frac{\pi d}{\ell} - \frac{\ell}{16\pi} \sin\frac{2\pi d}{\ell} \qquad (5.38)$$

This same problem but without the dead weight effect of the deck was previously modeled in *Example Problem 2.8* by equation (2.47), from which the equation for undamped, free vibrations is deduced as

$$[3\bar{m}f_1 d + 3\bar{m}_0(\ell - d)f_1 + m_d] \ddot{v} + 36\frac{EI}{\ell^3} v = 0 \qquad (5.39)$$

When equation (5.2) is used with the last equation, the resulting frequency is

$$\omega_0 = \sqrt{\frac{k_1}{m}} = \left[\frac{36EI/\ell^3}{3\bar{m}f_1 d + 3\bar{m}_0(\ell - d)f_1 + m_d}\right]^{1/2} \qquad (5.40)$$

In the last two equations, $\bar{m}$ is the virtual mass per unit length of each submerged leg, which is the sum of its actual mass per unit length $\bar{m}_0$ and its added mass per unit length, or

$$\bar{m} = \bar{m}_0 + C_A \rho \frac{\pi}{4} D^2 \qquad (5.41)$$

Several important features about the two system models can be deduced from their respective frequencies as given by equations (5.36) and (5.40). Those features are summarized as follows:

1. The Euler buckling load for this three-legged structure is $P_E = 3\pi^2 EI/\ell^2$. The Rayleigh model frequency of equation (5.36) shows that increasing the deck load reduces the effective bending stiffness of the legs. In fact, the structure buckles if the deck load is sufficiently high, a critical condition for which $\omega_0 = 0$ and $m_d g = P_E$.

2. If $P_E \gg m_d g$, then the bending stiffnesses for the two models are nearly the same, a result deduced by comparing the numerators of equations (5.36) and (5.40). That is, the coefficients of EI/ℓ^3 in these respective equations are $3\pi^4/8 = 36.5$ and 36.

3. If $P_E \gg m_d g$ and the deck mass is much greater than that of the legs, then the system mass is simply m_d for both models. However, such a design is probably unrealistic.

4. If the deck mass is of comparable magnitude to that of the legs, then the Rayleigh model shows that the virtual mass of the legs has a significant influence on ω_0. For instance, in the special case where $\ell \simeq d$, then the parameters of equations (5.37) and (5.38) become: $\ell' = 3\ell/8$, $\ell'' = 0$, and the virtual mass of the legs becomes $m = 3\bar{m}f_1\ell$. When the denominators of equations (5.36) and (5.40) are compared, then $3\bar{m}\ell' = 9\bar{m}\ell/8 = 3\bar{m}f_1\ell$, or $f_1 = 0.375$. This justifies the assumption in *Example Problem 2.8* in which 37.5 percent of the virtual mass for the submerged legs, together with 37.5 percent of the actual leg mass above water, was lumped with the total mass of the deck to formulate the single degree of freedom model. The recommended frequency expression for this case is thus

$$\omega_0 = \left[\frac{\left(3\pi^2 EI/\ell^2 - m_d g\right)\pi^2/(8\ell)}{3(0.375)\bar{m}\ell + m_d} \right]^{1/2}, \qquad \ell = d \qquad (5.42)$$

in which the virtual mass for each of the three legs is given by equation (5.41).

Realistic numerical data for a jackup platform with three steel pipes for legs are listed in Table 5.2. Listed also is the system frequency computed from equation (5.36): $\omega_0 = 1.36$ rad/sec or $f_0 = 0.217$ Hz. From the numerical values in the numerator of ω_0, it is observed that the deck weight has a small effect on the stiffness bending stiffness. However, numerical values in the denominator indicate that the leg mass is a significant portion of the system mass. In conclusion, it is noted that the fundamental period of oscillation of this platform is $T_0 = 1/f_0 = 4.61$ sec, which is well below the 12 to 15 sec periods of the highest energy offshore waves in a typical sea state.

5.2 FREQUENCIES FOR NONLINEAR STRUCTURES

In the last section it was demonstrated that for linear structural models with small motion, the free vibration frequency was independent of the amplitude of motion. Using first order perturbation theory, it will now be shown that for structures with nonlinear restoring forces, the free vibration frequency will be dependent on the amplitude of motion. In both the linear and nonlinear systems considered here, this amplitude of free vibration is imposed as an initial condition for the motion.

Consider the free, undamped motion of a virtual mass m with the independent coordinate v and with a nonlinear restoring force given by equation (2.69). The corresponding equation of motion is

$$m\ddot{v} + k_1 v + k_3 v^3 = 0 \qquad (5.43)$$

Recall that the restoring force term in the last equation was used to model the line stiffness of a spread moored ship. See *Example Problems 2.10* and *2.11*. Now rewrite this last nonlinear equation as

$$\ddot{v} + \omega_0^2 v + e v^3 = 0 \qquad (5.44)$$

in which $\omega_0^2 = k_1/m$ is the square of the natural frequency if $k_3 = 0$, and $e = k_3/m$. It is assumed that e is always small enough so that $ev^3 \ll \omega_0^2 v$ for all solutions $v = v(t)$.

An approximate solution to equation (5.44) can be derived as follows using classical perturbation theory (Cunningham, 1964). At time $t = 0$ the mass is displaced by the amplitude A and then released. At the instant of release, the initial velocity is zero. These two initial conditions are expressed as

$$v(0) = A; \qquad \dot{v}(0) = 0 \qquad (5.45)$$

Now assume that the solution to equation (5.44) and its corresponding frequency $\tilde{\omega}$ are

$$v = v_0(t) + e v_1(t) \qquad (5.46)$$

$$\tilde{\omega}^2 = \omega_0^2 + e f(A) \qquad (5.47)$$

In equation (5.46), the quantity $v_0(t)$ is a time-dependent displacement and should not be confused with the symbol v_0 used for the constant displacement amplitude in Section 5.1. When the quantity $v_1(t)$ and the amplitude function $f(A)$ are combined with the governing equation (5.43) and the result is regrouped in ascending powers of e, the result is

$$(\ddot{v}_0 + \tilde{\omega}^2 v_0)e^0 + [\ddot{v}_1 + \tilde{\omega}^2 v_1 - f(A)v_0 + v_0^3]e^1 + o(e^2) = 0 \qquad (5.48)$$

The coefficients of e^0 and e^1 are each equated to zero and terms of higher in e, or $o(e^2)$, are assumed to be small enough to be neglected. The result is the following two linear differential equations:

$$\ddot{v}_0 + \tilde{\omega}^2 v_0 = 0 \qquad (5.49)$$

$$\ddot{v}_1 + \tilde{\omega}^2 v_1 = f(A)v_0 - v_0^3 \tag{5.50}$$

A particular solution to equation (5.49) that satisfies the initial conditions of equations (5.45) is

$$v_0 = A \cos \tilde{\omega} t \tag{5.51}$$

This latter solution is then inserted into the right side of equation (5.50) and the following identity is used:

$$\cos^3 \tilde{\omega} t = \frac{3}{4} \cos \tilde{\omega} t + \frac{1}{4} \cos 3\tilde{\omega} t \tag{5.52}$$

This procedure leads to the general solution to equation (5.50), or

$$v_1 = B_1 \cos \tilde{\omega} t + B_2 \sin \tilde{\omega} t + \frac{A^3}{32\tilde{\omega}^2} \cos 3\tilde{\omega} t + \frac{t}{2\tilde{\omega}} \left[A f(A) - \frac{3}{4} A^3 \right] \tag{5.53}$$

where B_1 and B_2 are constants. In this solution, the last term on the right is called the *secular* term, and this term is observed to become unbounded as t becomes large. Unboundedness is not physically possible for this system so restrained, and therefore the coefficient of t in the secular term must vanish. Excluding the trivial case for $A = 0$, this reasoning leads to

$$f(A) = \frac{3}{4} A^2 \tag{5.54}$$

When this amplitude function is substituted into equation (5.47), with $e = k_3/m$, then

$$\tilde{\omega} = \left(\omega_0^2 + \frac{3k_3}{4m} A^2 \right)^{1/2} \tag{5.55}$$

This result clearly shows that the frequency of the nonlinear system increases due to the addition of the cubic restoring force term, assuming that $k_3 > 0$, and that this increase depends on the initial system displacement A. However, if $k_3 < 0$, then $\tilde{\omega} < \omega_0$.

Next v_1 is calculated from equation (5.53) by imposing the zero initial conditions $v_1(0) = \dot{v}_1(0) = 0$, with which the constants are evaluated as $B_1 = -A/(32\tilde{\omega}^2)$ and $B_2 = 0$. When this result and the solution for v_0, equation (5.51), are used with equation (5.46), the first-order perturbation solution becomes

$$v = A \cos \tilde{\omega} t - \frac{k_3 A^3}{32m\tilde{\omega}^2} (\cos \tilde{\omega} t - \cos 3\tilde{\omega} t) \tag{5.56}$$

This result shows that the free oscillation amplitude for the counterpart linear system ($k_3 = 0$) is distorted by a frequency component $3\tilde{\omega}$ when $k_3 > 0$.

Example Problem 5.5. Calculate ω_0 and $\tilde{\omega}$ for both surge and sway for the spread-moored ship described in *Example Problem 2.10*. Assume that the

ship is displaced first in the surge direction by an amplitude of 3.0 ft from
its equilibrium state and then is released from rest. Then repeat the same
calculation for the sway state. The numerical values of m, k_1, and k_3 for each
motion are given as the coefficients of equations (2.72) and (2.73). The results
for surge motion are

$$\omega_0 = \sqrt{\frac{k_1}{m}} = \sqrt{\frac{20,300}{3.52 \times 10^5}} = 0.240 \text{ rad/sec}$$

$$\tilde{\omega} = \sqrt{0.240^2 + \frac{3(400)3^2}{4(3.52 \times 10^5)}} = 0.255 \text{ rad/sec}$$

The results for sway motion are

$$\omega_0 = \sqrt{\frac{12,700}{6.12 \times 10^5}} = 0.144 \text{ rad/sec}$$

$$\tilde{\omega} = \sqrt{0.144^2 + \frac{3(950)3^2}{4(6.12 \times 10^5)}} = 0.177 \text{ rad/sec}$$

These results show that, although the surge frequency is increased by only
about 6 percent by the nonlinear restoring force term, the sway frequency is
increased by a significant amount, 22.9 percent. The calculation of the displacement responses using equation (5.56) is straightforward.

Example Problem 5.6.   Investigate the general behavior in free oscillations
of the nonlinear system modeled by equation (5.43). Consider two classes of
restraints: the *hardening* restraint for which the slope of the restoring force
$q(v) = k_1 v + k_3 v^3$ increases with v and $k_3 > 0$; and the *softening* restraint for
which the slope of $q(v)$ decreases with v and $k_3 < 0$. Softening restraint occurs
in a mooring line with clumped weights attached to it at the sea floor, where
these weights lift off the sea floor when the side displacement of the tethered
structure is sufficiently large (Wilson and Orgill, 1984). In particular, choose
several fixed ratios k_3/k_1, both positive and negative, and plot the free vibration
amplitude A vs. the frequency ratio $\tilde{\omega}/\omega_0$. Use ratios of k_3/k_1 in the range
given for the moored LST in *Example Problem 2.10* where k_3/k_1 ranges from
0.02 ft^{-2} to 0.1 ft^{-2}.

For ease of plotting and discussion, the governing equation (5.55) that relates
frequency to amplitude is recast in the following form:

$$A = \sqrt{\frac{4}{3}\frac{k_1}{k_3}\left(\frac{\tilde{\omega}^2}{\omega_0^2} - 1\right)} \tag{5.57}$$

The latter equation shows that for a *hard* restraint with $k_3/k_1 = 0.1$ and 0.02,
then A is a real number only for $\tilde{\omega}/\omega_0 \geq 1$. These results are shown in the right

half of Figure 5.3. For a *soft* restraint with $k_3/k_1 = -0.1$ and -0.02, then A is a real number only for $\tilde{\omega}/\omega_0 \leqq 1$. These results are shown in the left half of Figure 5.3.

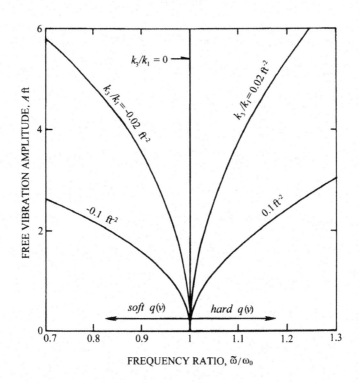

Figure 5.3 Free vibration frequency-amplitude behavior for a nonlinear structure.

There are several general conclusions for this study. First, for the same absolute value of k_3/k_1, and the same percent change in $\tilde{\omega}/\omega_0$ from unity, the results show that the soft restraint allows for smaller amplitudes than the hard restraint. Second, as $\tilde{\omega}/\omega_0$ extends beyond the ranges shown, the results become increasingly inaccurate because the perturbation solution begins to break down. Strictly speaking, these solutions are valid only for $\tilde{\omega}$ arbitrarily close to ω_0. Third, the vertical line at $\tilde{\omega}/\omega_0 = 1$ in Figure 5.3 is the condition reached as $|k_3/k_1|$ approaches zero. Here the amplitude of oscillation becomes independent of frequency, as is characteristic of linear systems.

5.3 RESPONSE FUNCTIONS FOR LINEAR STRUCTURES

Sections 5.1 and 5.2 illustrated methods for calculating the fundamental, free vibration frequency of several one degree of freedom structural models, characterized by either linear or nonlinear restoring forces. The remainder of this chapter makes use of these frequencies to compute the solutions $v = v(t)$ of these models in response to applied environmental loads $p_1(t)$. The governing equation considered now is of the linear form

$$m\ddot{v} + c_1\dot{v} + k_1 v = p_1(t) \tag{5.58}$$

where it is recalled that the subscript (1) on the parameters is the designation for a single degree of freedom system. In this section, $v(t)$ is computed for a harmonic load and an impulsive load. Then the response function is derived for a general loading function.

Harmonic Response Function

The harmonic response function $H(\omega)$, sometimes called the complex frequency response function, is derived from equation (5.58) in the following way. Choose a harmonic loading function of constant magnitude p_0 and frequency ω, or

$$p_1(t) = p_0 e^{j\omega t} \tag{5.59}$$

Then choose a harmonic solution of equation (5.58) in the form

$$v(t) = \frac{p_0}{k_1} H(\omega) e^{j\omega t} \tag{5.60}$$

After substituting equations (5.59) and (5.60) into equation (5.58), it follows that

$$(j^2 m\omega^2 + jc_1\omega + k_1)e^{j\omega t} H(\omega) = k_1 e^{j\omega t} \tag{5.61}$$

Since $j^2 = -1$, the harmonic response function is

$$H(\omega) = k_1(-m\omega^2 + jc_1\omega + k_1)^{-1} \tag{5.62}$$

and its modulus is

$$|H(\omega)| = [H(\omega)H^*(\omega)]^{1/2} = k_1[(k_1 - m\omega^2)^2 + c_1^2\omega^2]^{-1/2} \tag{5.63}$$

Note that for a linear structure in which the independent coordinate is the rotation θ instead of v, the frequency response function is similar in form to equation (5.62), but the constants would have a different meaning (J_0 would replace m, $\ddot{\theta}$ would replace $\ddot{v}$ and so on).

In some applications it is convenient to recast $H(\omega)$ and its modulus in terms of the system's undamped natural frequency ω_0 and the damping ratio ζ, defined by

$$\omega_0 = \sqrt{\frac{k_1}{m}}; \qquad \zeta = \frac{c_1}{2\sqrt{k_1 m}} \tag{5.64}$$

In terms of these latter definitions, the modulus of equation (5.63) becomes

$$|H(\omega)| = \left\{ \left[1 - \left(\frac{\omega}{\omega_0}\right)^2\right]^2 + 4\zeta^2 \left(\frac{\omega}{\omega_0}\right)^2 \right\}^{-1/2} \tag{5.65}$$

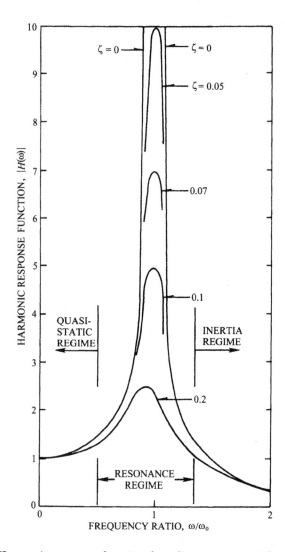

Figure 5.4   Harmonic response function for a linear system with small damping.

The utility of this form will become apparent later in the analysis of random motion.

Consider the four regimes of response as deduced from equation (5.65). Note that the behavior of $|H(\omega)|$ is the same as that for $|v(t)|$ of equation (5.60), since the latter quantities differ only by the multiple p_0/k_1. For the first regime, $0 \le \omega/\omega_0 \le 0.1$, the case in which $p_1(t)$ is applied very slowly. Then the system response is essentially static, $H(\omega) \to 1$, and

$$\frac{p_0}{k_1}\,|H(\omega)| \to \frac{p_0}{k_1} = \text{peak } v(t) \tag{5.66}$$

For the second regime, $0.1 \leq \omega/\omega_0 \leq 0.5$ and the response is quasi-static, or $v(t)$ is amplified only a little above its static value. For the third regime, $\omega/\omega_0 \to 1$, the response is in resonance and the peak response $v(t)$ is large compared to its static value, especially for ζ near zero. In the fourth regime, $\omega/\omega_0 \to \infty$ (> 10 for practical purposes) and the peak system response approaches zero for all damping ratios ζ. The latter three response regimes for $|H(\omega)|$ as a function of ω/ω_0 for several values of ζ are shown in Figure 5.4. The usual range for structural damping is $0.02 < \zeta < 0.10$.

For the case of structural excitation by a simple harmonic water wave, however, the response amplitude curves of Figure 5.4 become distorted and thus should not be used directly. This distortion occurs because the wave loading amplitude p_0 is not independent of the wave frequency ω. Recall from equation (4.32) that $p_0 = H \cdot G_0$ where H is the given wave height and G_0 is the transfer function which is dependent on ω.

Impulse Response Function

The impulse response function $h(t)$ is the solution to equation (5.58) for an impulsive load such as a swift kick or a hammer blow. Define the impulsive load as

$$p_1(t) = C\delta(t) \tag{5.67}$$

where $C = 1$ lb· sec or 1 N·s. The Dirac delta function $\delta(t)$ is defined as zero for all time except at $t = 0$; and in addition has the integral property

$$\int_{0^-}^{0^+} \delta(t)dt = 1 \tag{5.68}$$

Using this loading, the impulse response function is derived as follows. Since $h(t)$ is a solution to equation (5.58) for $p_1(t) = \delta(t)$, then

$$m\ddot{h}(t) + c_1\dot{h}(t) + k_1 h(t) = \delta(t) \tag{5.69}$$

For $t > 0$, then $\delta(t) = 0$ by definition, and the general solution to the homogeneous form of equation (5.69) is

$$h(t) = (C_3 \sin \omega_d t + C_4 \cos \omega_d t)e^{-\zeta\omega_0 t} \tag{5.70}$$

where C_3 and C_4 are arbitrary constants. The damped frequency in terms of ω_0 and ζ of equations (5.64) is

$$\omega_d = \omega_0(1 - \zeta^2)^{1/2} \tag{5.71}$$

One can verify by substitution using equations (5.64) and (5.71) that equation (5.70) is a solution to equation (5.69). The system is not displaced at $t = 0$, so $h(0) = 0$. With equation (5.70), then $C_4 = 0$. Next $\dot{h}(0)$ is found by differentiating equation (5.70) and setting $t = 0$. The result is

$$\dot{h}(0) = \omega_d C_3 \tag{5.72}$$

This value of $\dot{h}(0)$ must be compatible with equations (5.69) and (5.68). This is accomplished by integrating equation (5.69) term by term over the impulse time ε, the interval from $t = 0^-$ to $t = 0^+$, or

$$m\dot{h}(\varepsilon) + c_1 h(\varepsilon) + k_1 \int_{0^-}^{0^+} h(t)dt = \int_{0^-}^{0^+} \delta(t)dt \qquad (5.73)$$

As $\varepsilon \to 0$, the right side of the last equation approaches unity by definition, equation (5.68); $h(\varepsilon) \to 0$ since the system is not displaced initially, and the integral on the left side vanishes. Thus $\dot{h} = 1/m$. With equation (5.72), $C_3 = 1/(m\omega_d)$, and with equation (5.70) the solution for the unit impulsive force becomes

$$h(t) = \frac{1}{m\omega_d} e^{-\zeta\omega_0 t} \sin \omega_d t \qquad (5.74)$$

Typical units for $h(t)$ are inches per pound-second or meters per newton-second.

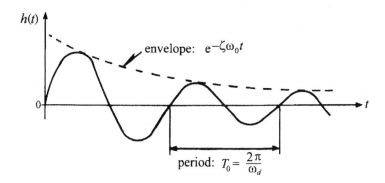

Figure 5.5   Impulse response function for a linear, lightly damped system.

Shown in Figure 5.5 is a sketch of $h(t)$ for light damping or $0 < \zeta \leq 0.1$. The corresponding period of free vibration is given by $T_0 = 2\pi/\omega_d$. For $\zeta \leq 0.1$, the frequency ratio lies in the range $0.994 \leq \omega_d/\omega_0 \leq 1.0$, indicating that $\omega_0 \simeq \omega_d$ is a good approximation for lightly damped linear systems.

Convolution Integral

Consider the response or solution to equation (5.58) for an arbitrary loading $p_1(t)$. At a particular time $t = \tau$, apply an impulsive load of magnitude $p_1(\tau)$ over the time interval $d\tau$. This impulsive load is depicted as a shaded portion of the function $p_1(t)$ in Figure 5.6. The response dv observed at time t is simply a multiple of the unit impulse response function over $d\tau$, which is

$$dv = p_1(\tau)h(t - \tau)d\tau \qquad (5.75)$$

For the last equation, note that the time between the application of the unit impulse and observation of its response is $(t - \tau)$ instead of t as in equation (5.74), where the impulse was applied at $\tau = 0$. Define the arbitrary loading $p_1(t)$ as zero for all times less than zero. Thus the total response due to all impulses applied in the interval $0 \leq \tau \leq t$ is given by integrating equation (5.75) over this time interval. Using equation (5.74), the result is

$$v(t) = \frac{1}{m\omega_d} \int_0^t p_1(\tau)e^{-\zeta\omega_0(t-\tau)} \sin \omega_d(t - \tau)d\tau \qquad (5.76)$$

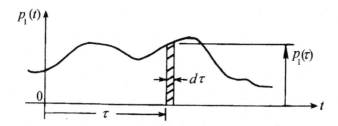

Figure 5.6   Representation of an arbitrary load history.

This is a particular solution to equation (5.58) and is the form used to evaluate the responses of offshore structures in the present text. It is one form of the convolution integral, sometimes called the Duhamel integral. Equation (5.76) does not include the two independent solutions to the homogeneous form of equation (5.58) for two reasons. First, their inclusion would necessitate the use of initial conditions $v(0)$ and $\dot{v}(0)$ to obtain the complete response solution, and those initial conditions are rarely if ever known for an offshore structure. Second, those homogeneous solutions die out rather quickly due to structural and external damping. It is appropriate to employ the Duhamel integral as given by equation (5.76) in computing steady state dynamic responses of offshore structures to the highly variable environmental loading conditions. The Duhamel integral in the above form will be used in the next section for earthquake motion analysis and in later chapters for random motion analysis.

5.4   RESPONSE OF LINEAR STRUCTURES TO EARTHQUAKE LOADING

Consider the motion of the fixed-legged platform shown in Figure 2.15. The equation of motion for this platform with horizontal base or sea floor earthquake excitation was derived in Chapter 2 as equation (2.36), which is repeated here with the negative sign on the right side omitted:

$$m\ddot{v} + c_1\dot{v} + k_1v = m\ddot{v}_g \qquad (2.36)$$

In this simplified model, the horizontal ground acceleration $\ddot{v}_g$ gives rise to an exciting force of magnitude $p_1(t) = m\ddot{v}_q$. Assume that the dynamic response is given to sufficient accuracy (without consideration to initial conditions) by equation (5.76), or

$$v(t) = \frac{1}{\omega_d} \int_0^t \ddot{v}_g(\tau) e^{-\zeta\omega_0(t-\tau)} \sin \omega_d(t-\tau) d\tau \qquad (5.77)$$

For a structure with known characteristics ω_0 and ζ, then ω_d is calculated from equation (5.71). Given an earthquake measurement of $\ddot{v}_g$, the response of this structure can then be calculated from the last equation by numerical integration. Such results are available in the open literature (Wiegel, 1970). For instance, the measured time history of $\ddot{v}_g$ for the SOOE component of the 1940 El Centro earthquake shown in Figure 2.14 was used with equation (5.77) to calculate several response histories $v(t)$, one for each fixed pair (ζ, ω_0). For each response history there is a maximum value of structural displacement $v_{\max}$ which occurs once during the time of excitation. Such a result is shown in Figure 5.7.

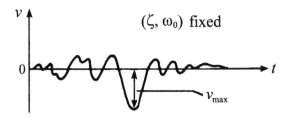

Figure 5.7   Typical time history of structural displacement resulting from horizontal ground motion (an earthquake).

Since ground velocity is commonly used as a measure of structural damage (Wiss, 1981), a peak ground velocity based on $v_{\max}$ is defined for this purpose. This is the pseudovelocity S_v given by

$$S_v = \omega_0 v_{\max} \qquad (5.78)$$

A typical plot of the pseudovelocity is shown in Figure 5.8 as a function of the structure's period, $T_0 = 2\pi/\omega_0$. It is noted that S_v is strongly dependent on the structural damping ζ. From plots such as these, the peak horizontal shear load $f_{\max}$ at the sea floor, and the peak overturning moment $M_{\max}$ can be estimated from

$$f_{\max} = k_1 v_{\max} \qquad (5.79)$$

$$M_{\max} = h_0 f_{\max} = h_0 k_1 v_{\max} \qquad (5.80)$$

where h_0 is the height of the platform.

Other factors involved in designing offshore structures for earthquake resistance not included in this simplified approach are inelastic structural behavior, ground-structural interactions, and the interactions of vertical and horizontal motions. Key references on these effects are included in the brief discussions of Gould and Abu-Sitta (1980) and in the more extensive analyses of Clough and Penzien (1993). There is some evidence, however, that the simplified approach just presented is conservative in that it often overpredicts the motion of the structure.

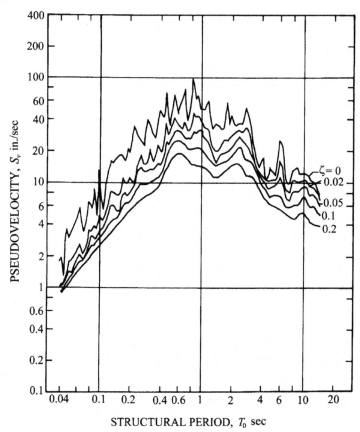

Figure 5.8   Pseudovelocity response of a single degree of freedom structure to the El Centro earthquake, south eastern component, May 18, 1940.

Example Problem 5.7.   For the jackup rig shown in Figure 2.17, estimate the critical responses to the horizontal motion of the 1940 El Centro earthquake for which $\ddot{v}_g$ is given in Figure 2.14. The structural data needed for this calculation are given in *Example Problems 2.8* and *5.3* as: $h_0 = 3180$ in., $\omega_0 = 1.36$ rad/sec, and $k_1 = 6.82 \times 10^4$ lb/in. The structural period is $T_0 = 4.61$ sec.

From Figure 5.8, the values of S_v for $\zeta = 0$ and 0.1 are about 12 and 8 in./sec, respectively. From equations (5.78)-(5.80), the corresponding upper and lower limits for maximum displacement, horizontal shear load, and overturning moment are thus

$$v_{\max} = 8.82 \text{ in. and } 5.88 \text{ in}$$
$$f_{\max} = 6.02 \times 10^5 \text{ lb and } 5.56 \times 10^5 \text{ lb}$$
$$M_{\max} = 1.19 \times 10^9 \text{ in.-lb and } 1.74 \times 10^9 \text{ in.-lb}$$

It is noted that these numerical results are quite sensitive to small changes in the structural period. That is, since T_0 lies in a trough, a 10 percent increase or decrease in T_0 would increase each of these calculated responses by about 50 percent. To avoid the many peaks and troughs that typify the S_v vs. T_0 curves calculated directly from earthquake data, such curves are often *smoothed* in a statistical sense before being used for design purposes.

5.5 RESPONSE CHARACTERISTICS OF NONLINEAR STRUCTURES

It was shown in Chapter 2 that offshore cable-stayed installations such as spread moored ships, floating platforms, and compliant towers all have nonlinear restraint forces. Under some special conditions these structures may be subjected to disturbing forces that are approximately harmonic. For instance, Wilson (1951) observed erratic surge oscillations of a ship moored in a harbor. The ship's excitation force was caused by harmonic harbor waves at a frequency close to the natural frequency of oscillation of the water within the harbor basin. The oscillation of the harbor waters was due to the waves in the adjacent sea.

For a cable-stayed offshore structure, neither the wave-induced exciting force amplitude p_0 nor the excitation frequency ω ever remains constant. However, to gain some physical insight about the response of the inherently nonlinear cable-stayed structures, the following undamped structural model is investigated in which p_0 and ω are constant.

$$m\ddot{v} + k_1 v + k_3 v^3 = p_0 \cos \omega t \qquad (5.81)$$

Responses to Harmonic Excitation

Solutions for this nonlinear structural model are now derived using the perturbation technique discussed in Section 5.2. Cast equation (5.81) as

$$\ddot{v} + \omega_0^2 v + e k_3 v^3 = e p_0 \cos \omega t \qquad (5.82)$$

where $e = 1/m$ is arbitrarily small. A steady-state solution to equation (5.82) is sought where this solution has the same frequency of oscillation as the wave excitation frequency ω. It is required that both this solution $v(t)$ and the frequency ω not differ greatly from their corresponding values values $v_0(t)$ and ω_0 of the linear system ($k_3 = 0$). That is

$$v = v_0(t) + e v_1(t) \qquad (5.83)$$

$$\omega^2 = \omega_0^2 + eg(\bar{a}) \tag{5.84}$$

Here $v_1(t)$ is the response correction function and $g(\bar{a})$ is the amplitude function, both arising from the presence of k_3. The procedure for calculating $v_1(t)$ and $g(\bar{a})$ follows that of Section 5.2, starting with equation (5.46). Also, the two initial conditions here are those of equations (5.45). Note that ω replaces the symbol $\tilde{\omega}$ of that previous formulation.

The procedure is summarized. When equations (5.83) and (5.84) are substituted into equation (5.82), the coefficients of e^0 and e^1 that result after some algebra are each equated to zero. When the trigonometric identity of equation (5.52) is applied and the higher order terms involving e^2, e^3, ... are ignored, the result is two linear differential equations in $v_0(t)$ and $v_1(t)$. A steady-state solution to each is found, they are superimposed according to equation (5.83), and the initial conditions $v(0) = \bar{a}$ and $\dot{v}(0) = 0$ are applied. The *secular* term in this solution, which is one of the terms of $v_1(t)$, is eliminated by equating to zero the coefficient multiplying the variable t. This procedure is physically appropriate since it preserves the boundedness of the solution as time increases. The result is

$$g(\bar{a}) = \frac{3}{4}k_3\bar{a}^2 - \frac{1}{\bar{a}}p_0 \tag{5.85}$$

With $e = 1/m$ and this last equation combined with equation (5.84), the sought-after relationship among the system parameters is derved as

$$m(\omega^2 - \omega_0^2)\bar{a} + p_0 = \frac{3}{4}k_3\bar{a}^3 \tag{5.86}$$

This last result can be compared to the result for free oscillations. For free oscillations, $\bar{a} = A$, $p_0 = 0$ and $\omega = \tilde{\omega}$, for which equation (5.86) is identical to equation (5.55), as it should be. Also it is not difficult to show that the forced vibration response $v(t)$ is given by equation (5.55), provided that ω is substituted for ω_0. Of course in the present case the amplitude $\bar{a}$ depends on the magnitude of the excitation p_0, as well as on the system's frequency and stiffness. As for the previous case of free oscillations, the forced oscillation response is distorted by a frequency component 3ω.

Equation (5.86), the key result needed to study the behavior of this nonlinear system, is recast in the following two forms using $\omega_0^2 = k_1/m$:

$$\bar{a}^3 - \frac{4}{3}\frac{k_1}{k_3}\left(\frac{\omega^2}{\omega_0^2} - 1\right)\bar{a} - \frac{4}{3}\frac{k_1}{k_3}\frac{p_0}{k_1} = 0 \tag{5.87}$$

$$KA^3 - (\Omega^2 - 1)A - 1 = 0 \tag{5.88}$$

The nondimensional parameters of the latter equation are

$$A = \frac{k_1}{p_0}\bar{a}; \qquad K = \frac{3}{4}\frac{k_3}{k_1}\left(\frac{p_0}{k_1}\right)^2; \qquad \Omega = \frac{\omega}{\omega_0} \tag{5.89}$$

In these nondimensional parameters, p_0/k_1 is the static deflection of its linear system counterpart for a constant excitation force p_0; and K is interpreted as the nonlinear stiffness parameter, assuming that k_1, p_0, and ω_0 are constant. For instance, when $K = 0$, the amplitude of motion is equal to that of the undamped linear system previously derived. That is, $|A|$ of equation (5.88) is equal to $|H(\omega)|$ of equation (5.65) for $K = \zeta = 0$.

Example Problem 5.8. In the San Diego harbor a ship moored with multiple lines was observed to oscillate or gallop in an erratic manner, even though the harbor waves were reasonably regular and were of normal height. The initially rather taut mooring lines were then tightened further, and after that this ship ceased to gallop and oscillated in the same regular manner as the ships moored nearby. Can the initially erratic ship motion and its subsequent correction be explained using a simple, nonlinear dynamic model?

Consider the simplified nonlinear dynamic model of the ship-wave system described by equation (5.81) in which a single incident harbor wave produced a hull force of amplitude p_0 at a excitation frequency ω. The connection of the ship's response amplitude $\bar{a}$ to ω and the restraint constants is equation (5.86). The task is to investigate the nature of the ship's response amplitude to the system parameters.

The mathematical question is: for what combination of system parameters does $\bar{a}$ of equation (5.86) have one or more real roots? If there is only *one* real root for $\bar{a}$, then the motion $v(t)$ is given by equation (5.56) with $\tilde{\omega} = \omega$, and is quite regular, without gallops. If *more than one* real root exists, then the motion could pass from one amplitude to another, giving rise to the observed erratic behavior of the moored ship. Fortunately, the nature of the roots for a third-order reduced cubic polynomial in $\bar{a}$, such as equation (5.87), was thoroughly studied early in the sixteenth century, and the results presented below are available in standard algebra texts (Rosenbach et al.,1958). Rewrite the latter equation in the form

$$\bar{a}^3 + \alpha\bar{a} + \beta = 0 \qquad (5.90)$$

where α and β are real, found by comparing equations (5.87) and (5.90). Define

$$R = \frac{1}{27}\alpha^3 + \frac{1}{4}\beta^2 \qquad (5.91)$$

where R determines the types of roots of equation (5.90). That is, there are three distinct real roots, at least two equal real roots, or a single real root for R negative, zero, or positive, respectively. With the values of α and β defined by equation (5.87), R can be cast in terms of the present system parameters. The three types of roots corresponding to R positive, zero, and negative are summarized. There are *three distinct real roots* if

$$\frac{1}{k_3}(\omega^2 - \omega_0^2)^3 > \frac{81}{16m^3}p_0^2 \qquad (5.92)$$

There are *at least two equal real roots* if

$$\frac{1}{k_3}(\omega^2 - \omega_0^2) = \frac{81}{16m^3}p_0^2 \qquad (5.93)$$

There is *only one real root* if

$$\frac{1}{k_3}(\omega^2 - \omega_0^2)^3 < \frac{81}{16m^3}p_0^2 \qquad (5.94)$$

To explain the ship's behavior, two reasonable assumptions are made. First, the wave excitation parameters p_0 and ω remained constant, or nearly so, from before to after the mooring lines were adjusted. Since the ship's mass m remained constant, then the right sides of equations (5.92)-(5.94) also remained constant. Second, the inequality given by equation (5.92) was true before the line adjustment since multiple real roots $\bar{a}$ lead to erratic motion, which is explained in more detail in the next section. This inequality (5.92) implies that $\omega > \omega_0$. The equality condition of equation (5.93) can be eliminated because such exactness is rarely possible in the physical world.

After tightening the mooring lines, k_1 definitely increased; and k_3 decreased somewhat because the static restraint stiffness curve $q(v)$ vs. v would tend to straighten out as the initial slope of this curve increased. For a small increase in k_1, the ω_0^2 term increased, causing the term in brackets on the left of equation (5.92) to decrease. Since this latter term is to the power three, its decrease more than offset the increase of $1/k_3$. The inequality of equation (5.92) with its multiple amplitudes $\bar{a}$ thus reverted to the inequality of equation (5.94), the condition of a single amplitude $\bar{a}$ with regular oscillations.

Jumps

The erratic changes or jumps in the response amplitudes for the nonlinear model described by equation (5.81) can be predicted from the $|A|$ vs. Ω behavior of equation (5.88). This behavior is shown in Figure 5.9 for two values of K. Several observations can be made about these curves. First, as K is decreased at constant Ω, as from the solid curves to the dashed curves, the response behavior approaches that of the undamped, linear system response $|H(\omega)|$ of Figure 5.4. Second, as K is increased at constant Ω, the response curves lean more to the right. Third, multiple amplitudes exist for values of $\Omega > \Omega_m$ where Ω_m is the value of Ω at the knee of each curve. Fourth, if linear damping had been included in the nonlinear model, equation (5.81), then that effect would have shown up as the amplitude-limiting, small dashed line shown as the upper knee for $K = 0.001$, around the point labeled $2'$ (Stoker, 1963).

Consider an example based on Figure 5.9 in which a particular set of conditions can lead to a jump in response amplitude. For a wave load of constant magnitude p_0 and for $K = 0.001$, suppose that the wave excitation frequency initially given by $\omega = 1.2\,\omega_0$ is decreased very slowly. The response amplitude follows the path from point 1 to point 2 on the lower knee of the solid curve, at which point the amplitude must almost double in magnitude by jumping to point 3 to allow for a subsequent decrease in ω. The response amplitude then decreases in a regular manner along the solid curve to point 4.

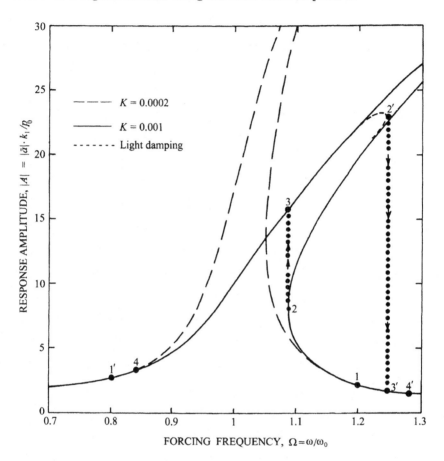

Figure 5.9 Response amplitudes and jumps for excursions in excitation frequency at constant p_0.

Again refer to Figure 5.9 and consider a second example for which p_0 is constant and $K = 0.001$. Assume light damping with an initial excitation frequency $\omega = 0.8\,\omega_0$. Then ω is increased very slowly from point $1'$ along this solid curve to point $2'$ at the upper knee of the damped amplitude curve. If there is a subsequent increase in ω, this must be accompanied by a decrease in amplitude of about a factor of ten, or a *downward* jump from point $2'$ to $3'$. Then smooth amplitude behavior persists along the same curve to point $4'$.

Because of system inertia it takes a finite time for a jump in amplitude to take place. If the total excursion time for the excitation frequency is much larger than $2\pi/\omega_0$, then the time required for a jump is of the order of $2\pi/\omega_0$. This is demonstrated in the problem of the oscillating buoy, which is discussed in Section 5.6.

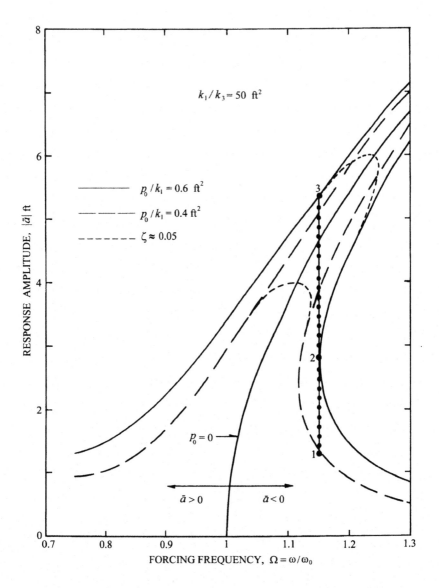

Figure 5.10   Jump responses for excursions in p_0 at constant excitation frequency.

Example Problem 5.9. Discuss the possibility of erratic jumps in the response amplitude of the moored LST described in *Example Problem 2.10.* Assume that $k_1/k_3 = 50$ ft^2 and that the amplitude of the wave load increases very slowly at a constant excitation frequency, where p_0/k_1 changes in the range of 0.4 to 0.6 ft. Let $\zeta = 0.05$.

Shown in Figure 5.10 are the amplitude-frequency plots for this ship, or the $|\bar{a}|$ vs. ω/ω_0 curves based on equation (5.87). The two curves shown, one for the smaller load and one for the larger load, were generated by calculating the real roots of $\bar{a}$ for $0.7 \leq \omega/\omega_0 \leq 1.3$. As in Figure 5.9, the response amplitude is negative or opposite in sign to p_0 to the right of the $p_0 = 0$ line and is positive to the left of this line. The two branches of each p_0/k_1 curve were connected by the broken lines to show the approximate effects of light damping, or $\zeta = 0.05$. The estimated peak amplitude of each was chosen as ten times the respective static response, the same amplification calculated for its linear counterpart, Figure 5.4. Damping is discussed in detail by Stoker (1963).

With the damping shown and with a value of $\omega/\omega_0 = 1.15$ for the lower load ratio of $p_0/k_1 = 0.4$ ft, the system oscillates initially with an amplitude of $\bar{a} = 1.3$ ft at point 1. For this lower load ratio and with system damping, the response amplitude is single-valued at this frequency. As the load ratio is increased to $p_0/k_1 = 0.6$ ft, the response amplitude increases to $\bar{a} = 2.85$ at the knee of the solid curve, or point 2. However, a response amplitude of $\bar{a} = 5.4$ ft could also exist on the upper branch of the solid curve. It follows that there can be an *upward* jump in amplitude response between points 2 and 3 as the excitation load ratio is increased. Furthermore, there can be a *downward* jump from points 3 to 2 and then a smooth transition to point 1 if the excitation load ratio were gradually decreased to its initially lower value. There is of course the possibility of jump behavior at load levels between those given, not only for $\omega/\omega_0 = 1.15$ but for larger values of the forcing frequency as well. The two load level ratios chosen, however, sufficiently illustrate the combination of system parameters that can lead to erratic jump responses in this nonlinear system.

Subharmonics

The preceding studies for the nonlinear structure modeled by equation (5.81) showed high amplitude responses or *resonance* for excitation frequencies ω near ω_0. This is not too surprising since resonance for ω near ω_0 is well-known in its linear system counterpart ($k_3 = 0$) and was shown in Figure 5.4. In nonlinear systems, however, resonance near ω_0 may sometimes occur when ω is not near ω_0. For instance, if high amplitude responses also exist near $\omega_0 = \omega/n$ where $n = 2, 3, \ldots$, then the response is defined as a *subharmonic* of order $1/n$. Such responses do not occur in linear systems, but exist under special conditions in nonlinear systems such as cable-stayed offshore structures in seas with regular, harmonic wave components.

The subharmonic response most commonly observed is for $n = 3$, or the one-third subharmonic (Wilson and Awadalla, 1973). There are several alternative, classical methods that can be used to show the existence of this and other subharmonics (Stoker, 1963). For instance for $\omega = 3\omega_s$, Cunningham (1964) applied first-order perturbation theory to equation (5.81) in the same, straightforward manner illustrated previously in this chapter. Cunningham's results are summarized as follows. The necessary relationship between the subharmonic response of amplitude a_s and its frequency of oscillation ω_s (close to ω_0), when the system of equation (5.81) is excited by the force $p_1(t) = p_0 \cos 3\omega_s t$, is

$$\omega_s^6 - \omega_0^2 \omega_s^4 = \frac{3k_3}{128m^3} p_0^2 \left(1 - \frac{4ma_s\omega_s^2}{p_0} + \frac{32a_s^2 m^2 \omega_s^4}{p_0^2} \right) \qquad (5.95)$$

Define the following three nondimensional parameters which will aid in investigating this last result relating the one-third subharmonic response amplitude to the excitation frequency $\omega = 3\omega_s$.

$$\Omega_s = \frac{\omega_s}{\omega_0}; \qquad K_s = \frac{3}{128}\frac{k_3}{k_1}\left(\frac{p_0}{k_1}\right)^2; \qquad A_s = \frac{k_1}{p_0}a_s \qquad (5.96)$$

Here, the frequency parameter Ω_s is of order unity and the stiffness parameter K_s is arbitrarily small (of order ε). When these nondimensional parameters are used with equation (5.95), the result is

$$\Omega_s^6 - (1 + 32K_sA_s^2)\Omega_s^4 + 4K_sA_s\Omega_s^2 - K_s = 0 \qquad (5.97)$$

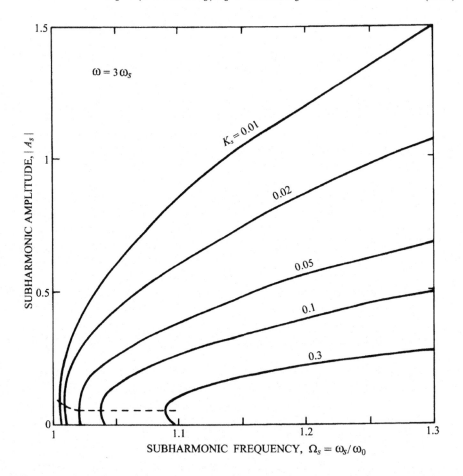

Figure 5.11   Effect of nonlinear stiffness on the amplitude and frequency of the one-third subharmonic response.

It is observed that the last equation is cubic in Ω_s^2 and quadratic in A_s and thus for a fixed value of K_s, the amplitude-frequency behavior can be deduced.

Such results are shown in Figure 5.11. For accuracy and consistency with first-order perturbation theory, neither K_s nor Ω_s should exceed the values shown. The knees in these curves indicate that for each K_s there is a minimum frequency below which the subharmonic cannot exist. This minimum, derived from equation (5.97) using $d\Omega_s/dA_s = 0$, yields the following equation from which the critical points (roots) $\Omega_s = \Omega_m$ can be computed:

$$\Omega_m^6 - \Omega_m^4 - \frac{7}{8}K_s = 0 \qquad (5.98)$$

With these roots, the critical amplitudes $A_s = A_m$ are computed from

$$A_m = \frac{1}{16\Omega_m^2} \qquad (5.99)$$

An approximate formula for the critical frequencies at the knee of each curve of constant K_s can be derived by substituting $\Omega_m = 1 + \varepsilon$ in equation (5.98), expanding the frequency terms, and ignoring terms in $\varepsilon^2, \varepsilon^3, \ldots$. The result is

$$\Omega_m \simeq 1 + \frac{7}{16}K_s \qquad (5.100)$$

This is an important result because it shows that, for hardening restraints where $k_3 > 0$, the minimum frequency for a one-third subharmonic to exist is just a little higher than ω_0.

Other important results by Cunningham (1964) are summarized. First, the subharmonic motion is stable only in the portions of the curves of Figure 5.11 that have positive slopes. No sustained subharmonic motion exists in the region below the broken line shown in this figure. Second, the effect of including linear damping of the form $c_1\dot{v}$ in the nonlinear model, equation (5.81), is to decrease A_s at constant K_s and Ω_s, and to eliminate subharmonic motion at some cutoff value of $\Omega_s = \Omega_c > \Omega_m > 1$. The value of Ω_c decreases as the damping increases. Third, for a subharmonic of order $1/n$ to exist for equation (5.81) where the excitation frequency is $n\omega_0$, the highest power of v in the restraining force polynominal must be at least of order n. Thus, a one-fifth subharmonic response cannot exist for equation (5.81), but a one-half subharmonic response can.

In summary, it is clear from these classical results for a nonlinear system that a *resonance* response amplitude A_s of frequency ω_s (near to but a little greater than ω_0) exists for excitation frequencies $\omega = 3\omega_s$. Experimental evidence for moored ship responses that support this analysis was presented by O'Brien and Muga (1964) and by Wilson and Awadalla (1973). This evidence will be discussed in Chapter 10.

5.6   NONLINEAR RESPONSES FOR A SALM BUOY

In practice, the responses to wave loading of offshore structures with nonlinear restraints are generally calculated numerically from the governing differential equations of structural motion. Some possible responses such as jumps and subharmonics were discussed above, based on closed form solutions to those equations. Such possible responses greatly aid in the interpretion of computer-derived numerical results, as the following example will demonstrate.

Many software packages are available to carry out numerical solutions. A popular package is Mathematica[R] (1999). For solutions to nonlinear dynamics problems, such software usually employs a step-by-step integration method, which was used in the following problem also. With this method, the response was evaluated at successive increments of time where the restraint and drag force terms were taken as constant during each interval and then updated at the end of each interval. Clough and Penzien (1993) and Paz (1980) discussed this basic method in the context of nonlinear structural dynamics.

The following problem is a rather comprehensive one. It begins with a physical description of the buoy, proceeds with a careful formulation of its dynamic model with its accompanying assumptions, and concludes with a discussion of numerical results relative to the closed form responses derived previously in this chapter.

Physical Description and Dynamic Model

A type of open sea mooring system used to load crude oil into tankers is the single anchor leg mooring (SALM) buoy. One of a wide variety of such buoy designs is shown in Figure 5.12: a buoy with a long, upright cylinder stabilized by cables to sea floor anchors and held in place by a short bottom chain to the pipeline end manifold (PLEM). This PLEM is anchored to the seafloor with pin piles. When the oil is not flowing through the flexible hoses to the ship, the tanks within the buoy are full, and the buoy has just sufficient buoyancy to keep it afloat. Then the tension in the bottom connecting chain is very slight. It is under these conditions that the rotational motion of the buoy is now analyzed. The chosen excitation is that of a single, harmonic, linear water wave. Responses to a range of excitation frequencies are sought.

The mathematical model of the upright buoy with its geometry and its loading are defined in Figure 5.13a. The buoy is treated as a rigid body rotating at angle θ about a frictionless pin at its base. The free body sketch of the buoy shown in Figure 5.13b identifies: the net cable tension force F_c at angle ϕ with the horizontal; the damping force F_d acting at the mass center G; and the total wave load $p_1(t)$. The center of buoyancy B is assumed to be coincident with G. The buoy weight and its buoyant force are equal and opposite, and the net restoring moment due to these forces is zero. The horizontal wave particle velocity and acceleration are assumed to be much larger than those for the buoy. When equation (2.3), the equation of plane motion for rotation of a rigid body about a fixed point 0, is applied to the free body sketch of Figure 5.13c, the result for small buoy rotations θ is

$$J_0\ddot{\theta} + h_c F_c \cos \phi - h_c \theta F_c \sin \phi + h_d F_d = - h_w p_1(t) \qquad (5.101)$$

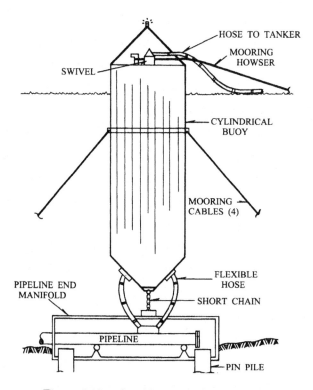

Figure 5.12   A cable-stayed SALM buoy.

Here, the external moments and the virtual mass moment of inertia J_0 are those with respect to the base point 0. The vertical dimensions h_c, h_d, and h_w locate the respective forces F_c, F_d, and $p_1(t)$. These latter three forces are now evaluated.

The cable tension force is based on a buoy with four identical, symmetrically arranged restraining cables where the wave force is in-line with two of these cables. For small in-plane rotations, the contributions of the two out-of-plane cables to the buoy's restoring force are negligible. The cables have nearly the same specific gravity as that of sea water so that their dead weight effects on the buoy are negligible. Each cable is a three-strand polypropylene line that is approximately straight and under a pretension force at $\theta = 0$. The increment in the tension force F_c required to stretch this line by an amount δ while its opposite line tends to go slack, is given in the form

$$F_c = a_0 + b_0 \delta^3 \qquad (5.102)$$

where a_0 and b_0 are material constants derived from experiments. If the buoy diameter is much smaller than the cable length, it follows from the geometry of the single stretched cable shown in Figure 5.13c that δ is approximately

$$\delta = h_c \theta \cos \phi \qquad (5.103)$$

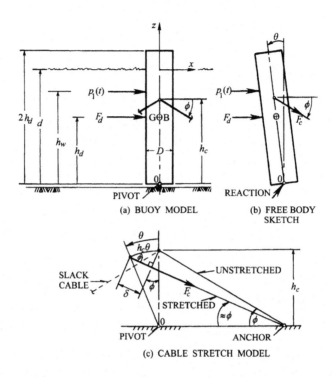

(a) BUOY MODEL

(b) FREE BODY SKETCH

(c) CABLE STRETCH MODEL

Figure 5.13   Mathematical models of SALM buoy and cable restraint.

Thus, if the variables on the right side of the latter equation are known, the cable force can be computed from equation (5.102).

The damping force F_d is modeled as linear-viscous and proportional to the horizontal buoy velocity $h_d\dot{\theta}$ at its mass center. That is

$$F_d = C'_D h_d \dot{\theta} \qquad (5.104)$$

where C'_D is the linear drag coefficient. For computations it is necessary to assign a realistic numerical value to C'_D so that the buoy oscillations are lightly damped. To do this, the homogeneous form of equation (5.101) is rewritten using equations (5.102)-(5.104) and then linearized. The terms involving $\theta^2, \theta^3, \ldots$ are neglected since θ is small. The result is the linear differential equation

$$\ddot{\theta} + 2\zeta\omega_0\dot{\theta} + \omega_0^2\theta = 0 \qquad (5.105)$$

for which the corresponding natural frequency for the undamped system is

$$\omega_0 = \sqrt{\frac{a_0 h_c^2 \cos^2\phi}{J_0}} \qquad (5.106)$$

The compatible value of the drag coefficient in terms of the damping factor and the other system constants is

$$C_D' = \frac{2\zeta}{h_d^2} \sqrt{a_0 h_c^2 J_0 \cos^2 \phi} \qquad (5.107)$$

For this study 5 percent damping was chosen, or $\zeta = 0.05$.

The last force component needed in equation (5.101) is that due to wave excitation, or $p_1(t)$. The assumptions used to model this force are summarized. The wave is a simple one based on linear theory with characteristics as described in Table 3.1. The compatibility equation relating the wave frequency ω, the wave number k, and water depth d is given by equation (3.16). The wave loading on the buoy is dominated by the inertia term $\bar{q}_I$ in Morison's equation, where the ratio of buoy diameter to wave height falls in the range $0.5 \leq D/H \leq 1$. The inertia coefficient has an average value of $C_M = 2$. The horizontal wave particle acceleration $\dot{u}$, evaluated at $x = 0$, is used to calculate the total horizontal wave load on the buoy. When $\bar{q}_I$ of equation (2.7) is integrated over the submerged height of the buoy, this load is calculated as

$$p_1(t) = \int_{-d}^{0} \bar{q}_I dz = \frac{\pi}{4} C_M \rho D^2 \int_{-d}^{0} \dot{u} dz = -\frac{\pi}{8k} C_M \rho D^2 \omega^2 H \sin \omega t \qquad (5.108)$$

The location of this load from the sea floor is

$$h_w = d + \bar{z} \qquad (5.109)$$

where its location from the still water line is

$$\bar{z} = \frac{\int_{-d}^{0} z \bar{q}_I \, dz}{\int_{-d}^{0} \bar{q} \, dz} = \frac{1 - \cosh kd}{k \sinh kd} \qquad (5.110)$$

It is noted that $\bar{z}$ is negative because it is below the still water line, and the coordinate z is measured positive upward from the still water line.

When equations (5.102)-(5.104), (5.108), and (5.109) are substituted into the original equation of motion (5.101), the result gives the final nonlinear model as

$$J_0 \ddot{\theta} + (a_0 h_c^2 \cos^2 \phi)\theta - (a_0 h_c^2 \sin \phi \cos \phi)\theta |\theta|$$

$$+(b_0 h_c^4 \cos^4 \phi)\theta^3 - b_0 h_c^4 \sin \phi \cos^3 \phi)\theta^3 |\theta|$$

$$+C_D' h_d^2 \dot{\theta} = (d + \bar{z})\frac{\pi}{8k} C_M \rho D^2 \omega^2 H \sin \omega t \qquad (5.111)$$

In this model, the absolute value sign was used on the even-order restoring force terms to preserve the proper sign of that force, or to assure that the restoring force is always antisymmetric about $\theta = 0$. For a fixed wave frequency and water depth, k is calculated from equation (3.16), and $\bar{z}$ is calculated from equation

(5.110). With a suitable choice of ζ, the damping constant is determined from equation (5.107). Then with a choice of wave height and C_M, together with the fixed system characteristics of geometry and inertia, all of the coefficients in equation (5.111) are known. The system parameters used to compute these coefficients are listed in Table 5.3.

Table 5.3   System Parameters for the SALM Buoy

Buoy diameter	$D = 240$ in.
Buoy height	$2h_d = 1320$ in.
Buoy moment of inertia	$J_0 = 5.57 \times 10^9$ lb-in.-sec^2
Cable angle	$\phi = 30$ deg
Cable location on buoy	$h_c = 960$ in.
Cable material parameter	$a_0 = 2.82 \times 10^4$ lb/in.
Cable material parameter	$b_0 = 1.34 \times 10^5$ lb/in.^{3}
Damping factor	$\zeta = 0.05$
Drag force location on buoy	$h_d = 660$ in.
Flow coefficient	$C_M = 2$
Water depth	$d = 1200$ in.
Water weight density	$\rho g = 0.0372$ lb/in.^{3}
Wave frequency range	$0 \leq \omega \leq 15$ rad/sec
Wave height	$H = 120$ in.

Numerical Results and Conclusions

Numerical solutions to equation (5.111) were obtained using a step-by-step Runge-Kutta integration procedure, subjected to the at-rest initial conditions of $\theta(0) = \dot{\theta}(0) = 0$. The discrete values of buoy rotation θ, as well as its angular velocity and angular acceleration, were calculated at a time step sufficiently small to show response behavior accurately. This time step was chosen as one-tenth of the smaller of either the natural period $T_0 = 2\pi/\omega_0$ of the equivalent linear system, or of the excitation period $T = 2\pi/\omega$.

Shown in Figure 5.14 as a function of the wave forcing frequency ω is the absolute value of the horizontal amplitude of displacement at the top of the buoy: $2h_a\theta_a$ where θ_a is the computed amplitude of rotation. This figure displays three important features.   First, the amplitude behavior is similar to the simpler nonlinear *hard* (cubic) restraint system, Figures 5.9 and 5.10: an amplitude curve that leans to the right. Second, the ratio of the peak dynamic response to the static response in Figure 5.14 is $5.4/0.5 = 10.8$, which compares favorably to the response ratio of 10 for a linear system with $\zeta = 0.05$, shown in Figure 5.4. Third, in the range of wave frequency $7.8 < \omega < 8.8$ rad/sec, the buoy's amplitude is multi-valued, indicating the occurrence of jumps.

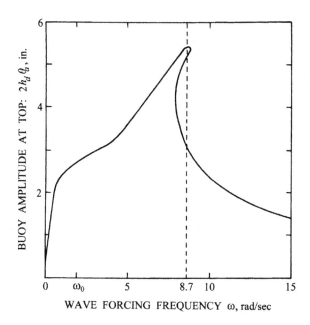

Figure 5.14   Response amplitude of the SALM buoy with harmonic excitation.

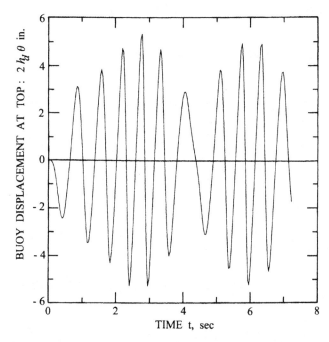

Figure 5.15   Time history of peak buoy displacement for a simple wave of height of
10 ft and frequency $\omega = 8.7$ rad/sec.

A typical time history of buoy response is shown in Figure 5.15. This response is for a wave excitation frequency of $\omega = 8.7$ rad/sec, an excitation frequency that produced three distinct amplitudes (*jumps*) as shown in Figure 5.14. The time history clearly shows an amplitude variation between the extremes of 5.4 in. and 3.2 in. However, the oscillation between these extremes is not so abrupt as the word *jump* implies because it takes two or three response cycles to effect an amplitude excursion. Further, the time history shows that although the excitation frequency is constant, the response frequency is not, since the time between crossings is not constant. This is characteristic of systems with nonlinear restoring forces.

PROBLEMS

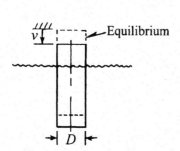

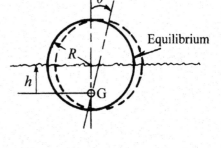

Figure 5.16   Cylindrical buoy.   Figure 5.17   Spherical buoy.

5.1   The cylindrical buoy shown in Figure 5.16 is weighted so that it has a low center of gravity. Its diameter is 2.5 ft, its weight is 2000 lb, and the density of the sea water is 64 lb/ft^3. Assume that the frictional resistance of the water is negligible, that $C_M = 0$, and that the surface of the water is relatively undisturbed while the buoy undergoes vertical oscillations. Let v be its displacement from vertical equilibrium and let the restoring force equal the weight of the water displaced by the motion v. Using its free body sketch, derive the equation of motion for the buoy in free oscillations. Then calculate its natural frequency and the time required for one full cycle of oscillation.

5.2   The spherical buoy shown in Figure 5.17 is a thin steel shell of mean radius R. It is weighted so that in static equilibrium it floats half out of the water, where its mass center is a distance $h > 3R/8$ below its geometric center. The mass center of the displaced water (the center of buoyancy) is at $h = 3R/8$; and J_G is the mass moment of inertia for the buoy with respect to its rotational axis through its mass center. Using its free body sketch, set up a dynamic model for small angles θ of rolling motion. Neglect the frictional resistance of the water. From this equation of motion, derive an equation for the buoy's natural frequency in roll. If $R = 3.5$ ft, $h = 1.75$ ft, and the period of oscillation is 2 sec, calculate J_G.

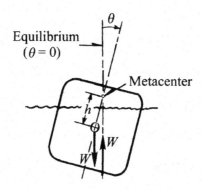

Figure 5.18   Cross section of a ship in roll motion.

5.3   A simplified model of a ship in roll motion is shown in Figure 5.18. The ship of actual mass m_0 is buoyed by a force $W = m_0 g$ equal to the weight of the water displaced. For small roll angles θ, the intersection of the line of action of the buoyant force with the ship's centerline defines the metacenter, or the roll axis at distance h from the mass center. Let J_0 be the ship's mass moment of inertia with respect to the roll axis. Neglecting the frictional resistance of the water, derive the equation of motion for roll, and from that deduce an expression for its roll frequency. The average cross section of the ship is approximately square with the dimensions 60 ft $\times$ 60 ft; and the ship has an approximate uniform distribution of mass. If $h = 3.5$ ft, calculate J_0 in terms of m_0. Then calculate the roll frequency and its period of oscillation.

5.4   Assume that the uniform beam of *Example Problem 5.3*, shown in Figure 2.16a, has hinged ends instead of clamped ends. Show that the following mode shape satisfies the geometric boundary conditions.

$$\psi(x) = \sin \frac{\pi x}{\ell}$$

Then calculate an expression for the natural frequency ω_0 similar in form to equation (5.30). By comparing this result to the frequency derived only from beam stiffness and the total virtual mass $m = f_1 \bar{m} \ell$, deduce a numerical value for the fraction f_1.

5.5   A three-legged platform is identical to that of *Example Problem 5.4* except that its legs are hinged instead of fixed to the deck. Using the Rayleigh Method and an appropriate mode shape, derive the expression for ω_0 in a form similar to equation (5.36). Using the data of Table 5.2, calculate a numerical value for f_1 and ω_0 for this platform, assuming $\ell \simeq d$. Discuss the meaning of your results in comparison to those computed in *Example Problem 5.4*, its more constrained counterpart.

5.6   The rotation θ of the monopod gravity platform shown in Figure 5.2 will result in an offset of the center of buoyancy B from its centerline to a new position B'. Examples of this shift are shown in Figures 2.13. Discuss qualitatively how this offset can affect the rocking frequency of the structure.

5.7   For the gravity platform of *Example Problem 5.2*, suppose that the ballast density is changed so that the center of mass and center of buoyancy are coincident. Discuss qualitatively how this coincidence can affect both the rocking frequency and the dynamic stability of this structure.

5.8   Suppose that the only motion for the gravity platform shown in Figure 5.2 is horizontal sliding, $v = v(t)$.

(a) From the free body sketch of this structure, derive the equation of motion for free, undamped horizontal sliding. Model the soil foundation stiffness k_1 according to equation (2.76).

(b) Assuming harmonic motion for the horizontal displacement in the form $v = v_0 \sin \omega_0 t$, derive an explicit expression for ω_0 similar in form to equation (5.11).

(c) Based on the soil and structural parameters given in Table 5.1, calculate ω_0 and its corresponding period T_0 for this platform. Compare your frequencies for both the lower bound and upper bound values for G_s given in Table 5.1 with the respective natural frequencies for rocking motion computed in *Example Problem 5.2*.

(d) If both rocking and sliding motion were taking place simultaneously, which of these two vibration modes do you think would dominate the structure's free vibrations? Explain.

5.9   Your task is to design, construct, and perform free vibration experiments on a desk-top model of a three-legged or four-legged jackup rig. The legs should be securely fixed to a solid base, and can be either hinged or fixed at the deck level.

(a) Design your model so that its two lateral periods of free vibration T_0, one in air and one in water, are sufficiently long to be measured accurately with a stop watch. Use any theoretical method you wish to predict T_0 for both cases.

(b) Construct your design.

(c) Perform free vibration experiments for the model in air and then with its legs mostly submerged in water. Measure T_0 in both cases as an average value over several oscillations.

(d) Compare your measured periods for the two types of experiments, and compare each to the theoretical values you calculated previously. Explain reasons for any discrepancies between your measured and predicted results.

5.10   Based on the numerical values given in *Example Problem 5.5* and the perturbation solution given by equation (5.56), plot the free vibration displacement $v(t)$ as a function of $\tilde{\omega}t$ for a ship in sway motion. Superimpose on this plot the harmonic displacement for its linear system counterpart ($k_3 = 0$). Discuss the distortion in the free vibration oscillations due to nonlinear cable restraints.

5.11   Derive the differential equation for the horizontal sliding motion only of the gravity platform shown in Figure 5.2. The structural geometry is given in the figure, other system parameters are summarized in Table 5.2, and the equations governing its soil foundation stiffness and damping are given by equations (2.76) and (2.77).

(a) Calculate an upper and lower bound for T_0, the undamped period of free horizontal vibration.

(b) Deduce a damping factor ζ for the upper and lower bound values of T_0.

(c) Assume horizontal ground excitation by the El Centro earthquake whose pseudovelocity response is given by Figure 5.8. Based on the numerical values of T_0 and ζ just calculated, estimate the bounds for the peak values of horizontal displacement, the horizontal shear load on the soil foundation, and the overturning moment. You may *smooth* the pseudovelocity curves to estimate these bounds.

(d) The preceding response calculations do not include the damping effects of the surrounding water. Deduce whether such additional damping would increase or decrease the system responses to this same earthquake excitation.

5.12   Use first order perturbation theory to derive the amplitude function $g(\bar{a})$ given by equation (5.85). Using this result, deduce equation (5.86).

5.13   Include linear, viscous damping in equation (5.82) and use first order perturbation theory to derive an algebraic result analogous to equation (5.86). Then predict the peak response amplitude for the two curves of Figure 5.10, for $\zeta = 0.05$.

5.14   Predict the excitation frequency $\omega = \omega_m$ below which there will be no amplitude jumps in the system modeled by equation (5.82). Do this in two ways: first by differentiating equation (5.88) to find a minimum Ω; and then by applying the *rule of roots* where $R = 0$ in equation (5.91).

5.15   Use first order perturbation theory to derive the one-third subharmonic relationship given by equation (5.95).

5.16   Derive equation (5.97) from (5.95). Then show that the frequency ratio Ω_m above which a one-third subharmonic response can exist is given by the lowest real root of equation (5.98).

5.17   For the nonlinear system modeled by equation (5.82), the minimum frequency ratio Ω_m below which a one-third subharmonic response cannot exist is predicted from equation (5.98) or approximately from equation (5.100). Which of these relationships predicts the lower value of Ω_m? Choose some realistic numerical values of K_s to validate your answer.

5.18   Assume a solution to the linear equation (5.105) of the form

$$\theta = \theta_0 e^{-\zeta \omega_0 t} \cos \omega_d t$$

where θ_0 is a constant and ω_d is given by equation (5.71). Use this solution to derive the following approximate expression for the damping ratio ζ :

$$\zeta \simeq \frac{1}{2\pi} \ln \frac{\theta_1}{\theta_2}$$

Here, θ_1 and θ_2 are any two consecutive peak ampitudes during free vibration.

5.19   Carry out the necessary integrations in equations (5.108) and (5.110) to verify the expressions given for the buoy wave loading $p_1(t)$ and its location $\bar{z}$.

5.20   Starting with equation (5.111), calculate an expression for the *quasi-static* response θ_{st} of this buoy, or the value of θ for $\omega \ll 1$. For the quasi-static load, use the amplitude of the wave loading given by the right side of this equation. Delete all terms involving $\dot{\theta}$ and $\ddot{\theta}$, and delete all terms containing θ with powers greater than one. Then solve the result for $\theta = \theta_{st}$. Use the numerical values of Table 5.3 to calculate θ_{st}, and compare your result with that obtained from Figure 5.14 for a small value of ω ($\omega = 0.05$ for instance). If these two results do not coincide, explain possible reasons for the discrepancy.

REFERENCES

Clough, R. W., and Penzien, J., *Dynamics of Structures*, second ed., McGraw-Hill, New York, 1993.

Cunningham, W. J., *Introduction to Nonlinear Analysis*, McGraw-Hill, New York, 1964.

Den Hartog, J. P., *Mechanical Vibrations*, third ed., McGraw-Hill, New York, 1947.

Forsythe, G. E., Malcolm, M. A., and Moler, C. B., *Computer Methods for Mathematical Computation*, Prentice-Hall, Englewood Cliffs, NJ, 1977.

Gould, P. L., and Abu-Sitta, S. H., *Dynamic Response of Structures to Wind and Earthquake Loading*, Wiley, New York, 1980.

Mathematica[R], version 4, Wolfram Media, Inc., Champaign, IL, 1999.

Nataraja, R., and Kirk, C. L., Dynamic Response of a Gravity Platform under Random Wave Forces, OTC-2904, *Proceedings of the Offshore Technology Conference*, 1977.

O'Brien, J. T., and Muga, B. J., Sea Tests on a Spread-Moored Landing Craft, *Proceedings of the Eighth Conference on Coastal Engineering*, Lisbon, Portugal, 1964.

Paz, M., *Structural Dynamics: Theory and Computation*, Van Nostrand Reinhold, New York, 1980.

Rosenbach, J. B., Whitman, E. A., Meserve, B. E., and Whitman, P. M., *College Algebra*, fourth ed., Ginn, Lexington, MA, 1958.

Stoker, J. J., *Nonlinear Vibrations in Mechanical and Electrical Systems*, Interscience, New York, 1963.

Wiegel, R. L., ed., *Earthquake Engineering*, Prentice-Hall, Englewood Cliffs, NJ, 1970.

Wilson, B. W., Ship Responses to Range Action in Harbor Basins, *ASCE Transactions* **116**, 1951.

Wilson, J. F., and Orgill, G., Optimal Cable Configurations for Passive Dynamic Control of Compliant Towers, *Journal of Dynamic Systems, Measurement, and Control* **106**, 1984.

Wilson, J. F., and Awadalla, N. G., Computer Simulation of Nonlinear Motion of Moored Ships, *Proceedings of the NATO Advanced Study Institute on Analytical Treatment of Problems in the Berthing and Mooring of Ships*, Wallingford, England, May 1973.

Wiss, J. F., Construction Vibrations: State-of-the-Art, *Journal of the Geotechnical Division, ASCE* **107** (GT2), February 1981.

6

Statistical Descriptions of Offshore Waves

Bruce J. Muga

Regular waves were treated in Chapter 3 in a classical, deterministic sense in which the analyses were based on nonviscous hydromechanic flow applied to either idealized surface wave forms or, in the case of the free-stream function theory, applied to a prescribed surface wave form of finite length. In this chapter, irregular waves based on experimental measures of surface wave heights are analyzed in a statistical sense to obtain wave spectra and other statistical parameters which will prove useful in predicting the response of offshore structures to irregular waves.

6.1 INTRODUCTION TO WAVE SPECTRA

A short time record representative of measured irregular surface wave forms offshore is shown in Figure 6.1. For such irregular forms, the wave periods and wave heights take on new meanings as statistical ways of describing these waves are now explored. Procedures are now defined to portray the important fluctuations and to condense these data into manageable forms.

Consider first the time interval at which the wave record is sampled at discrete points to give an adequate representation for $\eta(t)$. For instance, the record in Figure 6.1 can be marked off in equally spaced time intervals of $\Delta t = 1\,\text{sec}$ such that the sequence of ordinates, when connected by a smooth curve, reproduces the important details of the record. One suspects that there is an *optimum* time interval which is neither too fine nor too coarse and which best approximates the record. It is apparent that a 10-sec interval is too large, but also the choice of 100 points over the time interval 10 seconds is far too many. The Nyquist sampling theorem aids in the selection of an optimum interval Δt. This theorem states:

> The continuous time record $\eta(t)$ can be adequately represented by, and reconstituted from, a set of sample values $\eta_1, \eta_2, \cdots$, provided that f_s, the number of sample values per second, is at least twice the highest frequency $f_{\max}$ present in $\eta(t)$. That is, $f_s \geqq 2f_{\max}$.

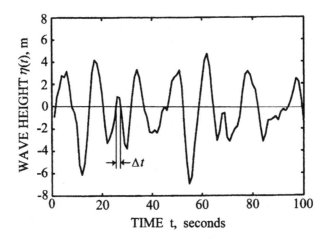

Figure 6.1   A 100 second section of a 17.07 minute record of surface wave height recorded at Macquaire Island in the Southern Ocean on May 13, 1989 (Young, 1999).

To apply the Nyquist theorem, let $f_{\max} = 1/(2\Delta t)$ where Δt is the shortest time between two consecutive crossings of the $\eta = 0$ axis. Thus, an appropriate sampling frequency is $f_s = 2/(2\Delta t) = 1/\Delta t$, and the corresponding time interval between sampling of $\eta(t)$ is simply Δt. For the 100-sec section of the wave record shown in Figure 6.1, $\Delta t = 2\,\text{sec}$. However, if the whole 17.07-min record were considered, then a shorter time would be apparent, or $\Delta t = 1\,\text{sec}$ (Young, 1999).

Consider a Fourier series representation of this whole 17.07 min or 1024 sec wave record. To reproduce this record, one would need $N = 1024$ sample points η. The Fourier series in this case would consist of 1024 terms involving a sum of sines and cosines, or

$$\eta(t) = 2\sum_{n=1}^{N/2} a_n \cos \frac{2\pi nt}{\tau_0} dt + 2\sum_{n=1}^{N/2} b_n \sin \frac{2\pi nt}{\tau_0} \qquad (6.1)$$

where τ_0 is the 1024-sec period and the mean value of $\eta(t)$ is zero ($a_0 = 0$). The coefficients in equation (6.1) are

$$a_n = \frac{1}{\tau_0} \int_0^{\tau_0} \eta(t) \cos \frac{2n\pi t}{\tau_0} dt$$

$$b_n = \frac{1}{\tau_0} \int_0^{\tau_0} \eta(t) \sin \frac{2\pi nt}{\tau_0} dt \qquad (6.2)$$

The total wave energy is proportional to the average of the squares of $\eta(t)$, and this energy is equal to the sum of the energy content of each of the individual wave components. This is shown with Parseval's theorem, which states that for any periodic function of time represented by equation (6.1) where the mean is zero ($a_0 = 0$), then the variance of $\eta(t)$ is

$$\frac{1}{\tau_0} \int_0^{\tau_0} [\eta(t)]^2 dt = 2 \sum_{n=1}^{n/2} (a_n^2 + b_n^2) = \sigma_\eta^2 \tag{6.3}$$

for which the root-mean-square (rms) value for $\eta(t)$ is σ_n.

Now define the wave spectral density, $S_\eta(\omega)$, as

$$\int_0^\infty S_\eta(\omega)\, d\omega = \sigma_\eta^2 \tag{6.4}$$

It is observed that this spectral function is the resolution of the variance of $\eta(t)$ into its frequency components. That is, the value of $S_\eta(\omega_n)$ at each discrete frequency ω_n can be approximated by the sum of the Fourier coefficients, or

$$S_\eta(\omega_n) \simeq \frac{2}{\Delta\omega}(a_n^2 + b_n^2) \tag{6.5a}$$

where

$$\omega_n = \frac{2n\pi}{\tau_0}; \qquad \Delta\omega = \frac{2\pi}{\tau_0} \tag{6.5b}$$

Here, $\Delta\omega$ is the spacing between adjacent harmonics. It is emphasized that the sampling interval Δt must satisfy the Nyquist sampling theorem to avoid falsely distorting the Fourier coefficients. In practice, an experimentally - based discrete energy spectrum is usually fit to a continuous function by *smoothing*. Examples of smoothed spectra are discussed later in this chapter.

The character of the function $S_\eta(\omega)$ depends on the surface wave elevation record. For example, $S_\eta(\omega)$ for unidirectional swell is a narrow-banded function centered about the dominant swell frequency. On the other hand, sea waves in general are most often characterized by a broader-banded spectral function than that for swell waves.

The interest in wave spectra arises from the need to select a representative design-loading history acting on the proposed offshore structure. The design wave spectrum, usually furnished by a specialist in hindcasting or forecasting, is chosen to represent the worst possible combination of waves which would lead to the highest loading for the structure. To complicate matters, there may be many different surface water wave elevation histories characterized by essentially the same spectrum. In other words, a one-to-one correspondence between the surface water wave elevation record and the spectrum does not generally exist. However, all records that yield identical spectra do have the same statistical properties, although details in the records may vary widely. The suggested procedure, then, is to select a number of water wave elevation records corresponding to the *design wave spectrum* and to analyze the proposed offfshore structure under these alternative conditions.

It is important that the mathematical model used to represent the sea surface (the time history of wave height) should conform as closely as possible to the actual waves. In practice, there are two ways to accomplish this. First, a purely deterministic method can be used in which the sea surface records are fitted to a harmonic or Fourier series. This representation is unaffected by nonlinear wave effects. Second, a statistical method can be used in which the sea surface records are fitted to a spectral function in the frequency domain. It is emphasized that the second method is based on the concept of superposition and is adequate only for linear seas. Various efforts have been made to include nonlinear phenomena in spectral descriptions but these methods are not widely utilized. For purposes of structural design, however, linear models usually provide adequate descriptions of the sea surface.

For further discussions of offshore waves as random processes, the reader is referred to the works of Longuet-Higgins (1963), Munk (1950), Kinsman (1965), Pierson and Neumann (1966), and the overview of Young (1999).

6.2   CONCEPT OF THE SIGNIFICANT WAVE

The concept of the significant wave was first published by Sverdrup and Munk (1947), and later discussed by Wiegel (1949), Bretschneider (1959), and Kinsman (1965). The significant wave height is a useful parameter for describing an irregular sea surface and its wave spectra.

What is the *significant wave height*? The significant wave height is not an identifiable wave form that propagates like a physical wave as in the classical wave theories. The significant wave height, H_s, is defined as the arithmetic average of the highest one-third of the waves in a wave record. Its associated statistical parameter, the significant period T_s, is defined as the average period of the highest one-third of the waves in this wave record. It is noted that the symbols $H_{1/3}$ and $T_{1/3}$ are also commonly used to denote H_s and T_s, respectively.

The significant wave height is an important parameter in statistical analysis of wave mechanics for a number of reasons. The statistical distribution of wave heights and most energy spectrum analyses are related to the significant height. In fact, the major portion of the wave energy of the spectrum surrounds the significant wave height. Thus, it has been the practice historically to report sea conditions in the form of significant wave heights. Also, the effects that irregular seas have on many types of fixed and floating objects and on shore processes such as littoral transport, have been related to significant wave heights, with an accuracy sufficient for these engineering applications.

In summary, the irregular sea can be described in an abbreviated format by two parameters: the significant wave height and the significant period. However, designers of offshore systems sometimes need to know other statistical features of the sea, such as the maximum wave height or the maximum water surface elevation. This led to a number of studies that attempted to establish further statistical correlations for offshore waves. A brief summary of these correlations is now presented.

Wave Height Distributions

Examinations of offshore wave data have shown that the surface wave heights H follow the Rayleigh (1880) probability density function, $p(H)$. The following exposition involving this probability follows that of Goda (1985) and Young (1999), originally proposed by Barber (1950) and Putz (1954). The particularly useful form of the Rayleigh probability density function is

$$p(H) = \frac{2H}{H_{rms}^2} e^{-H^2/H_{rms}^2} \tag{6.6}$$

in which H_{rms} is the root-mean-square (rms) wave height of a given record. The square of this latter quantity is defined as

$$H_{rms}^2 = \overline{H^2} = \int_0^\infty H^2 p(H)\, dH \tag{6.7}$$

As the notation $\sqrt{\overline{H^2}}$ implies, H_{rms} is formed by squaring the height of each wave in a given record, taking the arithmetic average of these quantities, and then taking the square root of the result. The mean or average wave height is defined by

$$H_0 = \int_0^\infty H p(H)\, dH \tag{6.8}$$

Based on the Rayleigh distribution, the average wave height, H_0, the significant wave height, H_s, and the 1/10 highest wave height, $H_{1/10}$, are deduced as

$$H_0 = 0.87 H_{rms}; \qquad H_s = 1.42 H_{rms}; \qquad H_{1/10} = 1.80 H_{rms} \tag{6.9}$$

These three wave heights are depicted in Figure 6.2, which is a plot of the Rayleigh distribution given by equation (6.6). Also shown is the height of the most probable wave, which is at the peak of the curve.

Further, the most probable maximum wave height, H_{max}, depends on the duration of the storm or length of the wave record. The following relationship is often used to approximate this maximum wave height:

$$H_{max} = 0.707 H_s \ln N \tag{6.10}$$

Here N is the number of waves in the record. When N is not known, a reasonable approximation to this maximum wave height is

$$H_{max} = 1.77 H_s \tag{6.11}$$

For very severe storm waves or for waves in very shallow water near the breaking point, equations (6.10) and (6.11) should be used with caution.

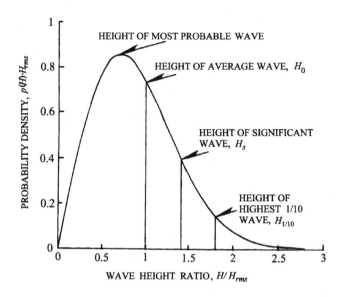

Figure 6.2 The Rayleigh distribution of wave heights.

A few selected results of wave height statistical correlations derived from field data, theoretical considerations, and from experiment are given in Table 6.1. From these data, it is seen that H_s/H_0 ranges from 1.35 to 1.63, and $H_{1/10}/H_s$ ranges from 1.27 to 1.29. These ratios compare favorably with the respective ratios derived above and given from equations (6.9), which are

$$H_s/H_0 = 1.63; \qquad H_{1/10}/H_s = 1.27 \tag{6.12}$$

Further, H_{max}/H_s in Table 6.1 ranges from 1.50 to 1.87, which compares favorably with the ratio of 1.77 of equation (6.11). One can conclude from these results that the Rayleigh probability distribution is a reasonable model for irregular surface waves offshore.

Experts in the field are in more agreement on wave height - distribution than on wave - period distribution. The average wave period $\overline{T}$ can be obtained from the energy spectrum. Sometimes the term *mean apparent period* is employed as a more precise definition to indicate the average elapsed time between successive crossings of the mean ordinate level of $\eta(t)$. Both expressions are related to the mean value of the instantaneous frequency which can be determined from a simple ratio of the first - and zero - order moments of the spectrum about its mean. In other words, the average period corresponds to the abscissa through which passes the line of action of the *center of gravity* of the spectrum. As indicated earlier, the significant period T_s is defined as the average period of the highest one-third of the waves in the record. Conceptually, it is closely identified with the period of maximum energy in the spectrum.

Table 6.1   A Compilation of Wave Height Statistical Correlations

Reference	Data Type	H_s/H_0	$H_{1/10}/H_s$	H_{max}/H_s
Munk (1944)	Field data	1.53	—	—
Seiwell (1949)	Field data	1.57	—	—
Wiegel (1949)	Field data	—	1.29	1.87
Barber (1950)	Theoretical	1.61	—	1.50
Putz (1950)	Field data	1.63	—	—
Longuet-Higgins (1952)	Theoretical	1.60	1.27	1.77
Putz (1952)	Theoretical	1.57	1.29	1.80
Darbyshire (1952)	Field data	1.60	—	1.50
Hamada et al. (1953)	Experimental	1.35	—	—

For typical irregular sea conditions where the average period ranges from 4 to 10 sec, studies have shown that T_s is nearly equal to the average period, $\overline{T}$. For longer average wave periods ($\gg$ 10 sec), it has been found that the average period is only 75 percent of the significant period. Based on a comprehensive study of wave conditions in many locations, the International Ship Structures Congress (ISSC) concludes that the average period can be taken as 90 percent of the significant period (Price and Bishop, 1974). It is emphasized that the average period as defined here does not correspond to the period of the average wave height.

Wave Height - Wave Spectrum Relationships

A historical advance in the description of irregular ocean surface waves was accomplished by Pierson (1952), who merged key concepts from classical mechanics and the theory of stochastic processes with the energy spectrum in order to predict the behavior of offshore waves (Kinsman, 1965). In its simplest form, the energy spectrum allocates the amount of energy of the sea surface according to frequency. As shown in Chapter 3, a small-amplitude sinusoidal wave has the form

$$\eta(x,t) = A\cos(kx - \omega t) \tag{6.13}$$

The total energy per unit surface area of this wave is

$$E = \frac{1}{2}\rho g A^2 = \frac{1}{8}\rho g H^2 \tag{6.14}$$

where H $(= 2A)$ is the wave height measured from crest to trough.

One of the fundamental premises of the spectral approach is that irregular waves are the result of the superposition of an infinite number of simple sine waves of small amplitudes that have a continuous frequency distribution. This process can be approximated with a finite number of small-amplitude sine waves having discrete frequencies. Under these conditions the mean total wave energy per unit surface area is given by

$$E = \frac{1}{8}\rho g(H_1^2 + H_2^2 + H_3^2 + \cdots + H_n^2 + \cdots) \tag{6.15}$$

where H_n is the wave height associated with the frequency ω_n.

Illustrated in Figure 6.3 is the distribution of wave spectral density as a function of wave frequency. The ordinate of each block is the spectral density $S_\eta(\omega)$ in units of (length)^2-time; and associated with each block is a wave height H_n of frequency ω_n, $n = 1, 2, \ldots$. Indications of height are intended merely to point out that the spectral density for a specified frequency corresponds to an identifiable wave. Wave records can be synthesized from the superposition of a finite number of discrete wave forms, as will be demonstrated at the end of this chapter. Amplitude spectra obtained from such synthetic records are indistinguishable from amplitude spectra derived from measured records of offshore waves. Such synthetic records are often quite adequate for engineering purposes.

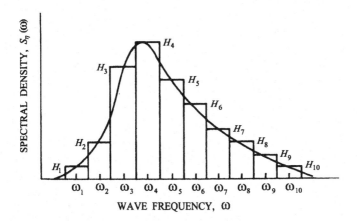

Figure 6.3   A representation of a wave spectrum.

The computation of spectra from wave records depends crucially on several factors, which include: the length of record, the sampling interval, the degree and type of filtering and smoothing, and the length of a statistical parameter called the autocovariance function. In general, some compromise between numerical stability, confidence, resolution, and the practical limitations of computers must be achieved. Historical analyses were developed by Blackman and Tukey (1959) and Borgman (1972). Present analyses include the use of computer packages such as Mathematica® (1999).

A general analytic form of the surface wave energy spectrum is

$$S_\eta(\omega) = A_0 \omega^{-m} e^{-B\omega^{-n}} \tag{6.16}$$

in which the empirical coefficients A_0, B, m, and n define the spectrum. In the most widely used forms, $m = 5$ and $n = 4$. Several empirical results for the coefficients A_0 and B, including their dependence on wind speed and significant wave height, are presented after the following alternative descriptions of wave spectra.

6.3   DESCRIPTIONS OF WAVE ENERGY SPECTRA

There are significant and sometimes subtle differences among the analytical descriptions of wave spectra as they appear in the open literature. Differences exist in terminology, notation, and even in the basic definition of $S_\eta(\omega)$. These alternative descriptions are depicted in the abscissa and ordinates of Figure 6.4.

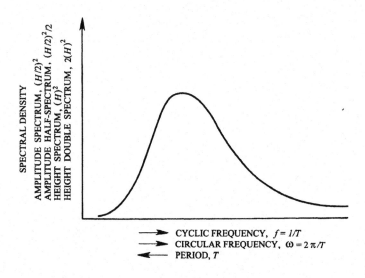

Figure 6.4   Alternative nomenclature for wave spectra.

Alternative Definitions

For the abscissa of the energy spectrum, the cyclic frequency $f = 1/T$ (Hz), the angular frequency $\omega = 2\pi/T$ (rad/sec), and the apparent period T are used. Experimentalists most often use f to present data; and those who expound theory use ω.

The ordinate is referred to as power spectral density, or simply spectral density. The average energy of the wave system is related to the sum of the square of the wave component heights H^2 or amplitudes $A^2 = (H/2)^2$. The following terminology was suggested by Michel (1967) for identifying the ordinate in terms of the significant wave height.

1. **Amplitude spectrum:** The following forms, used by Pierson (1952) and Pierson et al. (1955) define the spectral density as a function of the square of

the wave amplitude, for which the significant wave height is related to $(\text{area})_1$ under the spectral curve, or

$$H_s = 2.83\sqrt{2(\text{variance})} = 2.83\sqrt{(\text{area})_1} \qquad (6.17)$$

2. **Amplitude half-spectrum:** Rather than using twice the variance, some investigators use the variance directly. In the later case, the spectral density is related to one-half the square of the wave amplitude, and the significant wave height is related to the variance, the $(\text{area})_2$ under the spectral curve, or

$$H_s = 2.83\sqrt{2(\text{variance})} = 4\sqrt{\text{variance}} = 4\sqrt{(\text{area})_2} \qquad (6.18)$$

When equations (6.17) and (6.18) are compared, the results are: $(\text{area})_1 = 2(\text{variance})$ and $(\text{area})_2 = \text{variance}$.

3. **Height spectrum:** Other investigators choose to use wave heights, for which the area $(\text{area})_3$ under the spectral curve is four times the area obtained when wave amplitudes are used. This is because $(\text{wave height})^2 = 4(\text{wave amplitude})^2$. Therefore, the spectral density is a function of $(\text{wave height})^2$, and the relationships of the significant wave height to the areas under the three types of spectral curves are

$$H_s = 2.83\sqrt{2(\text{variance})} = 2.83\sqrt{(\text{area})_1}$$

$$= 1.414\sqrt{4(\text{area})_1} = 1.414\sqrt{(\text{area})_3} \qquad (6.19)$$

4. **Height double spectrum:** Investigators found that by taking twice the $(\text{height})^2$ rather than the $(\text{amplitude})^2$, the constant relating significant wave height and the square root of the area $(\text{area})_4$ under the spectral density curve could be made equal to unity. Accordingly, the area is eight times that given when amplitudes are used, and the significant wave height relationships are

$$H_s = 2.83\sqrt{(\text{area})_1} = \sqrt{8(\text{area})_1} = \sqrt{(\text{area})_4} \qquad (6.20)$$

The following precautions should be used when working with wave energy spectrum prepared by others. First, establish the units of the spectral density. Is this in Traditional English units of ft²-sec/rad or ft²/Hz, or is it in SI units of m²·s/rad or m²/Hz? Second, establish the frequency units of the abscissa. Is this in units of Hz for cyclic frequency f, or in units of rad/sec for circular frequency ω? Third, determine the specific formulation on which the ordinate is based. Sometimes this last step can be accomplished by simple examination of the symbolic notation; or the author of the data can be questioned. In any case, the Rayleigh distribution coefficient used to predict significant wave height (or wave amplitude) from the energy spectrum should be consistent with the nomenclature.

Empirical Forms

Particular forms of deepwater wave spectra in common use are those of Pierson and Moskowitz (1964) and Bretschneider (1959), which are based on theory and require sufficient data to fit the constants. The spectra of Pierson-Moskowitz are termed wind-speed spectra since wind speed is included directly in the spectral density function. The spectra of Bretschneider are termed height-period spectra since significant height H_s and the significant or mean apparent period $T_m = 2\pi/\omega_m$ are used directly in the spectral density function. It is emphasized that in deducing these spectra, use was made of the distribution functions originally derived theoretically by Longuet-Higgins (1952), and supported by the empirical relations based on the wave data of Putz (1952). These deepwater spectra have the form of equation (6.15) in which the constants (A_0, B) differ, depending on whether the height-double or amplitude-half spectrum assumption is made in reducing the data.

Another spectrum in common use is JONSWAP, which investigators have found appropriate for the design of offshore structures in the North Sea. Unlike the spectra in the form of equation (6.15), which are referred to as *fetch-unlimited* spectra, the JONSWAP spectrum includes the *fetch* in its formulation, and is called a *fetch-limited* spectrum. It is noted that the term fetch, denoted by X, is the distance from the shoreline of the wave field under consideration. A fetch-limited wave spectrum is one based on a wave field at distance X that has reached a steady state or time-independent condition. Such a condition can occur in deep water if the wind has blown for a sufficient length of time, out to sea, and in a direction perpendicular to a regular shoreline. These and other factors that affect the fetch, together with experimental measures of X appropriate for fetch-limited spectra, are thoroughly discussed by Young (1999). Useful forms for both fetch-unlimited and fetch-limited spectra are summarized.

Wind-Speed Spectra. The Pierson-Moskowitz fetch-unlimited spectra has the general form

$$S_\eta(\omega) = 0.0081 \frac{g^2}{\omega^5} e^{-B/\omega^4} \tag{6.21}$$

where g is the acceleration due to gravity. When the wind speed is known, then

$$B = 0.74 \left(\frac{g}{V}\right)^4 \tag{6.22}$$

where V is the wind speed at a height of 19.5 m above the still water level. Any consistent set of units for the quantities g, V, and ω are appropriate for equations (6.21) and (6.22), provided that ω is expressed in radians per unit time.

Significant Wave Height-Period Spectra. When the significant wave height rather than the wind speed is known, the constant B of equation (6.21), for H_s in units of m/s, is

$$B = \frac{3.11}{H_s^2} \tag{6.23}$$

The source of the latter relation is the International Towing Tank Conference (ITTC) (Price and Bishop, 1974), and the resulting spectrum is known as the ITTC spectra. Shown in Table 6.2 are the ITTC recommended values for significant wave height and average wave period for several wind speeds.

Table 6.2 ITTC Recommended Data for Equations (6.21)-(6.23)

V Wind Speed, m/s	$\overline{T}$, Average Wave Period, s	H_s, Significant Wave Height, m
5.14	2.7	---
10.3	5.3	3.1
15.4	8.0	5.1
20.6	10.7	8.1
25.7	13.4	11.0

Another form of the wave height-period spectra, developed by Ochi and Hubble (1976) from the work of Bretschneider (1959), is

$$S_\eta(\omega) = \frac{1.25}{4}\frac{\omega_m^4}{\omega^5}H_s^2\,e^{-1.25(\omega_m/\omega)^4} \tag{6.24}$$

in which ω_m is the frequency at the maximum of the spectrum, where the corresponding period is $T_m = 2\pi/\omega_m$. Listed in Table 6.3 are empirical equations for ω_m in terms of the significant wave height H_s and its probability of occurrence. Also listed in that table are values for ω_m and T_m for the special case of $H_s = 12.2$ m, a case considered later in *Example Problem 6.1.*

It is emphasized that the spectra expressed by equations (6.21) through (6.24) are amplitude half-spectra and that the square root of the area under each spectral diagram must be multiplied by 4 in order to obtain the corresponding significant wave height. That is, $H_s = 4\sqrt{(\text{area})_2}$.

Table 6.3 Spectral Model Data (Bretschneider, 1959 and Ochi, 1978)

Equation for ω_m, rad/s	For $H_s = 12.2$ m: ω_m, rad/s	T_m, s	Probability of Occurrence
0.048 (8.75− ln H_s)	0.30	20.9	0.0500
0.054 (8.44− ln H_s)	0.32	19.6	0.0500
0.061 (8.07− ln H_s)	0.34	18.5	0.0875
0.069 (7.77− ln H_s)	0.36	17.3	0.1875
0.079 (7.63− ln H_s)	0.41	15.5	0.2500
0.099 (6.87− ln H_s)	0.43	14.5	0.1875
0.111 (6.67− ln H_s)	0.46	13.6	0.0875
0.119 (6.65− ln H_s)	0.49	12.7	0.0500
0.134 (6.41− ln H_s)	0.52	12.0	0.0500

There are two general observations concerning significant wave height statistics. First, different wave time histories can have approximately the same significant wave height, but still have widely different spectral properties. Ochi and Whalen (1980) and Ochi (1981) offered a rational explanation and experimental documentation for such observations. Second, extensive global data for significant wave height based on nine years of satellite measurements taken in the 1990s were presented as contour maps by Young (1999). These data show mean monthly values of H_s, values which could be expected to be exceeded 10 percent, 20 percent, ... ,90 percent of the time. The highest value of H_s reported was 6 m with a probability of exceedence of 10 percent. Since confined events such as hurricanes affect such monthly averages very little, these satellite data have very limited use in the design of offshore structures.

Fetch-Limited Spectra.   The JONSWAP spectra has the general form

$$S_\eta(\omega) = \alpha g^2 \exp[-1.25(\omega/\omega_m)^4] \cdot \gamma^{\exp[-(\omega-\omega_m)^2/2\sigma^2\omega_m^2]} \qquad (6.25)$$

where

$$
\begin{aligned}
\gamma &= & &\text{3.3 for mean of selected JONSWAP data} \\
\gamma &= & &\text{7.0 for a very peaked spectrum} \\
\sigma &= & &0.07 \text{ for } \omega \le \omega_m \\
\sigma &= & &0.09 \text{ for } \omega > \omega_m \\
\omega_m &= & &2\pi(3.5)(g/V)(\overline{X})^{-0.33} \text{ peak frequency} \\
\alpha &= & &0.076(\overline{X})^{-0.22} \text{ or } \alpha = 0.0081 \\
\overline{X} &= & &gX/V^2 \\
X &= & &\text{fetch length;} \quad V = \text{wind speed}
\end{aligned}
$$

6.4   SELECTION OF DESIGN WAVE SPECTRA

The preceding spectra are best-fit curves of a number of individual spectra, each derived from actual wave records which are measured in generally similar sites and environments. They represent the mean of the points of the family of actual spectra and in no sense can be considered as *theoretical spectra.* For instance, a comparison of many individual spectra with formula spectra showed that the individual spectra height had a sigma variation of 30 percent or more from applicable formula spectrum (Hoffman, 1974).

This observation, which is consistent with the observations of other investigations, was explained by Ochi and Whalen (1980), who introduced the concept of families of wave spectra and provided a meaningful insight into this behavior. For example, using the two-parameter Bretschneider (1959) spectrum, equation (6.24), Ochi and Whalen presented a family of this spectra along with their probabilities of occurrence and their confidence limits. An example is shown in Figure 6.5, which will be discussed further in *Example Problem 6.1.* The important point is that each of the spectra represents the same significant wave height, although the frequency corresponding to each peak is different. This characteristic, which is partly reflected by a shape parameter γ, must also be considered in specifying design wave spectra. A shape parameter of $\gamma = 7$ gives a very sharply peaked JONSWAP spectrum, whereas by comparison, $\gamma = 3$ produces a lower peak and wider band about the peak. For a Pierson-Moskowitz spectrum, some investigators suggest that $\gamma = 1$. For further details, see Rye and Svee (1976) and Young (1999).

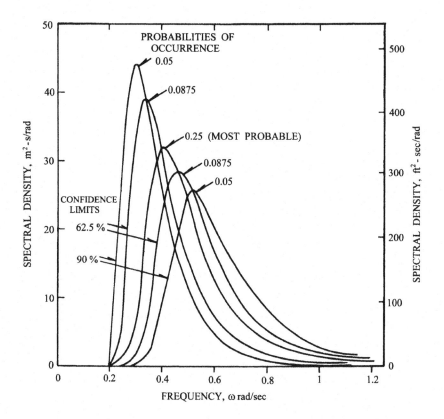

Figure 6.5   Illustration of a spectral family (Ochi, 1981).

The selection of wave spectra used to evaluate the design of a particular structure depends on several factors, including the risk criteria adopted by the owner of the structure. That is, the owner of the structure must decide what risk can be economically justified for a particular structure located at a given site. This presumes that all other design criteria are satisfied. In one case, for instance, the most economical overall design may be based on a storm occurring on the average once in 20 years, and in another case, once in 100 years. A comprehensive discussion of risk criteria was presented by Borgman (1963).

From another perspective, Freudenthal and Gaither (1969) suggested that the wave spectra eventually selected for design purposes should be based on the probability of occurrence of waves of a given intensity at the chosen site. This information is normally developed from hindcast studies in which the paths of all known storms that might have caused waves at the site are determined. From the data available on each of the storms, beginning usually with a knowledge of the wind field, estimates can be made of the intensity, duration, and direction of the wind-generated waves that reach the site. From this meteorological information, the lengths, periods, and heights of waves at the site can be determined. This approach is not exactly the way it is done in practice, but it does explain the basic methodology.

In practice, the way to forecast (or hindcast) waves is to forecast (or hindcast) the spectra of waves. From a spectrum one can obtain many of the properties of a wave such as height, average period, and average wavelength. Thus, one begins with a deepwater wave spectrum at a location that is not necessarily the same as the particular site for the structure. Then, using transformations which depend on the distance, direction, hydrography, and spectral and storm properties, one can determine the wave spectrum to be expected at the site under consideration. The well-known work by Pierson, Neumann, and James (1955) is highly recommended as an excellent introduction to the subject.

The significant wave heights to be expected at a known location depend on the historical climatological record. For instance, data for H_s at various water depths in the Norwegian North Sea, the Gulf of Mexico, and the Gulf of Alaska are shown in Table 6.4. At these sites, the significant wave heights are those generated by a storm with a recurrence interval of 100 years. These data were collected from a number of sources, each with a different database, and are presented only for comparison purposes. Alone, such data are insufficient for structural design since other special features such as subsea soil strength or floating ice at each geographic site must also be considered. However, these data illustrate that for a 100-year storm the significant wave heights of 12 m to 20 m are considerably higher then the mean values of 6 m (with the probability of exceedence of 10 percent) measured worldwide by satellite (Young, 1999).

Table 6.4 H_s at Three Sites, Based on a 100-Year Storm

Site	H_s, m	Water Depth, m
Norwegian North Sea	14 to 15	30
Gulf of Mexico	12 to 13	30 to 180
Gulf of Alaska	18 to 20	> 180

The selection of a design significant wave height can be done only after a hindcast study has been completed for the specific site. This hindcast study should yield both the significant wave heights corresponding to the selected storm recurrence intervals and the associated probability functions governing the modal frequencies ω_m. With this information and with a suitable spectral model, the family of wave spectra can be derived. The following example illustrates these ideas.

Example Problem 6.1.   The hindcast studies for a particular site in the Gulf of Mexico indicate that H_s for the 100-year storm is 12.2 m. Assume that the wave spectral family can be described by the two-parameter Bretschneider (1959) model, equation (6.24), and that the modal frequencies, ω_m, the corresponding periods T_m, the probabilities for ocurrence of H_s, for this family are those listed is Table 6.3. The results in Table 6.3 were derived by Ochi (1978) from selected wave data. Note that the sum of the probabilities for occurrence (or weighting factors) of H_s is equal to unity.

The equation for the most probable spectrum is obtained by substituting into equation (6.24) the frequency at the maximum of the spectrum: $\omega_m = 0.41$ rad/s from Table 6.3, for which the probability of occurrence is the maximum of 0.25. With $H_s = 12.2$ m, the spectrum of equation (6.24) becomes

$$S_\eta(\omega) = \frac{1.314}{\omega^5} e^{-0.03532/\omega^4} \quad \text{m}^2 \cdot \text{s/rad} \tag{6.26}$$

Equation (6.26), which is plotted in Figure 6.5, corresponds to the most probable spectrum associated with a storm of a 100-year recurrence interval. As such, this curve is not a design spectrum but a member of the family of spectra to be considered in design. Other members of the family are also shown in Figure 6.5, with selected frequencies ω_m, and the corresponding probabilities of occurrence as given in Table 6.3, together with their confidence limits.

The ideas illustrated in this example can be extended to more complex spectral formulations such as JONSWAP, which may be more appropriate for other sites such as the North Sea. The reader is encouraged to consult the most recent references so that the most up-to-date data can be used to generate the family of design spectra. Once these spectra are chosen, they can be employed for preliminary dynamic analyses of alternative structural designs, as will be illustrated in succeeding chapters.

6.5   SYNTHESIS OF TIME HISTORIES FROM SPECTRA

In the analysis of offshore structures, time domain solutions are frequently required. However, if the wave height excitation is expressed in the form of spectral density, it is necessary to transform this design spectra into an ensemble of representative time histories. This may be accomplished by utilizing Borgman's (1969) procedure for wave simulation. This is now presented with slight modifications.

The wave elevation $\eta(t)$ can be represented as

$$\eta(x,t) = \int_0^\infty \sin(kx - \omega t + \varepsilon)\sqrt{A^2(\omega)\,d\omega} \tag{6.27}$$

where $A^2(\omega)$ is the amplitude spectrum ordinate and ε is a random phase angle picked from a list of random numbers uniformly distributed over the interval from zero to 2π. The integral is discretized by partitioning the spectrum into equal portions (areas), instead of taking equidistant points on the frequency axis. This procedure avoids the presence of periodicities in the resulting time history.

Consider the partition

$$\omega_0 < \omega_1 < \omega_2 < \ldots < \omega_N = F \qquad (6.28)$$

where ω_0 is a small, positive value and F is a value beyond which the spectral ordinate is zero, for all practical purposes. Let

$$\Delta\omega_n = \omega_n - \omega_{n-1} \qquad (6.29)$$

$$\overline{\omega}_n = \frac{\omega_n + \omega_{n-1}}{2}, \quad n = 1, 2, \ldots, N \qquad (6.30)$$

The integral of equation (6.27) is approximated by the finite sum

$$\overline{\eta}(x,t) = \sum_{n=1}^{N} \sin(\overline{k}_n x - \overline{\omega}_n t + \varepsilon_n)\sqrt{A^2(\overline{\omega}_n)\,\Delta\omega_n} \qquad (6.31)$$

where $\overline{\omega}_n^2 = \overline{k}_n g$ and the overbar denotes the average value of the parameter. Let $S(\omega_n)$ represent the cumulative area under the spectral density curve, or

$$S(\omega_n) = \sum_{n=1}^{n} A^2(\omega_n)\,\Delta\omega_n \qquad (6.32)$$

Thus

$$A^2(\overline{\omega}_n)\,\Delta\omega_n \approx S(\omega_n) - S(\omega_{n-1}) = a^2 \qquad (6.33)$$

where a^2 is a constant. It follows that

$$\overline{\eta}(x,t) = \sum_{n=1}^{N} \sin(\overline{k}_n x - \overline{\omega}_n t + \varepsilon_n)\sqrt{S(\omega_n) - S(\omega_{n-1})} \qquad (6.34)$$

$$Na^2 = S(\omega_N) \approx S(\infty) = \int_0^{\infty} A^2(\omega)\,d\omega \qquad (6.35)$$

Recall the form for the Pierson-Moskowitz spectrum, equation (6.16), or

$$S_\eta(\omega) = A^2(\omega) = \frac{A_0}{\omega^5}\,e^{-B/\omega^4} \qquad (6.16)$$

Then from equations (6.16) and (6.32)

$$S(\omega) = \int \frac{A_0}{\omega^5} e^{-B/\omega^4} d\omega = \frac{A_0}{4B} e^{-B/\omega^4} \tag{6.36}$$

Equation (6.35) then gives

$$S(\infty) = \frac{A_0}{4B} \tag{6.37}$$

$$a^2 = \frac{A_0}{4BN} \tag{6.38}$$

Now determined are the positions of the partition frequencies $\omega_0, \omega_1, \omega_2, \ldots, \omega_N$. From equation (6.36) where $\omega = F$, it follows that

$$\frac{A_0}{4B} = e^{B/F^4} S(F) \tag{6.39}$$

Because of the equal area partition

$$S(\omega_n) = \frac{n}{N} S(F) = \frac{A_0}{4B} e^{-B/\omega_n^4} = S(F) e^{B/F^4} e^{-B/\omega_n^4} \tag{6.40}$$

It follows that

$$\frac{N}{n} e^{B/F^4} = e^{B/\omega_n^4} \tag{6.41}$$

When equation (6.41) is solved for ω_n, the result is

$$\omega_n = \left(\frac{B}{\ln(N/n) + B/F^4} \right)^{1/4}, \quad n = 1, 2, \ldots, N \tag{6.42}$$

which determines the partition frequencies.

The first-order simulation can be obtained by arbitrarily choosing the coordinate x equal to zero. Thus equation (6.34) becomes

$$\overline{\eta}(t) = a \sum_{n=1}^{N} \sin(-\overline{\omega}_n t + \varepsilon_n) \tag{6.43}$$

The random phase angles ε_n can be generated using one of the many available codes, but once selected, they are identified with a specific set of frequency components on a one-to-one basis. On the basis of some unpublished work by this writer, it has been found that an N value of at least 15 to 20 is necessary to produce time histories with statistical features similar to those predicted from low-order spectral moments. Time histories of other parameters such as water particle velocity, acceleration, and pressure can also be derived following the above approach.

A second order simulation of wave elevation for an irregular wave, derived by Longuet-Higgins and Stewart (1961), is

$$\overline{\eta}(t) = a \sum_{n=1}^{N} \sin(-\overline{\omega}_n t + \varepsilon_n) - \frac{1}{2}a^2 \sum_{n=1}^{N} \overline{k}_n \sin 2(-\overline{\omega}_n t + \varepsilon_n)$$

$$- \sum_{n=1}^{N} \sum_{n'=1}^{n-1} a^2 \overline{k}_{n'} \cos(-\overline{\omega}_{n'} t + \varepsilon_{n'}) \cos(-\overline{\omega}_n t + \varepsilon_n)$$

$$- \overline{k}_n \sin(-\omega_n t + \varepsilon_n) \sin(-\overline{\omega}_{n'} t + \varepsilon_{n'}) \tag{6.44}$$

In Chapter 10, this synthesis is illustrated for the case of surface wave excitation of a barge that deploys an OTEC pipeline.

REFERENCES

Barber, N. F., Ocean Waves and Swell, *Proceedings of the Institution of Civil Engineers, Marine and Waterways Division*, 1950.

Blackman, R. B., and Tukey, J. W., *The Measurement of Power Spectra*, Dover, New York, 1959.

Borgman, L. E., Risk Criteria, *Journal of the Waterways and Harbors Division, ASCE* **89** (WW3), 1963.

Borgman, L. E., Ocean Wave Simulation for Engineering Design, *Journal of the Waterways and Harbors Division, ASCE* **95** (WW4), 1969.

Borgman, L. E., Confidence Intervals for Ocean Wave Spectra, *Proceedings, Thirteenth Coastal Engineering Conference, ASCE,* Vancouver, B.C., 1972.

Bretschneider, C. L., *Wave Variability and Wave Spectra for Wind-Generated Gravity Waves,* Technical Memorandum No. 118, Beach Erosion Board, U.S. Army Corps of Engineers, 1959.

Darbyshire, J., The Generation of Waves by Wind, *Proceedings of the Royal Society Ser. A* **215**, 1952.

Freudenthal, A. M., and Gaither, W. S., Design Criteria for Fixed Offshore Structures, *First Offshore Technology Conference,* Houston, Texas, May 1969.

Goda, Y., *Random Seas and the Design of Marine Structures*, University of Tokyo Press, Japan, 1985.

Hamada, T., Mitsuyasu, H., and Hose, N., *An Experiemental Study of Wind Effect upon Water Surface,* Report of Transportation Technical Research Institute, Tokyo, June 1953.

Hoffman, D., *Proceedings of International Symposium on Dynamics of Marine Vehicles and Structures in Waves*, University College, London, 1974.

Kinsman, B., *Wind Waves, Their Generation and Propagation on the Ocean Surface*, Prentice-Hall, Englewood Cliffs, NJ, 1965.

Longuet-Higgins, M. S., On the Statistical Distribution of the Heights of Sea Waves, *Journal of Marine Research* **2** (3), 1952.

Longuet-Higgins, M. S., The Effect of Non-Linearities on Statistical Distributions in the Theory of Sea Waves, *Journal of Fluid Mechanics* **17** (3), 1963.

Longuet-Higgins, M. S., and Stewart, R. W., Changes in the Form of Short Gravity Waves on Long Waves and Tidal Currents, *Journal of Fluid Mechanics,* 1961.

Mathematica®, version 4, Wolfram Media, Inc., Champaign, IL, 1999.

Michel, W. H., *Sea Spectra Simplified*, Meeting of Gulf Section, Society of Naval Architects and Marine Engineers, April 1967; also in *How to Calculate Wave Forces and Their Effects*, Ocean Industry, May-June 1967.

Munk, W. H., *Proposed Uniform Procedure for Observing Waves and Interpreting Instrument Records*, Wave Report 26, Scripps Institute of Oceanography, 1944.

Munk, W. H., Origin and Generation of Waves, *Proceedings, First Conference on Coastal Engineering*, Council on Wave Research, Berkeley, CA, 1950.

Neumann, G., *On Wave Spectra and a New Method of Forecasting Wind Generated Seas*, Technical Memorandum No. 43, Beach Erosion Board, U.S. Army Corps of Engineers, 1953.

Ochi, M. K., Wave Statistics for the Design of Ships and Ocean Structures, *Transactions, Society of Naval Architects and Marine Engineers* **86**, 1978.

Ochi, M. K., *Waves for Mooring System Design*, Technical Note N1604, Civil Engineering Laboratory, Naval Construction Battalion Center, U.S. Navy, Port Hueneme, CA, March 1981.

Ochi, M. K., and Hubble, E. N., On a Six-Parameter Wave Spectra, *Proceedings, Fifteenth Coastal Engineering Conference* **1**, 1976.

Ochi, M. K., and Whalen, J. E., Prediction of the Severest Significant Wave Height, *Proceedings, Sixteenth Coastal Engineering Conference,* 1980.

Pierson, W. J., *A Unified Mathematical Theory for the Analysis, Propagation and Refraction of Storm Generated Ocean Surface Waves*, Research Division, College of Engineering, Department of Meteorology, New York University, Part 1, March 1, 1952; Part 2, July 1, 1952.

Pierson, W. J., and Moskowitz, L., A Proposed Spectral Form for Fully Developed Wind Seas Based on the Similarity Theory of Kitaigorodskii, *Journal of Geophysical Research* **69**, 1964.

Pierson, W. J., and Neumann, G., *Principals of Physical Oceanography*, Prentice-Hall, Englewood Cliffs, NJ, 1966.

Pierson, W. J., Neumann, G., and James, R. W., *Practical Methods for Observing and Forecasting Ocean Waves by Means of Wave Spectra and Statistics*, Publication No. 601, U.S. Navy Hydrographic Office, 1955. Reprinted 1960.

Price, W. G., and Bishop, R. E. D., *Probabalistic Theory of Ship Dynamics*, Chapman and Hall, London, 1974.

Putz, R. R., *Wave-Height Variability; Prediction of Distribution Function*, Technical Report No. HE 116-318, Series No. 3, Issue No. 318, Institute of Engineering Research, University of California, Berkeley, CA, 1950.

Putz, R. R., Statistical Distribution for Ocean Waves, *Transactions, American Geophysical Union* **33** (5), 1952.

Putz, R. R., Statistical Analysis of Wave Records, *Proceedings, Fourth Coastal Engineering Conference*, The Engineering Foundation Council on Wave Research, Berkeley, CA, 1954.

Rayleigh, Lord, On the Resultant of a Large Number of Vibrations of the Same Pitch and of Arbitrary Phase, *Philosophical Magazine* **10**, 1880.

Rye, H., and Svee, R., Parametric Representation of a Wave-Wind Field, *Proceedings, 14th Conference on Coastal Engineering*, Honolulu, Hawaii, 1976.

Seiwell, H. R., Sea Surface Roughness Measurements in Theory and Practice, *Annals, New York Academy of Science* **51** (3), 1949.

Sverdrup, H. U., and Munk, W. H., *Wind, Sea and Swell, Theory of Relations for Forecasting*, Publication No. 601, Hydrographic Office, U.S. Department of the Navy, 1947.

Wiegel, R. L., An Analysis of Data from Wave Recorders on the Pacific Coast of the United States, *Transactions, American Geophysical Union* **30**, 1949.

Young, I. R., *Wind Generated Waves*, Elsevier Science Ltd., Oxford, UK, 1999.

7

Statistical Responses for Single Degree of Freedom Linear Structures

James F. Wilson

The three commonly used methods for computing the dynamic responses of off-shore structures to waves are shown schematically in Figure 7.1. The first two methods involve the time domain and have been discussed in previous chapters. Here, the time history of structural response is computed either for a single design wave (the first method), or for multiple waves deduced from the partitioning of measured sea wave spectra (the second method). The third method, the topic of this chapter, involves the frequency domain. That is, for a single wave height spectrum and a transfer function relating wave height to structural loading, the statistical responses of the structure are calculated directly.

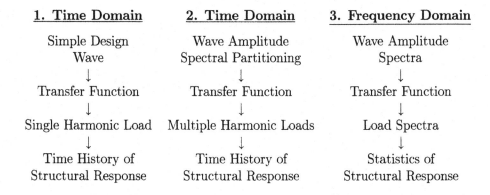

1. Time Domain	2. Time Domain	3. Frequency Domain
Simple Design Wave	Wave Amplitude Spectral Partitioning	Wave Amplitude Spectra
↓	↓	↓
Transfer Function	Transfer Function	Transfer Function
↓	↓	↓
Single Harmonic Load	Multiple Harmonic Loads	Load Spectra
↓	↓	↓
Time History of Structural Response	Time History of Structural Response	Statistics of Structural Response

Figure 7.1 Three approaches to the dynamic analysis of offshore structures.

The statistical structural responses for the single degree of freedom stuctures considered in this chapter are those for a single, nondirectional wave spectrum, and for *stationary* wave excitation. The derived structural responses are analogous to the *steady - state* time domain responses considered previously. Closed form solutions are derived by two statistical methods: the classical stationary stochastic analysis of Khintchine (1934) and Weiner (1949), and covariance propagation analysis originally applied by Bryson and Hu (1975) to electrical control systems. In both methods, the structural response spectrum is predicted in terms of the wave force spectrum on the structure, and the result leads to the root-mean-square (rms) structural displacement response and its probability of exceedence. These statistical ideas lay the mathematical framework for Chapters 9 and 10 in which statistical responses are deduced for multi-degree of freedom structures and for continuous structural elements.

7.1 AVERAGES AND PROBABILITIES

It is assumed *a priori* that the surface wave height $\eta(t)$, the wave load $p_1(t)$ and the structural displacement $v(t)$ or rotation $\theta(t)$ all have time histories of the general form shown in Figure 7.2. The variable $y = y(t)$ is used to denote such a general time history, which is defined as a random, stationary process of zero mean. A random, stationary process looks essentially the same over a time interval τ_0, no matter where this interval starts or stops. The interval τ_0 is of sufficient duration to capture the essential character of $y(t)$, and there is no startup, shutdown, or transient behavior for $y(t)$. A more precise definition of stationary will be given later in this section. The condition of zero mean is expressed as

$$E[y] = \frac{1}{\tau_0} \int_{-\tau_0/2}^{\tau_0/2} y(t)dt = 0 \tag{7.1}$$

In place of the expectation symbol $E[y]$, other notations commonly used to denote the time average of a function $y = y(t)$ are $\bar{y}$ and $< y(t) >$.

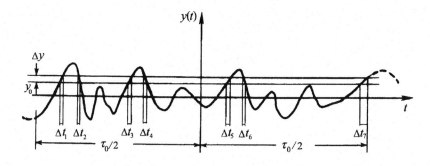

Figure 7.2 Typical time history of a stationary random variable of zero mean.

From the viewpoint of structural dynamics, equation (7.1) implies that the fluctuations are about a static, zero mean. For such cases the *variance* σ_y, defined as the mean square value of $y(t)$, is often used as a measure of these fluctuations. That is,

$$E[y^2] = \frac{1}{\tau_0} \int_{-\tau_0/2}^{\tau_0/2} y^2(t)dt = \sigma_y^2 \tag{7.2}$$

When equation (7.1) is satisfied, the *standard deviation* or the root-mean-square (rms) of $y(t)$ is defined as the positive square root of equation (7.2), or σ_y.

Once σ_y is calculated, its value needs to be interpreted in a statistical sense to be useful to the structural design engineer. For instance, if the rms value of deflection $y = v = v(t)$ is $\sigma_v = 2$ m, a practical question is: What are the chances that v will be smaller or larger than 2 m? Also, if the structure will fail statically for $v > 5$ m, what are the chances that $2.5\sigma_v = 5$ m will be exceeded? Thus, the *probability distribution function*, $P(v)$, and the *probability density function*, $p(v)$, expressed in terms of σ_v, are defined to answer these important questions.

Stripped of mathematical elegance, the increment of the probability distribution function, ΔP, is defined symbolically at $y = y_0$ as

$$\Delta P = P[y_0 \leq y \leq (y_0 + \Delta y)] \tag{7.3}$$

This is the probability that y is between a fixed value y_0 and $(y_0 + \Delta y)$. Specifically, ΔP is defined as the fraction of the total time τ_0 that $y(t)$ is in this bandwidth Δy. The vertical strips of Figure 7.2 show the typical time increments $\Delta t_1, \Delta t_2, \ldots$ within this bandwidth. Thus

$$\Delta P = \frac{1}{\tau_0}(\Delta t_1 + \Delta t_2 + \ldots) \tag{7.4}$$

The probability density function is defined as a limiting process, or

$$p(y) = \lim(\Delta y \longrightarrow 0) \frac{\Delta P}{\Delta y} = \frac{dP}{dy} \tag{7.5}$$

where y_0 is replaced by y for generality. Thus it is possible to obtain a plot of $p(y)$ versus y by dividing a sample trace $y(t)$ of duration τ_0 into a sufficient number of levels $y = y_0$, measuring the time spent at each level in each time band, and forming the ratio $\Delta P/\Delta y \simeq p(y)$ where ΔP is given by equation (7.4). With automated digital sampling (see Problem 7.3), this is accomplished in an efficient way.

In the statistical response calculations involving structures, $p(y)$ is rarely calculated. Instead it is assumed *a priori* that $p(y)$ has a *Gaussian* or *normal* distribution of zero mean, given in terms of the variance σ_y, or

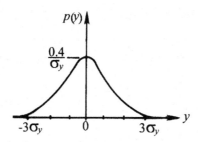

Figure 7.3   Gaussian probability density function of zero mean.

$$p(y) = \frac{1}{\sqrt{2\pi\sigma_y^2}} \exp\left(-\frac{y^2}{2\sigma_y^2}\right) \tag{7.6}$$

Equation (7.6) is depicted in Figure 7.3. In addition it is often assumed that the probability density function for the *peaks* of $y(t)$, with the amplitude a, is given by

$$p(a) = \frac{a}{\sigma_a^2} \exp\left(-\frac{a^2}{2\sigma_a^2}\right), \qquad 0 \le a \le \infty \tag{7.7}$$

Here σ_a^2 is the variance of the amplitude. Equation (7.7) is known as the Rayleigh probability density, which was presented in a somewhat different form for wave height in Chapter 6, or equations (6.6), (6.7), and Figure 6.2.

It is apparent from the above definitions that once the probability density function and the variance are established for a particular process, the probability of occurrence within prescribed limits of that process variable ($y(t)$ or its amplitude a) can then be calculated. These ideas are now illustrated.

Example Problem 7.1.   If $y(t)$ is a Gaussian process, what is the probability that $y(t)$ lies within the $\pm 3\sigma_y$ limits? Also out of 100 peaks of $y(t)$, how many would one expect to exceed $3\sigma_a$? To answer the first question, write the probability symbolically and then calculate the needed result by integrating equation (7.5) using equation (7.6). Thus

$$P[-3\sigma_y \le y \le 3\sigma_y] = \int_{-3\sigma_y}^{3\sigma_y} p(y)dy$$

$$= \frac{1}{\sqrt{2\pi\sigma_y^2}} \int_{-3\sigma_y}^{3\sigma_y} \exp\left(\frac{-y^2}{2\sigma_y}\right) dy = 0.9974 \tag{7.8}$$

It follows that the probability of y occurring outside the $\pm 3\sigma_y$ limits is (1 - 0.9974) = 0.0026, or only 0.26 percent. This is why "safe" design limits for the structural deflection, for $y = v$, for instance, are often chosen as $\pm 3\sigma_v$.

To answer the second question, consider the probability that the *peak* value of $y(t)$ lies within the zero to $3\sigma_a$ limits. Thus

$$P[0 \le a \le 3\sigma_a] = \int_0^{3\sigma_a} p(a)da = 0.989 \qquad (7.9)$$

which is based on the Rayleigh distribution, equation (7.7). It follows that the probability that any peak chosen at random *exceeds* $3\sigma_a$ is

$$P[|a| > 3\sigma_a] = 1 - 0.989 = 0.011 \qquad (7.10)$$

This shows that about one peak in 100 exceeds $3\sigma_a$. Further, it is not difficult to show that, for both the Gaussian and Rayleigh probability distribution functions, P is unity when the respective variable y or a extends throughout its whole range, or

$$P[-\infty \le (y \text{ or } a) \le \infty] = 1 \qquad (7.11)$$

Presented in the following sections of this chapter are two methods for evaluating the standard deviation σ_v or σ_θ for a structural displacement coordinate v or rotational coordinate θ. For the first method, it will be shown that solutions for σ_v and σ_θ can be based on a single measured surface wave height spectrum $S_\eta(\omega)$, the wave-to-structure load transfer function $G(\omega)$, and the structural response functions $h(t)$ and $H(\omega)$. For the second method, which is the newer of the two and based on control theory for electrical systems, the standard deviation is computed in closed form in terms of a three parameter spectral density representation of the excitation force. With the standard deviations computed by either method, appropriate probability density functions can then be used, as in the examples just presented, to evaluate probabilities of occurrence of structural displacements and rotations under wave loading.

7.2 STATIONARY AND ERGODIC HYPOTHESES

With the assumption that $y(t)$ is both *stationary* and *ergodic*, the statistical calculations leading to the structural standard deviation in displacement are simplified enormously. Although these assumptions are rarely checked in practice, it is nonetheless illuminating to elaborate on these two hypotheses from an "experimental" viewpoint, as suggested by Muga and Wilson (1970). To do this, cut a sample record $y(t)$ such as in Figure 7.2 into J equal charts, each of time duration τ_0. Again, τ_0 is of sufficient duration that it captures the essential character of $y(t)$. This ensemble of J charts is denoted as

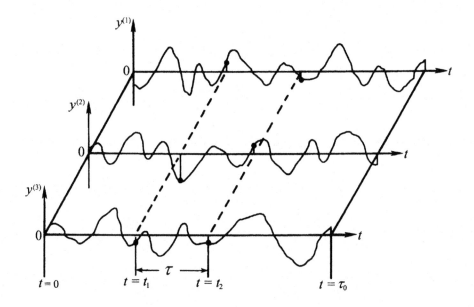

Figure 7.4　Ensemble of traces cut from Figure 7.2.

$$y^{(1)}, y^{(2)}, \ldots, y^{(i)}, \ldots, y^{(J)}$$

where i denotes a typical one. Figure 7.4 shows three typical charts of such an ensemble where each trace begins at time zero. The two broken lines across the ensemble correspond to arbitrary times t_1 and t_2. The *stationary* hypothesis is warranted if both of the following criteria are substantially met:

1. At a fixed time t_1, measure $y(t) = y^{(i)}(t_1)$ at $t = t_1$ on each chart. The average $y(t)$ is

$$\frac{1}{J} \sum_{i=1}^{J} y^{(i)}(t_1) \tag{7.12}$$

The numerical values of equation (7.12) should be about the same for any value of t_1 one chooses, where the value of J is a very large number.

2. Pick a constant time interval τ. For a time t_1 and another time t_2, where $(t_2 - t_1) = \tau$, measure the values of $y^{(i)}(t_1)$ and $y^{(i)}(t_2)$ on each chart i and form the sum

$$\frac{1}{J} \sum_{i=1}^{J} y^{(i)}(t_1) \, y^{(i)}(t_2) \tag{7.13}$$

The numerical values of equation (7.13) should be about the same for *any* value of t_1 and t_2, as long as $(t_2 - t_1) = \tau$ and J is very large.

The *ergodic* hypothesis is in two parts. The first part of this hypothesis states: the average value of y as given by equation (7.12), based on a constant t_1 over an ensemble of charts, is equal to the *time* average of y over just one typical chart. This must be true for all values of t_1. Suppose that the typical chart, which is representative of the ensemble of charts, is called $y^{(k)}(t)$. Suppose further that the chosen time interval over which $y^{(k)}(t)$ is considered is of sufficient duration that it represents the behavior of y reasonably well. Then the time average of $y^{(k)}(t)$ can be approximated as

$$\frac{1}{N} \sum_{n=1}^{N} y^{(k)}(t_n) \tag{7.14}$$

where $y^{(k)}(t_n)$ is the value of y at time t_n in the time interval τ_0. The values of t_n are equally spaced, and the total number of measurements is N. If expressions (7.12) and (7.14) are nearly equal, the first part of the ergodic hypothesis is approximately satisfied, or

$$\frac{1}{J} \sum_{i=1}^{J} y^{(i)}(t_1) \simeq \frac{1}{N} \sum_{n=1}^{N} y^{(k)}(t_n) \tag{7.15}$$

where J and N are both large numbers. In other words, the first part of the ergodic hypothesis is satisfied if the ensemble average of y for J charts is equal to the time average of y from a typical chart.

For the second part of the ergodic hypothesis another type of average is defined by

$$\frac{1}{N} \sum_{n=1}^{N} y^{(k)}(t_n)\, y^{(k)}(t_n + \tau) \tag{7.16}$$

where $y^{(k)}(t)$ is measured at N discrete times $t = t_n$ and $t = t_n + \tau$ along the one typical chart. Here τ is constant. If equations (7.13) and (7.16) are nearly equal, the second part of the ergodic hypothesis is approximately satisfied, or

$$\frac{1}{J} \sum_{i=1}^{J} y^{(i)}(t_1)\, y^{(i)}(t_2) \simeq \frac{1}{N} \sum_{n=1}^{N} y^{(k)}(t_n)\, y^{(k)}(t_n + \tau) \tag{7.17}$$

Based on this "experimental" viewpoint, these definitions are summarized. To the extent that the averages of equations (7.12) and (7.13) remain constant, $y(t)$ is stationary. To the extent that equality holds in equations (7.15) and (7.17), $y(t)$ is ergodic. It is observed that the ergodic condition implies a stationary condition, but $y(t)$ may conceivably be stationary without having ergodic properties.

7.3 AUTOCORRELATION AND SPECTRAL DENSITY

On the basis of the preceeding developments, two statistical parameters are defined. The first is the autocorrelation function, which can be thought of as the time average value of the product $y(t) \cdot y(t+r)$. Dropping the (k) superscript, this average is given approximately by equation (7.16), or

$$R_y(\tau) \simeq \frac{1}{N} \sum_{n=1}^{N} y(t_n)\, y(t_n + \tau) \qquad (7.18)$$

which is independent of time but not the time difference τ, provided that $y(t)$ is stationary. More precisely, $R_y(\tau)$ is defined by replacing the finite sum in equation (7.18) by the integral over the duration τ_0 where $N = \tau_0/\Delta t$ and $\Delta t \to dt$ in the limit. That is,

$$R_y(\tau) = \lim(\tau_0 \to \infty)\frac{1}{\tau_0} \int_{-\tau_0/2}^{\tau_0/2} y(t)\, y(t + \tau)dt = E[y(t)\, y(t + \tau)] \qquad (7.19)$$

It is noted that $R_y(\tau)$ is sometimes referred to as the autocovariance function of $y(t)$. As $\tau_0 \to \infty$, $y(t)$ fluctuates between positive and negative values and $R_y(\tau) \to 0$. Since $R_y(\tau)$ depends only on τ and not on absolute time t, then $R_y(\tau)$ is symmetric about $\tau = 0$, or $R_y(\tau) = R_y(-\tau)$.

The second important statistical parameter needed in this analysis is the *power spectral density function* or simply the *spectral density*, $S_y(\omega)$. This function was introduced in Chapter 6 to characterize wave height, where $S_\eta(\omega)$ was approximated by equations (6.5), and the Fourier coefficients were calculated from equation (6.2) for a given wave chart $\eta(t)$. Actually, $S_y(\omega)$ is defined precisely as the Fourier transform of $R_y(\tau)$, or

$$S_y(\omega) = \frac{1}{2\pi} \int_{-\infty}^{\infty} R_y(\tau)\, e^{-j\omega\tau} d\tau \qquad (7.20)$$

The reciprocal or inverse relationship is

$$R_y(\tau) = \int_{-\infty}^{\infty} S_y(\omega)\, e^{j\omega\tau} d\omega \qquad (7.21)$$

For $\tau = 0$, it is observed that $R_y(0) = \sigma_y^2$, the variance as defined by equation (7.2). Using the symmetry property, it follows that

$$\sigma_y^2 = E[y^2] = \int_{-\infty}^{\infty} S_y(\omega)\, d\omega = 2 \int_0^{\infty} S_y(\omega)d\omega \qquad (7.22)$$

In principle $R_y(\tau)$ can be calculated from equation (7.19) for a given $y(t)$, provided that τ_0 is sufficiently large, and $S_y(\omega)$ and σ_y^2 can be subsequently calculated from equations (7.20) and (7.22), respectively. The software package Mathematica® (1999) offers a convenient method for calculating the Fourier transform and the inverse Fourier transform from digitized data. In practice, however, it is usually more efficient to calculate $S_y(\omega)$ directly from experimental time histories (for $\eta(t)$, for instance) by means of an electronic instrument called a frequency analyzer such as described in Problem 7.4, or by means of computer-aided Fast Fourier Transform (FFT) methods to evaluate the Fourier coefficients of equation (6.2) for use in equations (6.5). Newland (1975) elaborated on these methodologies, and Cooley and Tukey (1965) presented efficient algorithms for calculating the Fourier coefficients.

Example Problem 7.2. A linear, single degree of freedom flexible stucture is subjected to a total wave force $p_1(t)$, in line with the motion of the structure. Assume a distribution of simple, linear waves for which the wave height $\eta(t)$ is stationary, ergodic, and Gaussian, with a zero mean and with a spectral density $S_\eta(\omega)$. Starting from basic definitions, relate the spectral density of the wave load $S_{p1}(\omega)$ to $S_\eta(\omega)$ through a known transfer function $G(\omega)$ defined by equation (4.24).

First, rewrite the structural load-wave height relationship as

$$p_1(t) = \eta(t)\,|G(\omega)| \tag{7.23}$$

Then rewrite equation (7.20) twice using equation (7.19), first substituting η for y and then p_1 for y. The results for the respective spectral densities are

$$S_\eta(\omega) = \frac{1}{2\pi} \int_{-\infty}^{\infty} E[\eta(t)\,\eta(t+\tau)]e^{-j\omega\tau}\,d\tau \tag{7.24}$$

$$S_{p1}(\omega) = \frac{1}{2\pi} \int_{-\infty}^{\infty} E[p_1(t)\,p_1(t+\tau)]e^{-j\omega\tau}\,d\tau \tag{7.25}$$

When equation (7.23) is substituted into equation (7.25) and this result is compared to equation (7.24), the required relationship is deduced as

$$S_{p1}(\omega) = |G(\omega)|^2 S_\eta(\omega) \tag{7.26}$$

7.4   STRUCTURAL RESPONSE STATISTICS: PART I

The derivation of response statistics that follow are based on a linear model in the form of equation (2.43), written in terms of the displacement coordinate v. However, similar results can be obtained based on a linear model in the form of equation (2.81), written in terms of the rotational coordinate θ. After going through the following derivation for the displacement model, the reader is encouraged to then derive the response statistics for the rotational model.

Given the linear structural model

$$m\ddot{v} + c_1\dot{v} + k_1 v = p_1(t) \tag{7.27}$$

for which the load spectral density $S_{p1}(\omega)$ is known through equation (7.26), the task now is to calculate the response spectral density $S_v(\omega)$ and the variance σ_v^2. Because these results are so important in applications, all of the assumptions and mathematical details needed for this derivation are included here. Because two particular response functions for equation (7.27) are required eventually, these are repeated for convenience. One is the harmonic response function given by equation (5.62), or

$$H(\omega) = k_1(-m\omega^2 + jc_1\omega + k_1)^{-1} \tag{7.28}$$

The other is the impulse response function given by equation (5.74), or

$$h(t) = \frac{1}{m\omega_d}e^{-\zeta\omega_0 t}\sin\omega_d t \tag{7.29}$$

The Fourier Transform

A preliminary step is to relate $H(\omega)$ to $h(t)$ through the Fourier transform. To do this, the steady-state solution to equation (7.27), as given by the convolution integral in equation (5.76), is first written as

$$v = \int_{-\infty}^{t} p_1(\tau)\,h(t-\tau)d\tau \tag{7.30}$$

Here the lower limit $\tau = 0$ was replaced by $\tau = -\infty$ since $p_1(t)$ vanishes for $\tau < 0$, leaving the value of the integral unchanged. Further it is recalled that $h(t-\tau)$ is the response to a unit impulse at $(t-\tau) = 0$. For $(t-\tau) < 0$, the response v is zero because the unit impulse has not yet come into existence. Thus for $\tau < t$, $h(t-\tau) = 0$, and the upper limit $t = \tau$ may be extended to $t = \infty$ without changing the value of this integral. That is

$$v = \int_{-\infty}^{\infty} p_1(\tau)\,h(t-\tau)d\tau \tag{7.31}$$

Now define a variable change: $\theta = t - \tau$, where $d\tau = -d\theta$. The lower limit $\tau = -\infty$ now changes to $\theta = \infty$, and the upper limit $\tau = \infty$ changes to $\theta = -\infty$. It follows that

$$v = \int_{\infty}^{-\infty} p_1(t-\theta)h(\theta)(-d\theta) \tag{7.32}$$

Change the sign of the integral and reverse the limits of integration, or

$$v = \int_{-\infty}^{\infty} p_1(t-\theta)h(\theta)d\theta \tag{7.33}$$

Now rename the dummy variable of integration, or let $\theta = \tau$, which gives

$$v = \int_{-\infty}^{\infty} p_1(t-\tau)h(\tau)d\tau \tag{7.34}$$

To show the relationship between $h(t)$ and $H(\omega)$, the solution v must now be expressed in terms of $H(\omega)$. To do this, set

$$p_1(t) = p_0 e^{j\omega t} \tag{7.35}$$

which can be rewritten as

$$p_1(t - \tau) = p_0 e^{j\omega t} e^{-j\omega \tau} \tag{7.36}$$

With this last result, the solution to equation (7.34) becomes

$$v = p_0 e^{j\omega t} \int_{-\infty}^{\infty} h(\tau) e^{-j\omega \tau} d\tau \tag{7.37}$$

It is recalled that equation (5.60) is the solution to equation (7.27) compatible with equations (7.28) and (7.35). That is,

$$v = \frac{p_0}{k_1} H(\omega) e^{j\omega t} \tag{7.38}$$

When the last two results are equated and the dummy variable is changed to t, the connection between $h(t)$ and $H(\omega)$ is established as

$$\frac{1}{k_1} H(\omega) = \int_{-\infty}^{\infty} h(t) e^{-j\omega t} dt \tag{7.39}$$

The function $H(\omega)/k_1$ is the Fourier transform of $(2\pi)h(t)$. The inverse Fourier transform of the latter yields

$$h(t) = \frac{1}{2\pi} \int_{-\infty}^{\infty} \frac{1}{k_1} H(\omega) e^{j\omega t} d\omega \tag{7.40}$$

The Autocorrelation Functions

From the definition of the autocorrelation of response $R_v(\tau)$ given by equation (7.19) and the solution v given by equation (7.31), it follows that

$$R_v(\tau) = E[v(t)v(t + \tau)]$$

$$= E\left[\int_{-\infty}^{\infty} h(\theta_1) p_1(t - \theta_1) d\theta_1 \int_{-\infty}^{\infty} h(\theta_2) p_1(t + \tau - \theta_2) d\theta_2 \right] \tag{7.41}$$

where θ_1 and θ_2 replace τ to avoid confusion. Assume that $v(t)$ is stable and that these integrals converge. The term $R_v(\tau)$ can then be written as a double integral where the order of averaging and integration is interchanged. That is

$$R_v(\tau) = E\left[\int_{-\infty}^{\infty} \int_{-\infty}^{\infty} h(\theta_1) h(\theta_2) p_1(t - \theta_1) p_1(t + \tau - \theta_2) d\theta_1 d\theta_2 \right]$$

$$= \int_{-\infty}^{\infty} \int_{-\infty}^{\infty} h(\theta_1)h(\theta_2)E[p_1(t - \theta_1)p_1(t + \tau - \theta_2)]d\theta_1 d\theta_2 \qquad (7.42)$$

Next assume that $p_1(t)$ is stationary and ergodic. The autocorrelation function for this loading is then independent of time t and that portion of the integrand of equation (7.42) involving $p_1(t)$ can be written as

$$E[p_1(t - \theta_1)p_1(t + \tau - \theta_2)] = R_{p1}(\tau - \theta_2 + \theta_1) \qquad (7.43)$$

Equation (7.43) is simply the autocorrelation function for $p_1(\tau)$ with a time lag of $(-\theta_2 + \theta_1)$. The last two equations are combined to give

$$R_v(\tau) = \int_{-\infty}^{\infty} \int_{-\infty}^{\infty} h(\theta_1)h(\theta_2)R_{p1}(\tau - \theta_2 + \theta_1)d\theta_1 d\theta_2 \qquad (7.44)$$

Response Parameters

The spectral density of the response v is defined as the Fourier transform of $R_v(\tau)$. That is, from equation (7.20),

$$S_v(\omega) = \frac{1}{2\pi} \int_{-\infty}^{\infty} R_v(\tau)e^{-j\omega\tau}d\tau \qquad (7.45)$$

When the last two equations are combined, then

$$S_v(\omega) = \frac{1}{2\pi} \int_{-\infty}^{\infty} e^{-j\omega\tau}d\tau \int_{-\infty}^{\infty} \int_{-\infty}^{\infty} h(\theta_1)h(\theta_2)R_{p1}(\tau - \theta_2 + \theta_1)d\theta_1 d\theta_2 \quad (7.46)$$

After interchanging the order of integration in equation (7.46) and inserting the following identity in the integrands,

$$e^{j\omega\theta_1}e^{-j\omega\theta_2}e^{-j\omega(\theta_1 - \theta_2)} = 1 \qquad (7.47)$$

the result is a product of three integrals given by

$$S_v(\omega) = \int_{-\infty}^{\infty} h(\theta_1)e^{j\omega\theta_1}d\theta_1 \cdot \int_{-\infty}^{\infty} h(\theta_2)e^{-j\omega\theta_2}d\theta_2$$

$$\cdot \frac{1}{2\pi} \int_{-\infty}^{\infty} R_{p1}(\tau - \theta_2 + \theta_1)e^{-j\omega(\tau - \theta_2 + \theta_1)}d\tau \qquad (7.48)$$

When compared with equation (7.39), it is observed that the first two integrals on the right of the last equation are $H(-\omega)/k_1$ and $H(\omega)/k_1$, respectively. With equation (7.20), the last integral in equation (7.48) is identified as the power spectral density of $p_1(t)$ with a time shift of $(-\theta_2 + \theta_1)$. The product of these first two integrals is

$$\frac{1}{k_1^2}H(-\omega) \cdot H(\omega) = \frac{1}{k_1^2}|H(\omega)|^2 \qquad (7.49)$$

With this last result, equation (7.48) becomes

$$S_v(\omega) = \frac{1}{k_1^2}|H(\omega)|^2 S_{p1}(\omega) \tag{7.50}$$

This last remarkably simple and useful result relates the power spectral density of $p_1(t)$ to the power spectral density of $v(t)$ through the complex frequency response function. As shown in *Example Problem 7.2*, $S_{p1}(\omega)$ is given by equation (7.26) in terms of $S_\eta(\omega)$ and $G(\omega)$. With this and the modulus of the harmonic response function given by equation (5.63), the response spectral density of equation (7.50) for the linear structural model of equation (7.27) is deduced as

$$S_v(\omega) = \frac{|G(\omega)|^2 S_\eta(\omega)}{(k_1 - m\omega^2)^2 + c_1^2\omega^2} \tag{7.51}$$

The variance of the response is then

$$\sigma_v^2 = 2\int_0^\infty \frac{|G(\omega)|^2 S_\eta(\omega)}{(k_1 - m\omega^2)^2 + c_1^2\omega^2}\,d\omega \tag{7.52}$$

which follows from its definition given by equations (7.22) and (7.51).

These results are summarized. To calculate σ_v, the model parameters k_1, c_1, and m are identified. Then $G(\omega)$ is calculated by the methods discussed in Chapter 4 where linear small-amplitude wave theory is assumed. After a design wave height spectrum is chosen, such as the Pierson-Moskowitz form of equation (6.21) or the JONSWAP form of equation (6.24), then σ_v is calculated from equation (7.52) in which the limits of integration $(0, \infty)$ can be replaced for practical purposes by $(0.16, 1.60)$ rad/sec.

It is important to note that for a linear structure, if $\eta(t)$ and $p_1(t)$ are Gaussian, then the response $v(t)$ is also Gaussian (Newland, 1975). Thus, with the assumption that the excitation is Gaussian, then the probability that $v(t)$ will exceed the calculated $\pm 3\sigma_v$ limits is only 0.26 percent. Thus if a *static* displacement of $v = \pm 3\sigma_v$ yields peak stresses within the allowable limits, the structure is a practical one from the general viewpoint of structural dynamics. However, this type of calculation does not exclude the possibility of local material failure by fatigue.

The following comprehensive example brings together many of the basic ideas elaborated on in this and preceding chapters. In working through such problems, the reader is reminded of the many and sometimes subtle assumptions involved in this analysis and is cautioned to temper the interpretation of numerical results accordingly.

Example Problem 7.3. The free, undamped lateral motion of the three-legged jackup rig shown in Figure 2.17 has already been investigated in *Example Problem 5.4*. Now include light damping, and subject this structure to steady, unidirectional linear waves with a significant wave weight $H_s = 15$ m and with a distribution given by the Pierson-Moskowitz spectrum, equation (6.21). Based on the result given by equation (7.52) for a single degree of freedom structural model, calculate the variance σ_v^2 for the horizontal displacement of the deck. Then assuming that the wave height distribution is Gaussian, calculate and interpret the following quantities based on $\pm 3\sigma_v$: the horizontal deck displacement, the horizontal shear force in each leg, and the overturning moment. State any further assumptions needed to make these calculations.

The first step is to clarify the mathematical model of equation (7.27). Identify the coordinate v, the constants for the stiffness k_1, the mass m, and the damping c_1. Let v be the absolute horizontal displacement of the deck where v is on the average much smaller than the horizontal wave particle velocity u. From *Example Problem 5.4*, the undamped natural frequency ω_0 for this structure is given as the last item in Table 5.2, or

$$\omega_0 = \sqrt{\frac{k_1}{m}} = \sqrt{\frac{6.82 \times 10^4 \text{ lb/in.}}{3.68 \times 10^4 \text{ lb-sec}^2/\text{in.}}} = 1.36 \text{ rad/sec} \qquad (7.53)$$

In this calculation, the equivalent stiffness and the virtual mass are: $k_1 = 8.18 \times 10^5$ lb/ft and $m = 4.42 \times 10^5$ slugs (or lb-sec^2/ft). Lacking the required data for c_1, assume $\zeta = 0.05$. With equation (5.64), it follows that $c_1 = 2\zeta\sqrt{k_1 m} = 6.01 \times 10^4$ lb-sec/ft.

The second step is to choose a reasonable wave theory and then calculate $G(\omega)$, the corresponding transfer function for wave loading of the legs. Assume that the inertial flow regime dominates. Then $p_1(t)$, or the total horizontal load on all three legs, is found by setting $C_D = 0$ and integrating $\bar{q}$ given by Morison's equation (2.14) over the range from $z = -d$ to $z = 0$. That is,

$$p_1(t) = 3C_M \frac{\pi}{4}\rho D^2 \int_{z=-d}^{0} \dot{u}\, dz \qquad (7.54)$$

This last form implies that the wave forces on the legs are assumed to be statistically independent of each other so that the horizontal wave particle acceleration $\dot{u}$ is essentially the same on all three legs. Let $x = 0$ be the location of the water particles on each leg so that the value of $\dot{u}$ for a single wave as given in Table 3.1 can be employed, or

$$\dot{u} = -\frac{H}{2}\omega^2 \frac{\cosh k(z+d)}{\sinh kd} \sin \omega t \qquad (7.55)$$

where the wave period T was replaced by $2\pi/\omega$ and the wave amplitude A was replaced by $H/2$. With this last result, equation (7.54) can be integrated to give

$$\frac{1}{H}p_1(t) = -\frac{3\pi}{8}\frac{\omega^2}{k}\rho D^2 C_M \sin \omega t \qquad (7.56)$$

Now assume deepwater waves so that the wave number-frequency relationship of equation (3.16) reduces to $\omega^2/k = g$. Making this substitution in the righthand side of equation (7.56) and then expressing the result in complex notation leads to the required transfer function, or

$$G(\omega) = j\frac{3\pi}{8}\rho g D^2 C_M e^{j\omega t} \tag{7.57}$$

It follows that

$$|G(\omega)|^2 = \left(\frac{3\pi}{8}\rho g D^2 C_M\right)^2 \tag{7.58}$$

Since it is tacitly assumed in the mathematical model that $p_1(t)$ is applied at the *deck level*, this transfer function is conservative; that is, the predicted value of σ_v^2 will be on the high side.

The third step is to specify the constants g and B of the Pierson-Moskowitz wave height spectrum of equation (6.21), or

$$S_\eta(\omega) = 0.0081\frac{g^2}{\omega^5}e^{-B/\omega^4} \tag{7.59}$$

In this example, traditional English units are used so that the numerical coefficient of the exponential term above is $(0.0081)(32.2)^2 = 8.40$ ft^2/sec^4. From equation (6.23), $B = 3.11/H_s^2 = 3.11/15^2 = 0.0138$. Note that the constant 3.11

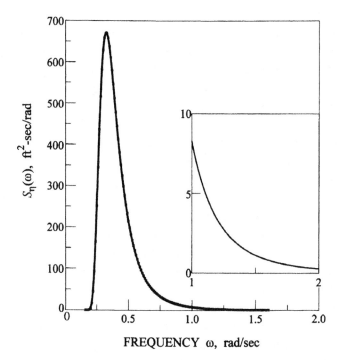

Figure 7.5 Pierson-Moskowitz wave height spectrum for $H_s = 15$ m.

in this calculation requires that H_s be expressed in meters. The latter equation thus becomes

$$S_\eta(\omega) = \frac{8.40}{\omega^5} e^{-0.0138/\omega^4} \text{ ft}^2\text{-sec/rad} \tag{7.60}$$

The units displayed for equation (7.60) are appropriate only if ω is given in rad/sec. A plot of equation (7.60), given in Figure 7.5, shows that for all practical purposes the wave energy is confined to frequency range of 0.16 to 1.6 rad/sec, the limits of integration in the variance integral. With equations (7.58) and (7.60), the variance equation (7.52) is thus

$$\sigma_v^2 = \left(\frac{3\pi}{8}\rho g D^2 C_M\right)^2 2 \int_{0.16}^{1.6} \frac{8.40\, e^{-0.0138/\omega^4}}{\omega^5[(k_1 - m\omega^2)^2 + c_1^2\omega^2]} d\omega \tag{7.61}$$

where

$$
\begin{aligned}
\rho g &= 64.3 \text{ lb/ft}^3, \text{ water density} \\
D &= 12 \text{ ft, single leg diameter} \\
C_M &= 2.0, \text{ assumed inertia coefficient} \\
m &= 4.42 \times 10^5 \text{ slugs (lb-sec}^2\text{/ft), equivalent mass} \\
k_1 &= 8.18 \times 10^5 \text{ lb/ft, bending stiffness} \\
c_1 &= 6.01 \times 10^4 \text{ lb-sec/ft, damping}
\end{aligned}
$$

When equation (7.61) was evaluated numerically, the variance was $\sigma_v^2 = 0.276$ ft^2, giving $\sigma_v = 0.526$ ft for the *rms* deck deflection. For a Gaussian process, the probability of exceeding the following dynamic responses is 0.26 percent:

deck displacement: $v = 3\sigma_v = 1.58$ ft
horizontal shear load per leg: $f_{max} = k_1 v/3 = 4.30 \times 10^5$ lb
overturning moment: $M_{max} = \ell f_{max} = 1.14 \times 10^8$ ft-lb,
(leg height: $\ell = 265$ ft)

Approximate Responses

In some applications, the spectral density of the loading $S_{p1}(\omega)$ can be approximated as a constant S_0 over a frequency band between the limits of ω_1 and ω_2 and as zero outside that frequency band. That is

$$S_{p1}(\omega) = |G(\omega)|^2 S_\eta(\omega) = S_0 = \text{const.}, \quad \omega_1 \le (\omega, \omega_0) \le \omega_2 \tag{7.62a}$$

$$S_{p1}(\omega) = 0, \quad \omega < \omega_1 \text{ and } \omega > \omega_2 \tag{7.62b}$$

Equations (7.62) define *band-limited white noise*. The designation *white noise* originated with the description of white light for which the spectrum is nearly uniform over the range of frequency for visible light. In the present applications, it is noted that S_0 is a one-sided spectrum since it is based on the one-sided experimental wave spectrum $S_\eta(\omega)$. Further, it is assumed that the undamped frequency ω_0 of the single degree of freedom system is within the defined frequency band of equations (7.62). With these equations it follows that the variance of the deflection response, equation (7.52), has the following form:

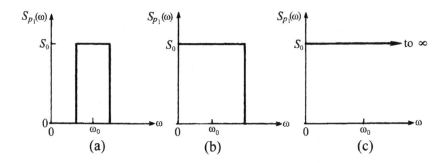

Figure 7.6   Three idealizations of the load excitation spectrum: (a) band-limited white noise; (b) white noise with a cut-off frequency; and (c) ideal white noise.

$$\sigma_v^2 = 2S_0 \int_{\omega_1}^{\omega_2} \frac{d\omega}{(k_1 - m\omega^2)^2 + c_1^2\omega^2} \tag{7.63}$$

The solution to the latter equation, given by Crandall and Mark (1963), is

$$\sigma_v^2 = \frac{\pi S_0}{k_1 c_1}[I(\omega_2/\omega_0) - I(\omega_1/\omega_0)], \quad \omega_1 \le \omega_0 \le \omega_2 \tag{7.64}$$

in which the integrals $I(\omega_i/\omega_0)$, $i = 1, 2$, are computed from

$$I(\omega_i/\omega_0) = \frac{1}{\pi}\tan^{-1}\frac{2\zeta(\omega_i/\omega_0)}{1 - (\omega_i/\omega_0)^2}$$

$$+ \frac{\zeta}{2\pi\sqrt{1-\zeta^2}}\ln\frac{1 + (\omega_i/\omega_0)^2 + 2(\omega_i/\omega_0)\sqrt{1-\zeta^2}}{1 + (\omega_i/\omega_0)^2 - 2(\omega_i/\omega_0)\sqrt{1-\zeta^2}} \tag{7.65}$$

This solution includes, in addition to band-limited white noise, two other special cases: white noise with a cut-off frequency ω_c, and ideal white noise. These three cases are depicted in Figure 7.6. For white noise S_0 with a cut-off frequency ω_c, then $\omega_1 = 0$, $\omega_2 = \omega_c$ and the integral term of equation (7.64) becomes

$$I(\omega_c/\omega_0) - I(0) = I(\omega_c/\omega_0) \tag{7.66}$$

For ideal white noise where S_0 is uniform for $0 \le \omega \le \infty$, the integral term is

$$I(\omega_2/\omega_0) - I(0) = I(\infty) = 1 \tag{7.67}$$

The latter result follows from the behavior of the arctangent term in equation (7.65): as $\omega_2 \to \infty$, its argument is large and negative, and $\tan^{-1}(-\infty) \to \pi$.

Also, as $\omega_2 \to \infty$, the argument of the natural logrithmic term approaches unity and $\ln (1) = 0$. Thus, for ideal white noise, S_0 is uniform for all frequencies and

$$\sigma_v^2 = \frac{\pi S_0}{k_1 c_1} \tag{7.68}$$

Example Problem 7.4. For the jackup rig described in *Example Problems 5.4* and *7.3*, compute the upper bound for the *rms* deck deflection, σ_v, for each of the three idealized loading spectra shown in Figure 7.6. In all three cases, choose

$$S_0 = |G(\omega)|^2 S_\eta(\omega) = (4.76 \times 10^8)(336.3) = 1.60 \times 10^{11} \text{ lb}^2\text{-sec/rad}$$

Here the quantity 336.3 ft^2-sec/rad is arbitrarily chosen as one-half the peak value of the wave height spectrum shown in Figure 7.5, the value that corresponds to the frequency of $\omega = 0.324$ rad/sec. The results are summarized.

(a) For the *band-limited* spectrum of Figure 7.6a, choose the same frequency limits that were used in the direct integration of equation (7.61), or $\omega_1 = 0.16$ rad/sec and $\omega_2 = 1.6$ rad/sec. Compute the variance by evaluating equations (7.64)-(7.66), from which

$$\sigma_v = 3.1066 \text{ ft}$$

(b) For the *cut-off frequency* spectrum of Figure 7.6b, choose $\omega_1 = 0$ and $\omega_2 = \omega_c = 1.6$ rad/sec. Compute the variance by evaluating equations (7.64) and (7.65) using (7.67). The result is

$$\sigma_v = 3.1079 \text{ ft}$$

(c) For *ideal white noise*, use equation (7.68) to give

$$\sigma_v = 3.1976 \text{ ft}$$

These results show a small but progressive increase in the *rms* deflection as the band width increases. However, these idealized approximations all led to a gross overestimate in σ_v by about an order of magnitude, compared to the value of 0.372 ft computed by direct integration of equation (7.61). The conclusion is that direct integration gives the best answer, at least for this type of problem in which the structure's fundamental frequency ω_0 is at the very low end of the wave height spectrum.

Extensions

There are several possible extensions to the classical statistical results obtained thus far in this chapter. These extensions involve the relaxation of certain restrictive assumptions upon which the variance of the response given by equation (7.52) was based. Two of those assumptions were: there exists a single, stationary excitation wave force of a known spectrum $S_{p1}(\omega)$ and the wave excitation force has a zero mean value. This assumption of zero mean was investigated by Tung (1974), who showed how a single, stationary wave excitation spectrum is modified when the mean is not zero, which corresponds to the case of steady current forces and wave forces acting simultaneously on the structure. In typical examples, however, Tung showed that such simultaneous action had a definite but relatively small effect on the structural response. Thus the practice of superimposing the effects of current as a static loading on the structure under wave excitation seems to be justified, providing that all loading due to vortex shedding in the current field is negligible.

Another extension is the case of two or more stationary wave excitation forces that are uncorrelated or statistically independent, and all with a zero mean value. Suppose that the N such spectral densities are known and denoted by $S_i(\omega)$, $i = 1, 2, ..., N$. From lengthy but straightforward calculations, the structural response analogous to equation (7.52) then becomes

$$S_v(\omega) = \sum_{i=1}^{N} \frac{1}{k_i^2} |H_i(\omega)|^2 S_i(\omega) \tag{7.69}$$

where $H_i(\omega)/k_i$ is the ith harmonic response function.

Statistical response theories and numerical results for linear and occasionally for nonlinear systems, subjected to stationary and nonstationary excitation, appear from time to time in engineering and applied mathematics journals. These analyses, although rarely lacking in elegance, do require experimentally derived wave data (which are lacking) to be useful in applications to offshore structures. For further expositions, the reader can consult the works of Gould and Abu-Sitta (1980), Lutes and Sarkani (1997), Newland (1975), and Yang (1986), all of whom include many source references. Of particular interest may be the incorporation of a time lag in excitation such as discussed by Hedrick and Firouztash (1974), an analysis applicable to response calculations for structures whose components (legs, braces, etc.) are sufficiently close so that there is a correlation of the wave forces among the components.

Presented up to this point was the classical statistical response analysis for linear structures subjected to stationary excitation, an analysis that forms the basis for similar studies of the multi-degree of freedom and of continuous linear structures in Chapters 9 and 10. Presented now is an introduction of an alternative to this classical statistical analysis.

7.5 STRUCTURAL RESPONSE STATISTICS: PART II

Modern control theory, developed mainly after 1960 for use by electrical and mechanical engineers, offers some powerful techniques of dynamic statistical analysis that are applicable to offshore structures. In this analysis, the linear differential equations representing structural motion the excitation forces are cast in state-variable form, those forms are transformed to a statistical representation called covariance propagation, and the latter result is then solved to obtain the statistical responses. These ideas are now discussed and illustrated using a single degree of freedom linear structure subjected to stationary wave excitation. The theory is freely drawn from the expositions of Bryson and Hu (1975), Lin (1967), and Hedrick (1984).

State Variable Form

Equation (7.27), governing the structural motion $v = v(t)$, is rewritten as

$$\ddot{v} + 2\zeta\omega_0\dot{v} + \omega_0^2 v = \frac{1}{m}p_1(t) \tag{7.70}$$

$$\omega_0 = \sqrt{\frac{k_1}{m}}; \qquad \zeta = \frac{c_1}{2\sqrt{k_1 m}} \tag{7.71}$$

The spectral density of a stationary excitation force, equation (7.26), is

$$S_{p1}(\omega) = |G(\omega)|^2 S_\eta(\omega) \tag{7.72}$$

where $G(\omega)$ is the load transfer function and $S_\eta(\omega)$ is the surface wave height spectral density. This force excitation spectral density is now arbitrarily fitted to the following equation:

$$S_{p1}(\omega) \simeq \hat{\alpha}\left[\left(1 - \frac{\omega^2}{\hat{\omega}^2}\right)^2 + \left(2\hat{\zeta}\frac{\omega}{\hat{\omega}}\right)^2\right]^{-1} \tag{7.73}$$

Here, the constants $\hat{\alpha}$, $\hat{\omega}$, and $\hat{\zeta}$ are picked to give a *best fit* to the right side of equation (7.72). There are two reasons for picking the latter form. First, when $|G(\omega)|^2$ is constant, then $S_\eta(\omega)$ has this general shape of equation (7.73). Second, that form is precisely the spectral density obtained by passing white noise $w(t)$ through a linear filter given by

$$\ddot{p}_1(t) + 2\hat{\zeta}\hat{\omega}\,p_1(t) + \hat{\omega}^2 p_1(t) = \hat{\omega}^2\hat{\alpha}^{1/2}w(t) \tag{7.74}$$

where the white noise has zero mean, or

$$E[w(t)] = 0 \tag{7.75}$$

$$E[w(t)\,w(t+\tau)] = Q\delta(\tau) \tag{7.76}$$

and the intensity of the white noise Q is unity.

Equations (7.70) and (7.74) are now expressed in the *state variable* form, or four first order differential equations in the following matrix form. That is

$$\dot{\mathbf{z}} = \mathbf{F}\mathbf{z} + \Gamma\mathbf{w} \tag{7.77}$$

where $\mathbf{F}$ and Γ are constant matrices and for brevity the argument t is omitted from the variables $\mathbf{z}$, $\mathbf{w}$ and their components. The state variables are defined from the two governing differential equations (7.70) and (7.74). These and the other matrices of equation (7.77) are as follows:

$$\mathbf{z} = [z_1 \quad z_2 \quad z_3 \quad z_4]^T = \left[v \quad \dot{v} \quad \frac{p_1}{m} \quad \frac{\dot{p}_1}{m} \right]^T \tag{7.78}$$

$$\dot{z}_1 = \dot{v} = z_2 \tag{7.79a}$$

$$\ddot{z}_2 = \ddot{v} = -2\zeta\omega_0 z_2 - \omega_0^2 z_1 + z_3 \tag{7.79b}$$

$$\dot{z}_3 = \frac{\dot{p}_1}{m} = z_4 \tag{7.79c}$$

$$\dot{z}_4 = -2\hat{\zeta}\hat{\omega} z_4 - \hat{\omega}^2 z_3 + \frac{1}{m}\hat{\omega}^2\hat{\alpha}^{1/2}w \tag{7.79d}$$

$$\mathbf{F} = \begin{bmatrix} 0 & 1 & 0 & 0 \\ -\omega_0^2 & -2\zeta\omega_0 & 1 & 0 \\ 0 & 0 & 0 & 1 \\ 0 & 0 & -\hat{\omega}^2 & -2\hat{\zeta}\hat{\omega} \end{bmatrix} \tag{7.80}$$

$$\mathbf{\Gamma w} = \left[0 \quad 0 \quad 0 \quad \frac{1}{m}\hat{\omega}^2\hat{\alpha}^{1/2}w \right]^T \tag{7.81}$$

In this case, the only nonzero term of the 4 by 4 matrix Γ is $\Gamma_{4,4}$ which is $\hat{\omega}^2\hat{\alpha}^{1/2}/m$.

Covariance Propagation Equation: Derivation

The next task is to cast the state variable form equation (7.77) in its statistical counterpart, the *covariance propagation equation*, also in state variable form. It will be shown that solutions to this latter matrix equation yield statistical responses, which include the variance of displacement σ_v^2 for a single degree of freedom structure. The derivation that follows, based on the expositions of Hedrick (1984) and Lin (1967), is general in that the results are applicable to multi-degree of freedom linear structures also. However, zero mean is assumed for both the state variable $\mathbf{z}$ and the white noise $\mathbf{w}$, or

$$E[\mathbf{z}] = E[\mathbf{w}] = \mathbf{0} \tag{7.82}$$

Define the covariance propagation matrix $\mathbf{Z}$ for zero mean. Let

$$\mathbf{Z}(t) = \mathbf{Z} = E[\mathbf{z}\,\mathbf{z}^T] \tag{7.83}$$

The time derivative of the last equation is

$$\frac{d}{dt}\mathbf{Z} = \dot{\mathbf{Z}} = E[\dot{\mathbf{z}}\mathbf{z}^T + \mathbf{z}\dot{\mathbf{z}}^T] \tag{7.84}$$

Substitute $\dot{\mathbf{z}}$ of equation (7.77) into the last equation:

$$\dot{\mathbf{Z}} = E[(\mathbf{Fz} + \mathbf{\Gamma w})\mathbf{z}^T + \mathbf{z}(\mathbf{Fz} + \mathbf{\Gamma w})^T] \tag{7.85}$$

Taking advantage of the linearity of the expectation operator E, the last equation becomes

$$\dot{\mathbf{Z}} = \mathbf{F}E[\mathbf{z}\mathbf{z}^T] + E[\mathbf{z}\mathbf{z}^T]\mathbf{F}^T + \mathbf{\Gamma}E[\mathbf{w}\mathbf{z}^T] + E[\mathbf{z}\mathbf{w}^T]\mathbf{\Gamma}^T$$

$$= \mathbf{FZ} + \mathbf{ZF}^T + \mathbf{\Gamma}E[\mathbf{w}\mathbf{z}^T] + E[\mathbf{z}\mathbf{w}^T]\mathbf{\Gamma}^T \tag{7.86}$$

What remains is to evaluate the last two terms on the right of the latter equation. Begin by defining the state transition matrix $\phi(t, \tau)$ with the following properties:

$$\frac{d}{dt}\phi(t, \tau) = \mathbf{F}\phi(t, \tau), \qquad \phi(t, t) = \mathbf{I} \tag{7.87}$$

Since the system is linear, it follows that

$$\mathbf{z}(t) = \phi(t, \tau_0)\mathbf{z}(t_0) + \int_{t_0}^{t} \phi(t, \tau)\mathbf{\Gamma w}(\tau)d\tau \tag{7.88}$$

Define the expectation as

$$E[\mathbf{w}(t)\mathbf{z}(t)] = E\left[\mathbf{w}(t)\left(\phi(t, t_0)\mathbf{z}(t_0) + \int_{t_0}^{t} \phi(t, \tau)\mathbf{\Gamma w}(\tau)d\tau\right)^T\right] \tag{7.89}$$

Assume that the following expectation is valid:

$$E[\mathbf{z}(t_0)\mathbf{w}^T(t)] = 0 \tag{7.90}$$

Use the last result and then invoke linearity to put the expectation operator E under the integral of equation (7.89). Since $\mathbf{w}(t)$ is independent of the integration variable τ, include $\mathbf{w}(t)$ under that integral as well. Thus, equation (7.89) becomes

$$E[\mathbf{w}(t)\mathbf{z}^T(t)] = \int_{t_0}^{t} E[\mathbf{w}(t)\mathbf{w}^T(\tau)]\mathbf{\Gamma}^T\phi(t, \tau)d\tau \tag{7.91}$$

Assume white noise of the form

$$E[\mathbf{w}(t)\mathbf{w}^T(\tau)] = \mathbf{Q}\delta(t - \tau) \tag{7.92}$$

with which equation (7.91) becomes

$$E[\mathbf{w}(t)\mathbf{z}^T(t)] = \int_{t_0}^t \mathbf{Q}\delta(t-\tau)\mathbf{\Gamma}^T\phi(t,\tau)d\tau \qquad (7.93)$$

With the identity relation of equation (7.87), the last result becomes

$$E[\mathbf{w}(t)\mathbf{z}^T(t)] = \left(\int_{t_0}^t \mathbf{Q}\delta(t-\tau)d\tau\right)\mathbf{\Gamma}^T \qquad (7.94)$$

The problem in evaluating the last integral is that the impulse occurs at the *end point* of the time interval. Use the symmetry property of the white noise autocorrelation function, or $R_w(\tau) = R_w(-\tau)$. The impulse can be approximated as the magnitude of any symmetric function whose time duration ε approaches zero in the limit. Choose a rectangular pulse at $\tau = t$ of duration ε and of magnitude Q/ε. The symmetry property leads to the evaluation of the integral of equation (7.94) as

$$\int_{t_0}^t \mathbf{Q}\delta(t-\tau)d\tau = \int_{t-\varepsilon/2}^{t+\varepsilon/2} \mathbf{Q}\delta(t-\tau)d\tau = \frac{1}{2}\mathbf{Q} \qquad (7.95)$$

With this last result and equation (7.94), the third term on the right of equation (7.86) is determined as $\mathbf{\Gamma}\mathbf{Q}\mathbf{\Gamma}^T/2$, which, after some algebra, turns out to be identical to the last term on the right of equation (7.86). Thus, the covariant propagation equation (7.86) becomes

$$\dot{\mathbf{Z}} = \mathbf{F}\mathbf{Z} + \mathbf{Z}\mathbf{F}^T + \mathbf{B}, \qquad \mathbf{B} = \mathbf{\Gamma}\mathbf{Q}\mathbf{\Gamma}^T \qquad (7.96)$$

where the vector for the initial conditions $\mathbf{Z}(0)$ is given.

For the particular example of a single degree of freedom structure modeled by equations (7.70)-(7.81), note that $\mathbf{Z} = E[\mathbf{z}\mathbf{z}^T]$ is a 4 by 4 covariance matrix whose diagonal elements are the variances of the corresponding state variable $\mathbf{z}$. Thus $\mathbf{Z}_{1,1} = \sigma_v^2$ is the variance of displacement. In this case $\mathbf{B}$ is the 4 by 4 matrix all of whose elements are zero except the element $B_{4,4} = \hat{\omega}^4\hat{\alpha}/m^2$. Since $\mathbf{F}$ is constant and $\mathbf{B}$ is statistically stationary, the steady-state solution for $\mathbf{Z}$ is found by setting $\dot{\mathbf{Z}} = \mathbf{0}$.

Two general methods for obtaining steady state solutions to equation (7.96) are discussed next: a numerical method and the Laplacian method.

Steady State Solutions

There are several numerical algorithms available for solving the steady - state covariance equation (7.96) for $\mathbf{Z}$ where

$$\mathbf{F}\mathbf{Z} + \mathbf{Z}\mathbf{F}^T + \mathbf{B} = 0 \qquad (7.97)$$

Davison and Man (1968), R. Smith (1968), and P. G. Smith (1971) discussed such methodologies. In a typical procedure, the eigenvalues of $\mathbf{F}$ are first calculated, an arbitrary scalar parameter β is chosen as two and one-half times the real part of the largest eigenvalue, and successive matrix solutions are generated by the following recursion formula:

$$\mathbf{Z}_{n+1} = \mathbf{Z}_n + \mathbf{V}^{2n}\mathbf{Z}_n(\mathbf{V}^{2n})^T \tag{7.98}$$

The quantities of equation (7.98) are defined as follows, in which $\mathbf{I}$ is the identity matrix:

$$\mathbf{U} = (\beta\mathbf{I} - \mathbf{F})^{-1} \tag{7.99a}$$

$$\mathbf{V} = \mathbf{U}(\beta\mathbf{I} + \mathbf{F}) \tag{7.99b}$$

$$\mathbf{W} = 2\beta\mathbf{U}\mathbf{B}\mathbf{U}^T \tag{7.99c}$$

$$\mathbf{Z}_1 = \mathbf{W} + \mathbf{V}\mathbf{W}\mathbf{V}^T \tag{7.99d}$$

The generation of the successive terms in the series is aborted when the successive changes in partial sums becomes sufficiently small, or less than one percent.

If the system is not too large, then the following closed form solution to equation (7.97), based on Laplacian transforms, can be used (Lin, 1967). That is,

$$\mathbf{Z} = \int_0^\infty \exp(\mathbf{F} \cdot t)\mathbf{B}\exp(\mathbf{F}^T \cdot t)dt \tag{7.100a}$$

$$\exp(\mathbf{F} \cdot t) = \mathcal{L}^{-1}(s\mathbf{I} - \mathbf{F})^{-1} \tag{7.100b}$$

$$\exp(\mathbf{F} \cdot t) = \mathcal{L}^{-1}(s\mathbf{I} - \mathbf{F}^T)^{-1} \tag{7.100c}$$

where $\mathcal{L}^{-1}$ is the inverse Laplace transform and s is the Laplace operator.

A Closed Form Solution

The solution to the covariance propagation equation for the single degree of freedom system described by equations (7.70)-(7.81) was calculated using the integral solution of equations (7.100). One result is an expression for the variance of the displacement in terms of the three structural constants (m, k_1, c_1) and the three load excitation parameters $(\hat{a}, \hat{\omega}, \hat{\zeta})$. That is

$$\sigma_v^2 = Z_{1,1} = \sum_{n=1}^{4} \frac{e^2\gamma_n^2}{2\alpha_n} + \frac{2e^2\gamma_1\gamma_2}{\alpha_1 + \alpha_2} + \frac{2e^2\gamma_3\gamma_4}{\alpha_3 + \alpha_4}$$

$$+\frac{2e^2\gamma_1\gamma_3}{\alpha_1 + \alpha_3} + \frac{2e^2\gamma_1\gamma_4}{\alpha_1 + \alpha_4} + \frac{2e^2\gamma_2\gamma_3}{\alpha_2 + \alpha_3} + \frac{2e^2\gamma_3\gamma_4}{\alpha_3 + \alpha_4} \tag{7.101}$$

where

$$a = -k_1/m; \qquad b = -c_1/m; \qquad c = -\hat{\omega}^2$$

$$d = -2\hat{\zeta}\hat{\omega}; \qquad e = \hat{\omega}^2\hat{\alpha}^{1/2}/m$$

$$2s_1 = d + (d^2 + 4c)^{1/2}; \qquad\qquad 2s_2 = d - (d^2 + 4c)^{1/2}$$

$$2s_3 = b + (b^2 + 4a)^{1/2}; \qquad\qquad 2s_4 = b - (b^2 + 4a)^{1/2}$$

$$C_1 = d(d-b)/[c(a-c)(d-b) - (bc - ad)(c + d^2 - bd - a)]$$

$$C_2 = -C_1(c + d^2 - bd - a)/(d - b)$$

$$C_3 = -C_1; \qquad C_4 = -(1 + aC_2)/c$$

$$\gamma_1 = (s_1C_1 + C_2)/(s_1 - s_2); \qquad \gamma_2 = -(s_2C_1 + C_2)/(s_1 - s_2)$$

$$\gamma_3 = (s_3C_3 + C_4)/(s_3 - s_4); \qquad \gamma_4 = -(s_4C_3 + C_4)/(s_3 - s_4)$$

$$\alpha_i = -s_i, \quad i = 1, 2, 3, 4$$

Although this integral solution is exact, it is recalled that the results are an approximation to reality because of the assumptions inherent in the mathematical model. In addition to the single degree of freedom approximation for the structure, the assumed form for $S_{p1}(\omega)$, equation (7.73), does not exactly replicate $S_\eta(\omega)$. Note that as $\omega \to 0$, $S_{p1}(\omega) \to \hat{\alpha}$, whereas $S_\eta(\omega) \to 0$. However, these two spectra have similar behavior otherwise: both form a single peak and both approach zero as ω becomes large.

Example Problem 7.5. Consider the same jackup rig of *Example Problem 7.3*, for which the characteristic constants m, k_1, and c_1 are given just after equation (7.51). Based on the theory of covariant propagation, or the closed form response results obtained in equation (7.101), compute numerically the *rms* deck deflection σ_v. Use the Peirson-Moskowitz wave height spectrum with a significant wave height of 15 m as the basis for the wave force excitation of this structure.

The first task is to achieve an approximate fit for $\hat{\alpha}$, $\hat{\zeta}$, and $\hat{\omega}$ when the two forms for the wave force excitation spectrum are equated. From equations (7.72) and (7.73), then

$$\hat{\alpha}\left[\left(1 - \frac{\omega^2}{\hat{\omega}^2}\right)^2 + \left(2\hat{\zeta}\frac{\omega}{\hat{\omega}}\right)^2\right]^{-1} \simeq |G(\omega)|^2 S_\eta(\omega) \qquad (7.102)$$

The two terms on the right side of the last equation are known from *Example Problem 7.3*: $|G(\omega)|^2 = 4.76 \times 10^8$ lb^2/ft^2, and $S_\eta(\omega)$ of equation (7.60), which has its peak at 672.5 ft^2-sec/rad at the frequency $\omega = 0.324$ rad/sec. Now match the peaks for each side of equation (7.102) at $\omega = \hat{\omega} = 0.324$ rad/sec, which leads to

$$\hat{\alpha} = (2\hat{\zeta})^2(4.76 \times 10^8)(672.5) = 1.28 \times 10^{12}\hat{\zeta}^2 \text{ lb}^2\text{-sec/rad}$$

Arbitrarily choose $\hat{\zeta} = 0.1$. Using the constants in equation (7.101), Mathematica®
(1999) gave the following result: $\sigma_v = 0.135$ ft. Note that this result is about
one-fourth the value calculated by a different theory in *Example Problem 7.3*. A
closer agreement between the two results can be obtained by choosing $\hat{\zeta} = 0.2$,
for which $\sigma_v = 0.196$ ft. A least squares fit of the constants $\hat{a}$, $\hat{\zeta}$, $\hat{\omega}$ based on
equation (7.102) may lead to even closer agreement between the results for σ_v
obtained in the two statistical methods. This exercise is left to the reader.

PROBLEMS

7.1   If $y(t)$ is Gaussian with zero mean and has a variance of σ_y, calculate
by numerical integration the probability that $y(t)$ is outside the levels $y = \pm\sigma_y$
and $y = \pm2\sigma_y$. Check your results using Gaussian probability tables such as
found in a statistics reference book.

7.2   Suppose that $y(t)$ is Gaussian, exists only over a narrow band of frequencies, and is a smooth function of time. Also assume that each cycle crosses
the mean level $y(t) = 0$ so that the maxima always occur for $y(t) > 0$ and the
minima always occur for $y(t) < 0$.

(a) Sketch a function $y(t)$ which behaves as defined.

(b) The probability distribution for the peaks of $y(t)$ so defined is given by
equation (7.7), the well-known Rayleigh distribution. Sketch $p(a)$ as a function
of the amplitude a of $y(t)$. Prove that the maximum value of $p(a)$ occurs when
the amplitude is equal to the standard deviation, or $a = \sigma_y$.

(c) Compute the probability that any peak of $y(t)$ exceeds these two values:
$2\sigma_y$, $3\sigma_y$. Note that the probability that any peak of $y(t)$ exceeds a is given by

$$\int_a^\infty p(a)da = \exp\left(-\frac{a^2}{2\sigma_y^2}\right)$$

7.3   A digital time history of $y(t)$ is available in the form of points $y(t_n)$ at
even increments of time, $n = 1, 2, \ldots, n$. Outline a computer-aided method that
will generate a probability density function $p(y)$ from these data. How would
you check $p(y)$ to see if it was Gaussian?

7.4   A schematic diagram of an instrument called a spectrum analyzer
is shown in Figure 7.7. The wave height input $\eta(t)$ is assumed to be a stationary ergodic random process. This input is filtered by a filter whose harmonic response function $H(\omega)$ is a constant H_0 in the narrow frequency band
$(\omega_0 - \Delta\omega/2) \leq \omega \leq (\omega_0 + \Delta\omega/2)$. The filter output $y(t)$ is squared, and its time
average $z(t)$ is calculated from

$$z(t) = E[y^2(t)] = \frac{1}{\tau_0}\int_{-\tau_0/2}^{\tau_0/2} y^2(t)dt$$

The mean level of $z(t)$, estimated from the output meter for sufficiently long
time periods τ_0, is

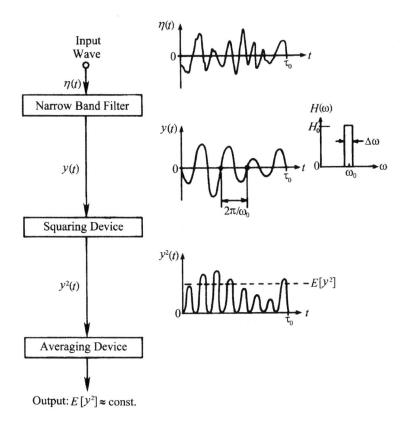

Figure 7.7   Schematic diagram of a spectrum analyzer.

$$E[z(t)] = \frac{1}{\tau_0} \int_{-\tau_0/2}^{\tau_0/2} E[y^2(t)]dt$$

(a) Based on this description and the definitions, deduce that

$$E[y^2(t)] = \int_{-\infty}^{\infty} |H(\omega)|^2 S_\eta(\omega)d\omega \simeq 2H_0^2 \, \Delta\omega \, S_\eta(\omega_0), \qquad \Delta\omega \ll \omega_0$$

(b) State the assumption that leads to $E[z(t)] = E[y^2(t)]$. Then show that the mean level $E[z(t)]$ of the output meter is a direct measure of the wave spectral density, or

$$S_\eta(\omega_0) \simeq \frac{E[z(t)]}{2H_0^2 \, \Delta\omega}$$

7.5   The Pierson-Moskowitz wave height spectrum for a significant wave height of $H_s = 15$ m is given by equation (7.60) and is plotted in Figure 7.5.

Use this equation and employ numerical integration, with reasonable limits for the integral, to compute the area A under the curve of Figure 7.5. Then verify the result discussed in Chapter 6 that $H_s = 4A^{1/2} = 15$ m.

7.6 In *Example Problem 7.4*, three rms displacement responses σ_v for the jackup rig were computed, each based on an idealized model: band-limited white noise, white noise with a cut-off frequency, and ideal white noise. Using the same parameters as for this example problem, solve equation (7.63) by numerical integration to obtain σ_v for each of these three idealized models. Compare your results to the corresponding results in the text that were derived from closed form solutions to the integral. Explain possible differences in the results. Also explain why these idealized models give results for σ_v that are about six times that obtained from numerical integration of equation (7.61).

7.7 Reconsider the jackup rig described in *Example Problem 5.4*, with the design parameters given in Table 5.2. The statistical responses to this same structure were discussed in *Example Problem 7.3*. It is proposed to add more equipment to the deck of this jackup rig so that the deck weight m_{dg} would increase from 1.02×10^7 lb to 5×10^7 lb.

(a) What percentage of the Euler buckling load is this new deck load? Use equation (5.36) to answer this and to explain whether the new deck weight will increase the chance of structural buckling.

(b) With the new deck weight, the other parameters of Table 5.2, and equation (5.36), calculate the following quantities: the equivalent bending stiffness k_1, the equivalent mass m, the equivalent damping constant c_1 based on $\zeta = c_1/\sqrt{4k_1 m} = 0.05$, and the undamped structural frequency ω_0.

(c) Based on the same parameters of *Example Problem 7.3*, except for the modified values of k_1, m, and c_1, use equation (7.61) and numerical integration to compute the variance σ_v^2. Would you expect σ_v to be higher or lower when compared to the original rig with the lower deck mass? Explain.

(d) Compute the response σ_v of the modified structure to ideal white noise. Compare this result to that obtained in part (c). What assumptions account for the differences in these two results?

7.8 Consider the concrete monotower shown in Figures 2.2 and 5.2. This structure is idealized as a rigid body consisting of a rectangular box caisson and a uniform leg. The structure rotates in the plane with angle θ about the base point 0. The foundation is assumed to be linearly elastic with a stiffness k_θ and damping c_θ, modeled respectively by equations (2.78) and (2.79), in which the frequency ω is identified as the fundamental frequency in free vibration, or ω_0. Let J_0 denote the virtual mass moment of the structure's inertia about point 0, and let $M(t)$ denote the moment on the structure induced by wave action. The governing equation (2.6) for rotational motion thus has the following form:

$$J_0\ddot{\theta} + c_\theta\dot{\theta} + k_\theta\theta = M(t)$$

(a) By comparing the symbols of the above equation to those for translational motion v, or equation (7.27), deduce for rotational motion the harmonic response function $H(\omega)$ in a form similar to equation (7.28).

(b) This monotower is subjected to a simple wave as described in Table 6.1. Derive the explicit equations for the two major components of the wave-induced moment $M(t)$: the moment on the uniform leg, and the moment on the caisson. HINTS: For the leg moment, use $\bar{q}$ of *Example Problem 4.4*, integrate $\bar{q}$ over the leg height to calculate the total horizontal wave load on the leg, compute the centroid of this load from point 0, and form the product of the last two quantities to give the first component of the moment. The second component of the moment, that on the caisson, was derived in *Example Problem 4.3*. Carry through those calculations, expressed symbolically by equation (4.22).

(c) Based on the results of part (b), derive the equation for the transfer function $G(\omega)$ corresponding to the total wave-induced moment $M(t)$ on the whole structure.

7.9   Assume that the wave field imposed on the monotower described by Problem 7.8 is stationary and ergodic.

(a) By comparing the symbols in the equation (7.27) to the equation of motion for rotation θ given in Problem 7.8, deduce without calculation the explicit forms of $S_\theta(\omega)$ and σ_θ^2 for rotational motion. These results will be analogous to equations (7.50), (7.51), and (7.52).

(b) For the quantities $H(\omega), G(\omega), S_\theta(\omega)$, and σ_θ^2 corresponding to rotational motion, write down a consistent set of units, first in the traditional English system, and then in the SI system.

7.10   The purpose of this problem is to obtain numerical results for the responses of the concrete monotower, for which the theoretical results were obtained in Problems 7.8 and 7.9. The system parameters for this tower are summarized in Table 5.1, which also lists two values for the free, undamped rocking frequency ω_0 based on two soil foundation stiffnesses.

(a) Based on the weaker soil foundation for which $G_s = 10$ MPa and $\omega = \omega_0 = 1.41$ rad/s, compute the soil foundation parameters k_θ and c_θ using equations (2.78) and (2.79).

(b) Compute the foundation damping factor $\zeta = c_\theta / \sqrt{4 k_\theta J_0}$. This damping factor is analogous to $\zeta = c_1 / \sqrt{4 k_1 m}$, the damping factor for equation (7.27).

(c) The monotower is subjected to steady, unidirectional waves with a significant wave height H_s of 15 m and with a distribution $S_\eta(\omega)$ given by the Pierson-Moskowitz spectrum, equation (7.60). With this spectrum, the numerical results of part (a), and the equation previously derived in Problem 7.9 for the variance, compute σ_θ^2 by numerical integration.

(d) Based on the $\pm 3\sigma_\theta$ limits, compute the angle of rotation, the horizontal displacement of the deck, and the horizontal shear force at the base, and the overturning moment for this 180 m high monotower.

REFERENCES

Bryson, A. E., and Hu, Y. C., *Applied Optimal Control: Optimization, Estimation, and Control*, Wiley, New York, 1975.

Cooley, J. W., and Tukey, J. W., An Algorithm for the Machine Calculation of Complex Fourier Series, *Mathematics of Computation* **19**, 1965.

Crandall, S. H., and Mark, W. D., *Random Vibration in Mechanical Systems*, Academic Press, New York, 1964.

Davison, E. J., and Man, F. T., The Numerical Solutions of $A^T Q + Q A = -C$, *IEEE Transactions on Automatic Control*, August 1968.

Gould, P. L., and Abu-Sitta, S. H., *Dynamic Response of Structures to Wind and Earthquake Loading*, Wiley, New York, 1980.

Hedrick, J, K., Mean Value and Covariance Propagation Equation, Appendix B, *Dynamics of Offshore Structures*, J. F. Wilson, ed., Wiley, New York, 1984.

Hedrick, J. K., and Firouztash, H., The Covariance Propagation Equation with Time-Delayed Inputs, *IEEE Transactions on Automatic Control*, October 1974.

Khintchine, A., Korrelations Theorie der Stationaren Stochastischen Prozesse, *Mathematische Annalen* **109**, 1934.

Lin, Y. K., *Probabilistic Theory of Structural Dynamics*, McGraw-Hill, New York, 1967.

Lutes, L. D., and Sarkani, S., *Stochastic Analysis of Structural and Mechanical Vibrations*, Prentice Hall, Upper Saddle River, NJ, 1997.

Mathematica®, version 4, Wolfram Media, Inc., Champaign, IL, 1999.

Muga, B. J., and Wilson, J. F., *Dynamic Analysis of Ocean Structures*, Plenum, New York, 1970.

Newland, D. E., *An Introduction to Random Vibrations and Spectral Analysis*, Longman, London, 1975.

Smith, P. G., Numerical Solution of the Matrix Equation $AX + XA^T + B = 0$, *IEEE Transactions on Automatic Control*, June 1971.

Smith, R., Matrix Equation $XA + BX = C$, *Siam Journal of Applied Mathematics* **16** (1), 1968.

Tung, C. C., Peak Distribution of Random Wave-Current Force, *Journal of the Engineering Mechanics Division, ASCE* **100** (EM5), October 1974.

Weiner, N., *Extrapolation, Interpolation, and Smoothing of Stationary Time Series*, Wiley, New York, 1949.

Yang, C. Y., *Random Vibrations of Structures*, Wiley, New York, 1986.

8

Multi-Degree of Freedom Linear Structures

James F. Wilson

The first approximation to determining the motion of a structure in the offshore environment is to model that structure as a single degree of freedom system in which the motion is described by a single coordinate. This was done in previous chapters. For instance, for an exploratory drilling rig rigidly fixed to the mat at the sea floor, the sway motion was computed in terms of a single displacement coordinate $v = v(t)$ at the top of the rig; and for a rigid monopod gravity platform on a compliant subsea soil foundation, the rotational motion was computed in terms of a single rotational coordinate $\theta = \theta(t)$.

In a refined dynamic analysis, several independent coordinates are used to describe structural motion, hence the term multi-degree of freedom system. For the monopod gravity platform, for instance, if the rigid deck had flexible connections to the rigid leg, then the rotation of the deck could be described by a coordinate θ_1 and the rotation of the legs by another coordinate θ_2. This dynamic model is a two degree of freedom system described by two ordinary, coupled differential equations of motion involving these two coordinates. Coupling occurs since one motion affects the other through the flexible deck-leg interface. More than two independent coordinates could be defined if one needed to account for the flexibility of the legs and the deck, and these coordinates could include horizontal displacement coordinates.

In this chapter, differential equations for multi-degree of freedom structural models are derived using both the Newtonian and the Lagrangian approach and solved using the popular normal mode method. These classical theories were freely drawn and condensed from the expositions of Chopra (2001), Clough and Penzien (1993), and Utku (1984), in which some changes in structural nomenclature were made to avoid redundancy with the common symbols of fluid mechanics. The basic assumptions used to formulate the structural models are: (1) the number of independent coordinates N chosen to describe the motion is equal to the number of degrees of freedom; (2) the restoring forces are linear functions of the chosen coordinates (linear structures); and (3) the damping is linear-viscous. Numerical examples illustrating these models are given in the following chapters.

8.1   EQUATIONS OF MOTION: GENERAL FORM

The general set of equations representing structural motion investigated in this chapter are defined by the following matrix form:

$$\mathbf{M}\ddot{\boldsymbol{\xi}} + \mathbf{C}\dot{\boldsymbol{\xi}} + \mathbf{K}\boldsymbol{\xi} = \mathbf{p} \tag{8.1}$$

Equation (8.1) represents a finite set of N ordinary, linear differential equations in N independent coordinates $(\xi_1, \xi_2, \ldots, \xi_i, \ldots, \xi_N)$, in which the coefficient matrices $\mathbf{M}, \mathbf{K}$, and $\mathbf{C}$ are constant. The single and double overdots of a coordinate represent the velocity and acceleration, respectively, of that coordinate. That is, $(\dot{\ })$ is the operator d/dt. The notation is that each upper case, boldfaced letter represents an $(N \times N)$ matrix, and that each lower case, boldfaced letter $(\boldsymbol{\xi}, \mathbf{p})$ represents a $(1 \times N)$ matrix or column vector of time-dependent elements. When $N = 1$, equation (8.1) reduces to a single degree of freedom structural model, equation (2.43), and the respective terms of these two equations have an analogous meaning. That is, the respective values of $\boldsymbol{\xi}$, $\mathbf{p}, \mathbf{M}, \mathbf{K}$, and $\mathbf{C}$ are: the coordinate vector, the loading vector, and the mass, stiffness, and damping matrices. Each of these terms is now discussed in general and evaluated for simple representations of offshore structures, for $N = 2$ and 3.

The Coordinate Vector, $\boldsymbol{\xi}$

The first step in modeling an offshore structure for dynamic analysis is to carefully define a set of N independent coordinates ξ_i, $i = 1, 2, \ldots, N$, that contain the dominant features of the structural motion. These coordinates are the elements of the column vector

$$\boldsymbol{\xi} = [\xi_1, \xi_2, \ldots, \xi_i, \ldots, \xi_N]^T \tag{8.2}$$

in which the superscript T denotes transpose (the interchange of the row elements shown to its defined column array). This vector can be composed of a mixture of displacement coordinates, designated in applications as $v_1, v_2, \ldots$, and rotational coordinates, designated as $\theta_1, \theta_2, \ldots$. Each chosen coordinate describes the motion of a *node point* on the structure, such as the mass center of a structural element or a junction point on a frame. Each chosen coordinate must then be tested as follows for independence. Freeze the motion of $(N - 1)$ node points and then check whether the lone remaining node can have motion when loaded. If and only if motion occurs at that lone node is its associated coordinate independent. Then repeat this mental test for each of the remaining nodes, one-by-one, to check for their independence.

Example Problem 8.1.   To illustrate the choice of independent coordinates, consider the simple representation of a jacket template structure shown in Figure 8.1a, a two-bay, four-legged tower fixed at the sea floor and in plane motion. This tower, which is symmetric about its vertical centerline, is a scaled-down version of the five-bay configuration discussed by Mansour and Millman (1974). Assume that, as the whole tower sways side-to-side with wave loading, the deck at location 1 and the horizontal bracings at location 2 have negligible rotations. Choose v_1 and v_2 as the horizontal displacements from the equilibrium state of levels 1 and 2, respectively. For convenience, these two coordinates locate node points 1 and 2 on the single stalk model shown in Figure 8.1b. To test whether these two coordinates are truly independent, mentally supress all motion of one coordinate ($v_1 = 0$) and check whether the structure can deflect at the other coordinate ($v_2 \neq 0$) for a horizontal load $p_2(t)$ applied at the latter coordinate. Then repeat this mental procedure, where the roles of level 1 and level 2 for coordinate supression and loading are reversed. Because of the structural geometry with its flexible leg sections separating v_1 and v_2, this structure passes these two mental tests and thus the two chosen coordinates are judged to be independent, and the coordinate vector is $\boldsymbol{\xi} = [\xi_1, \xi_2]^T = [v_1, v_2]^T$.

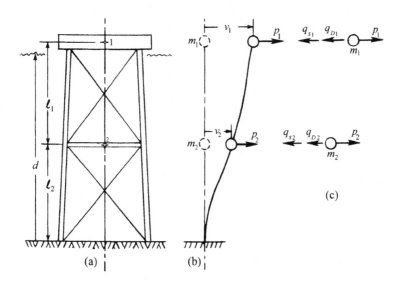

Figure 8.1   (a) A jacket template structure; (b) stalk model with external loads; (c) free body sketches of the two masses.

In engineering practice in which a design has progressed well beyond the conceptual stages, then the analysis for dynamic integrity will require more than two or three independent coordinates. Even for a tower in plane motion, with 10 bays in horizontal motion and with an additional degree of freedom to account for deck rotation, a value of $N = 11$ would be an appropriate choice for an initial analysis, but would be insufficient for a final design. With the use of computer-aided finite element techniques, local flexibilities of the deck, the legs, and the soil foundation, and out-of-plane motion, can be accounted for; and such models can have literally thousands of independent coordinates. However, the examples in this chapter, which limit N to 2 or 3, are of sufficient complexity to illustrate the basic methods of dynamic structural analysis, and with this understanding, extensions of such analyses to higher degrees of freedom systems become apparent.

The Loading Vector, p

The loading vector $\mathbf{p} = \mathbf{p}(t)$ is comprised of N loads p_i, $i = 1, 2, \dots, N$, where the load $p_i = p_i(t)$ is located at the respective node point i. The vector representation is

$$\mathbf{p} = [p_1, p_2, \dots, p_i, \dots, p_N]^T \tag{8.3}$$

Example Problem 8.2. To illustrate the formulation of a loading vector, refer to *Example Problem 8.1* and Figure 8.1. This jacket template structure is subjected to a single, deepwater harmonic wave of height H, wave number k, and frequency ω. To compute the wave loading, make the following assumptions: linear wave theory is applicable; the flow is predominantly in the inertia regime so that inertia loading of the four legs gives most of the loading; the drag forces on the smaller cross bracings are small by comparison to the inertial loading; and all four legs, each of diameter D, experience the same water particle acceleration $\dot{u}$ at any instant of time. The latter assumption is conservative and offsets somewhat the omission of drag loading on the cross bracings. Note that the largest amplitude of wave force will be transferred to the structure if $\dot{u}$ has the same phase for each leg, as is assumed in this case. For $x = 0$, then $\dot{u}$ is a maximum and it follows from Table 3.1 that

$$\dot{u} = -\frac{H}{2}\omega^2 \frac{\cosh k(z+d)}{\sinh kd}; \quad \omega^2 = gk \tanh kd \tag{8.4}$$

With this water particle acceleration, the components of the loading vector can be computed from Morison's formulation, or equation (2.14) with $C_D = 0$. The loading per unit length of all four legs is thus

$$\bar{q}(z,t) = 4C_M \frac{\pi}{4}\rho D^2 \dot{u} \tag{8.5}$$

The total load acting at each node is approximated by integrating $\bar{q}(z,t)$ over the appropriate portions of the four legs up to each node point. Recall that the coordinate z has its origin at the still water line and is measured positive upward. The results are

$$p_1 = \int_{-(d-\ell_2)}^{0} \bar{q}(z,t)dz = -\frac{\pi}{2}C_M \rho D^2 \frac{\omega^2}{k} H \left(1 - \frac{\sinh k\ell_2}{\sinh kd}\right) \sin \omega t \tag{8.6a}$$

$$p_2 = \int_{-d}^{-(d-\ell_2)} \bar{q}(z,t)dz = -\frac{\pi}{2}C_M \rho D^2 \frac{\omega^2}{k} H \frac{\sinh k\ell_2}{\sinh kd} \sin \omega t \tag{8.6b}$$

in which d is the water depth. For this case, the loading vector is expressed as

$$\mathbf{p} = [p_1, p_2]^T \tag{8.7}$$

The Mass Matrix, M

In the present analysis, $\mathbf{M}$ is assumed to be a diagonal matrix with elements $m_i > 0$, $i = 1, 2, \ldots, N$, in which the element's subscript is its associated node point. The notation is as follows:

$$\mathbf{M} = \mathrm{diag}(m_1, m_2, \ldots, m_N) = \begin{bmatrix} m_1 & 0 & \cdots & 0 \\ 0 & m_2 & \cdots & 0 \\ \vdots & \vdots & \ddots & \vdots \\ 0 & 0 & \cdots & m_N \end{bmatrix} \qquad (8.8)$$

The conditions under which $\mathbf{M}$ is diagonal are now discussed and typical calculations for $\mathbf{M}$ are illustrated.

Recall that each generalized coordinate represents either the displacement or the rotation of a portion of the structure's mass at a numbered node point. Sometimes a fixed point labeled 0 is considered a node point. The question remains: How does one determine the portion of the structural mass to be associated with each node point? Equivalently: How does one formulate the mass matrix?

The *mass lumping* method is probably the most popular method of discretizing the supporting framework and the rigid body portions of an offshore structure. For a framework, the mass lumping requires some experience on the part of the analyst. For flexural motion of structural frame elements, the analyst can use as a guideline two particular cases discussed in Chapter 5. For instance, it was calculated in *Example Problem 5.3* that, if 37 percent of the mass of a uniform beam clamped on both ends (a cross member of a supporting framework) is lumped at midspan of an equivalent massless beam of the same flexural stiffness, then the fundamental flexural frequencies of the two beam models are the same, for practical purposes. In an another case, the jackup rig of *Example Problem 5.4*, it was calculated that if 37.5 percent of a cantilevered beam's mass is lumped at the tip of its massless counterpart (an elastic beam of the same geometry, restraint, and bending stiffness), then both configurations have nearly the same fundamental flexural frequency. In such cases, the mass not accounted for is of no consequence; but for argument's sake it can be lumped at a fixed end node point 0. In these two cases, then, the criterion for lumping the mass is based on preserving the fundamental flexural frequency between the continuous element and its simple lumped mass counterpart, and this frequency equivalence is based on the conservation of potential and kinetic energy during motion.

In practical cases, however, the end constraints for an element of a supporting framework are not so simple as these two cases just discussed. Thus, without making further calculations, the choice of the fraction of element mass to be lumped at a node becomes quite subjective. Nevertheless, the experienced analyst knows that lumping to a node between 25 percent and 40 percent of the element mass surrounding that node usually leads to an adequate structural dynamic model with a diagonal mass matrix.

Lumping a portion of the structure modeled as a rigid body is not so subjective as that for a framework. For instance, for a rigid deck in plane rotation about a node at its mass center G, or for a concrete monopod in plane rotation about a fixed base point 0, *all* of the mass is used. For such a model, the rigid structural mass is assumed to be symmetric with respect to a vertical centerline. With this symmetry assumption, the diagonal element will be the mass moment of inertia with respect to the node point and all of the associated products of inertia terms due to antisymmetrical mass distribution disappear, leading to zero off-diagonal terms in the mass matrix. An example of a diagonal $\mathbf{M}$ for the coupled motion between a rigid deck and its flexible supporting structure is given later in this chapter.

A variation of the lumped mass method called the *consistent mass* theory can also be used to calculate $\mathbf{M}$. This theory, however, is usually quite tedious to implement and is beyond the scope of the present text. For further discussions, the reader is referred to the following expositions: Clough and Penzien (1993), who applied this theory to beams and frames; Chopra (2001), who illustrated the method for plane frames; and Utku (1984), who based his rigorous analysis on the principle that the sum of the kinetic energy for each discretized structural element of a plane frame or truss is equal to the kinetic energy for the whole structure. It is noted that the consistent mass matrix method leads to a banded, symmetric matrix with some non-zero and some negative off-diagonal terms. This writer has found that a judicious modeling of offshore structural supporting framwork using the lumped mass method first discussed usually leads to quite satisfactory results, whereas refinements achieved by employing the consistent mass matrix to the same model lead to results for dynamic responses that are nearly the same in many cases.

In applying the lumped mass methods, it is always very important to interpret the mass of all submerged components as *virtual* mass. It is this writer's experience that to neglect the use of virtual mass can lead to an error in a structure's fundamental frequency of 40 percent to 50 percent; and such an error can lead to comparable errors in the structure's dynamic responses.

Example Problem 8.3. Compute now the virtual mass for *Example Problem 8.1*, shown in Figure 8.1. Here, nodes 1 and 2 are on the vertical centerline of the structure, in line with the base point 0, and the system mass is symmetrically distributed with respect to this centerline. Choose node 1 at the mass center of the deck and its equipment, at the height $(\ell_1 + \ell_2)$ from 0 at the sea floor. Locate node 2 at the mass center of the horizontal members, at height ℓ_2 from 0. The masses lumped at nodes 1 and 2 are approximated as follows:

$$m_1 = m_p + 0.375\, m_{v1} \tag{8.9}$$

$$m_2 = m_{vh} + 0.375\, m_{v1} + 0.375\, m_{v2} \tag{8.10}$$

In the last equations, m_p is the total actual mass of the deck and its equipment (not submerged), and m_{vh} is the virtual mass of the horizontal members at node 2. The subscripts $v1$ and $v2$ for the other mass terms denote virtual mass for portions of the legs and nonhorizontal members (or supporting structure) surrounding nodes 1 and 2, respectively. For simplicity, assume that all members below the deck are fully submerged. The coefficient 0.375 is the proportion of the structure's mass that surrounds a particular node point, deduced from a somewhat similar cantilevered structure using energy considerations, *Example Problem 5.4*. The remaining mass of the whole system can be considered lumped at the fixed node point 0, and is of no consequence. The virtual masses for each node are approximated by

$$m_{vh} = (\text{actual mass of horizontal members at node 2}) + C_A \, \rho \mathcal{V}_h$$
$$m_{v1} = (\text{actual mass of legs and X-members in } \ell_1) + C_A \, \rho \mathcal{V}_1$$
$$m_{v2} = (\text{actual mass of legs and X-members in } \ell_2) + C_A \rho \mathcal{V}_2$$

Recall that the quantities in added mass terms are: the added mass coefficient C_A (which, lacking any experimental data for this geometry, can be approximated as unity); the mass density of water ρ; and the volumes of water $\mathcal{V}_h, \mathcal{V}_1$ and $\mathcal{V}_2$ displaced by the respective structural components: the submerged horizontal members, the submerged members surrounding node 1, and the submerged members surrounding node 2. The mass matrix for this example is thus

$$\mathbf{M} = \text{diag}(m_1, m_2) \tag{8.11}$$

where m_1 and m_2 are given by equations (8.9) and (8.10).

The Stiffness Matrix, K

The stiffness matrix for an N degree of freedom structure is a symmetric array of $N \times N$ elements in the following form:

$$\mathbf{K} = \begin{bmatrix} k_{11} & k_{12} & \cdots & k_{1N} \\ k_{21} & k_{22} & \cdots & k_{2N} \\ \vdots & \vdots & \ddots & \vdots \\ k_{N1} & k_{N2} & \cdots & k_{NN} \end{bmatrix} \tag{8.12}$$

The constants k_{ij}, where $i, j = 1, 2, \ldots, N$, are the static *stiffness influence coefficients*, which are generated from the actual structure after each node point and its associated coordinate is identified, but before any attempt is made to lump the mass. In this section, a displacement ξ_{si} means a generalized static displacement of node i: either a translation of node i or a rotation about node i. Also, a force q_{si} means a generalized static force at node i: a simple force causing a displacement of node i along the line of action of that force, or a moment causing a rotation about node i.

The constant k_{ij} is defined as the force that is required at node i to counteract a unit elastic displacement $\xi_{sj} = 1$ imposed at node j, under the condition that all displacements $\xi_{si} = 0$ for $i \neq j$. In this single displacement case, $q_{si} = k_{ij}(1) = k_{ij}$, $i \neq j$. If such a displacement condition is applied sequentially to each node, then the net force at each node j can be obtained by superposition. For instance, the elastic force q_{s1} at node 1 is a linear combination of the forces at node 1 resulting from the unit displacements at all nodes, or

$$q_{s1} = k_{11}\xi_{s1} + k_{12}\xi_{s2} + \cdots + k_{1N}\xi_{sN} \qquad (8.13)$$

Likewise, superposition gives the elastic force q_{s2} at node 2 as

$$q_{s2} = k_{21}\xi_{s1} + k_{22}\xi_{s2} + \cdots + k_{2N}\xi_{sN} \qquad (8.14)$$

By induction, the force q_{si} resulting from deflections at all nodes is

$$q_{si} = k_{i1}\xi_{s1} + k_{i2}\xi_{s2} + \cdots + k_{ij}\xi_{sj} + \cdots + k_{iN}\xi_{sN} = \sum_{j=1}^{N} k_{ij}\xi_{sj} \qquad (8.15)$$

Equation (8.15) can be expressed in matrix notation in the following two ways:

$$\begin{bmatrix} q_{s1} \\ q_{s2} \\ \vdots \\ q_{sN} \end{bmatrix} = \begin{bmatrix} k_{11} & k_{12} & \cdots & k_{1N} \\ k_{21} & k_{22} & \cdots & k_{2N} \\ \vdots & \vdots & \ddots & \vdots \\ k_{N1} & k_{N2} & \cdots & k_{NN} \end{bmatrix} \begin{bmatrix} \xi_{s1} \\ \xi_{s2} \\ \vdots \\ \xi_{sN} \end{bmatrix} \qquad (8.16)$$

$$\mathbf{q}_s = \mathbf{K}\,\boldsymbol{\xi}_s \qquad (8.17)$$

Example Problem 8.4. In practice, a reliable structural computer package (SAP 2000, for instance) is used to compute the $\mathbf{K}$ matrix. This general procedure is illustrated for the simple jacket template structure of *Example Problem 8.1*, shown in Figure 8.1. A plane frame, static force-displacement analysis is made in the two stages depicted in Figures 8.2. In the first stage, Figure 8.2a, a unit displacement $\xi_{s1} = 1$ is imposed at level 1, and $\xi_{s2} = 0$ is imposed at level 2. The corresponding forces k_{11} and k_{21} necessary to achieve those displacements are computed. In the second stage, Figure 8.2b, $\xi_{s1} = 0$ and $\xi_{s2} = 1$, for which the respective forces at levels 1 and 2 are k_{12} and k_{22}. Each such force is assumed to be positive in the direction of its coordinate. If a force is opposite to the direction of its coordinate, then that force is a negative number. In this problem, it is intuitive that k_{12} ($= k_{21}$) is a negative number. The stiffness matrix for this example is

$$\mathbf{K} = \begin{bmatrix} k_{11} & k_{12} \\ k_{21} & k_{22} \end{bmatrix} \qquad (8.18)$$

In the computer analysis of such a frame, a typical structural package requires such input data as the coordinate locations of the junctions between the basic beam elements, the junction fixity (in this case, welded joints), and the value of Young's modulus, the cross-sectional area, and the cross-sectional area moment of inertia for each beam element in the plane frame. In this example, the deck can be modeled as a rigid body, or a beam with high flexural stiffness compared to that of the supporting structural elements. Note that a computer analysis requires units for the geometry and material properties, and that any compatible set of units can be used. Thus, the unit displacements can be 1 m or 1 ft, for instance, depending on whether the SI or the traditional English units are employed.

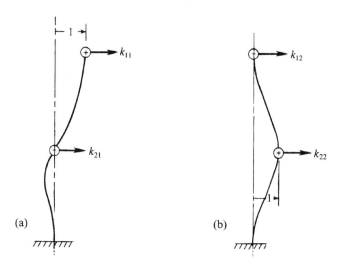

Figure 8.2   Definition of the stiffness influence coefficients for *Example Problem 8.1*.

The Damping Matrix, C

For a stucture with N degrees of freedom, the damping matrix is defined as a symmetric array of $N \times N$ constants c_{ij} in the following form:

$$
\mathbf{C} = \begin{bmatrix} c_{11} & c_{12} & \cdots & c_{1N} \\ c_{21} & c_{22} & \cdots & c_{2N} \\ \vdots & \vdots & \ddots & \vdots \\ c_{N1} & c_{N2} & \cdots & c_{NN} \end{bmatrix}
\tag{8.19}
$$

In this analysis, the damping force q_{Di} for the structural node coordinate ξ_i is assumed to be a linear combination of the generalized coordinate velocities $\dot{\xi}_i$, $i = 1, 2, \ldots, N$. The constants relating the nodal damping forces to the nodal velocities are called the *damping influence coefficients* c_{ij}; and the relationship is analogous to force-deflection relation given by equation (8.15). That is

$$
q_{Di} = c_{i1}\dot{\xi}_1 + c_{i2}\dot{\xi}_2 + \cdots + c_{ij}\dot{\xi}_j + \cdots + c_{iN}\dot{\xi}_N = \sum_{j=1}^{N} c_{ij}\dot{\xi}_j
\tag{8.20}
$$

Equation (8.20) can be expressed in matrix form in the following two ways:

$$
\begin{bmatrix} q_{D1} \\ q_{D2} \\ \vdots \\ q_{DN} \end{bmatrix} = \begin{bmatrix} c_{11} & c_{12} & \cdots & c_{1N} \\ c_{21} & c_{22} & \cdots & c_{2N} \\ \vdots & \vdots & \ddots & \vdots \\ c_{N1} & c_{N2} & \cdots & c_{NN} \end{bmatrix} \begin{bmatrix} \dot{\xi}_1 \\ \dot{\xi}_2 \\ \vdots \\ \dot{\xi}_N \end{bmatrix}
\tag{8.21}
$$

$$
\mathbf{q}_D = \mathbf{C}\,\dot{\boldsymbol{\xi}}
\tag{8.22}
$$

The damping matrix can be cast in several different specialized forms, each of which has the advantage of easily utilizing available experimental data to determine the elements c_{ij}. One such form is *Rayleigh damping* in which $\mathbf{C}$ is proportional to the system's mass and also the system's stiffness. That is

$$
\mathbf{C} = a_1 \mathbf{M} + a_2 \mathbf{K}
\tag{8.23}
$$

in which a_1 and a_2 are constants. A more explicit form for $\mathbf{C}$ based on Rayleigh damping will be presented later in this chapter.

Other specialized forms of $\mathbf{C}$ are beyond the scope of the present work. Those forms include *Caughey damping,* for which Rayleigh damping is a special case (Caughey and O'Kelly, 1965; Chopra, 2001); and *complex stiffness damping* (Clough and Penzien, 1993).

8.2 EQUATIONS OF MOTION: NEWTON'S METHOD

One method of formulating the differential equations of motion for a lumped mass structural model is to apply Newton's second law to the free body sketch of each discrete mass m_i. To illustrate, let each such mass be located by a coordinate ξ_i and have an absolute acceleration $\ddot{\xi}_i$. In these terms, Newton's second law is

$$
\sum F_{\xi_i} = m_i \ddot{\xi}_i
\tag{8.24}
$$

in which the sum on the left represents all forces applied to m_i in the direction of ξ_i. Those forces, which have lines of action acting through the mass center of each m_i, include the the net restoring force q_{si} due to structural stiffness, the net viscous damping force q_{Di}, and the lumped value of the environmental load p_i. Note that equation (8.24) is analogous to equation (2.1) for a single degree of freedom with one mass and one coordinate.

Suppose a portion of the structure can be modeled as a rigid mass m_i whose rotation is defined by the coordinate ξ_i. Define J_i as the mass moment of inertia of m_i about an axis perpendicular to the plane of motion and through its mass center. Then, the form of Newton's law of motion for this rigid body is

$$
\sum M_{\xi_i} = J_i \ddot{\xi}_i
\tag{8.25}
$$

Here, the sum on the left represents all moments applied to that mass, moments that are positive in the direction of ξ_i and that lead to an angular acceleration $\ddot{\xi}_i$. Such moments arise from forces whose lines of action are not through the center of mass of the body, moments due to structural stiffness, structural damping, and environmental loading. The mass center need not be fixed.

Example Problem 8.5. Shown in Figure 8.1c are the free body sketches for the two lumped masses of the jacket template structure. For each mass m_i, and for each corresponding coordinate $\xi_i = v_i$, there are three types of in-line loads: q_{si} and q_{Di}, which both oppose the positive direction of ξ_i, and the positively directed load p_i induced by wind, waves, and currents. When equation (8.19) is applied to mass m_i, the result is

$$p_i - q_{si} - q_{Di} = m_i \ddot{v}_i \qquad (8.26)$$

When the forces q_{si} and q_{Di} of equations (8.15) and (8.20) are combined with equations (8.26), the result is the set of two differential equations of motion, or

$$m_i \ddot{v}_i + c_{i1} \dot{v}_1 + c_{i2} \dot{v}_2 + k_{i1} v_1 + k_{i2} v_2 = p_i, \qquad i = 1, 2 \qquad (8.27)$$

This same set of differential equations expressed in matrix form is

$$\begin{bmatrix} m_1 & 0 \\ 0 & m_2 \end{bmatrix} \begin{bmatrix} \ddot{v}_1 \\ \ddot{v}_2 \end{bmatrix} + \begin{bmatrix} c_{11} & c_{12} \\ c_{21} & c_{22} \end{bmatrix} \begin{bmatrix} \dot{v}_1 \\ \dot{v}_2 \end{bmatrix}$$

$$+ \begin{bmatrix} k_{11} & k_{12} \\ k_{21} & k_{22} \end{bmatrix} \begin{bmatrix} v_1 \\ v_2 \end{bmatrix} = \begin{bmatrix} p_1 \\ p_2 \end{bmatrix} \qquad (8.28)$$

which has the form of the general matrix equation (8.1).

Example Problem 8.6. Newton's method is now used to derive the equations of motion for the three degree of freedom model of the jacket template structure shown in Figure 8.3a. This structure is modeled as the stalk configuration of Figure 8.3b with the mass lumped at nodes 1 and 2, as for *Example Problem 8.1*. The difference now is that the deck is allowed to rotate with angle θ about node 1 as nodes 1 and 2 undergo horizontal displacements v_1 and v_2. Thus, there are three degrees of freedom and the coordinate vector is

$$\boldsymbol{\xi} = [\xi_1, \xi_2, \xi_3]^T = [v_1, v_2, \theta]^T \qquad (8.29)$$

The loading vector, due to the external environmental forces and moments, is

$$\mathbf{p} = [p_1, p_2, M_d]^T \qquad (8.30)$$

in which p_1 and p_2 are the horizontal forces lumped at levels 1 and 2, and M_d is the net moment about the mass center G of the deck, due to wave slamming. The mass matrix is

$$\mathbf{M} = \text{diag}(m_1, m_2, J_d) \qquad (8.31)$$

where m_1 and m_2 are the same as for *Example Problem 8.1* and are given by equations (8.9) and (8.10); and J_d is the mass moment of inertia of the deck only, taken with respect to an axis perpendicular to the plane of motion and through G. Another way to express the latter quantity is $J_d = m_d r^2$ where m_d is the deck mass and r is the deck's radius of gyration.

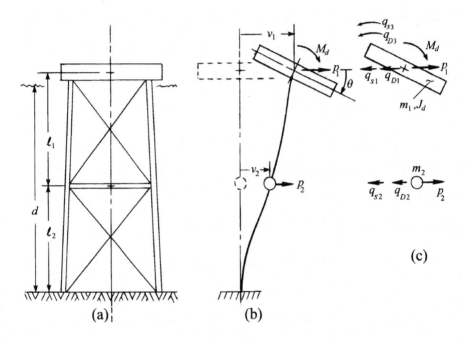

Figure 8.3 (a) Jacket template structure; (b) three degree of freedom model with external loads; (c) free body sketches.

The respective generalized stiffness and damping forces for node i are special cases of equations (8.15) and (8.20) for $N = 3$, or

$$q_{si} = k_{i1}v_1 + k_{i2}v_2 + k_{i3}\theta, \qquad i = 1, 2, 3 \qquad (8.32)$$

$$q_{Di} = c_{i1}\dot{v}_1 + c_{i2}\dot{v}_2 + c_{i3}\dot{\theta}, \qquad i = 1, 2, 3 \qquad (8.33)$$

Here, the stiffness influence coefficients k_{ij} for $i, j = 1, 2, 3$, are the forces and moments defined in Figure 8.4. These coefficients are computed from a structural program based on the three sets of displacement conditions shown, respectively, in Figures 8.4a, 8.4b, and 8.4c: $v_1 = 1$, $v_2 = \theta = 0$; $v_1 = \theta = 0$, $v_2 = 1$; and $v_1 = v_2 = 0$, $\theta = 1$. The damping influence coefficients will be determined later in this chapter.

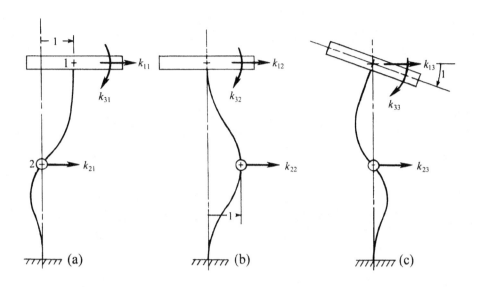

Figure 8.4   Definition of the stiffness influence coefficients for Example Problem 8.6.

Refer again to Figure 8.3 in which the free body sketch for each of the two masses is shown. For translational motion, the application of equation (8.24) to each mass gives

$$p_1 - q_{s1} - q_{D1} = m_1\ddot{v}_1 \tag{8.34a}$$

$$p_2 - q_{s2} - q_{D2} = m_2\ddot{v}_2 \tag{8.34b}$$

For rotational motion of m_1 (the total mass of the deck plus the lumped portion of the virtual mass of the upper legs, as previously defined), the application of equation (8.25) gives

$$M_d - q_{s3} - q_{D3} = J_d\ddot{\theta} \tag{8.34c}$$

When equations (8.32) and (8.33) are combined with (8.34), the results are the three equations of motion, or

$$m_1\ddot{v}_1 + c_{11}\dot{v}_1 + c_{12}\dot{v}_2 + c_{13}\dot{\theta} + k_{11}v_1 + k_{12}v_2 + k_{13}\theta = p_1 \tag{8.35a}$$

$$m_2\ddot{v}_2 + c_{21}\dot{v}_1 + c_{22}\dot{v}_2 + c_{23}\dot{\theta} + k_{21}v_1 + k_{22}v_2 + k_{23}\theta = p_2 \tag{8.35b}$$

$$J_d\ddot{\theta} + c_{31}\dot{v}_1 + c_{32}\dot{v}_2 + c_{33}\dot{\theta} + k_{31}v_1 + k_{32}v_2 + k_{33}\theta = M_d \tag{8.35c}$$

This same set of equations expressed in matrix form is

$$\begin{bmatrix} m_1 & 0 & 0 \\ 0 & m_2 & 0 \\ 0 & 0 & J_d \end{bmatrix} \begin{bmatrix} \ddot{v}_1 \\ \ddot{v}_2 \\ \ddot{\theta} \end{bmatrix} + \begin{bmatrix} c_{11} & c_{12} & c_{13} \\ c_{21} & c_{22} & c_{23} \\ c_{31} & c_{32} & c_{33} \end{bmatrix} \begin{bmatrix} \dot{v}_1 \\ \dot{v}_2 \\ \dot{\theta} \end{bmatrix}$$

$$+ \begin{bmatrix} k_{11} & k_{12} & k_{13} \\ k_{21} & k_{22} & k_{23} \\ k_{31} & k_{32} & k_{33} \end{bmatrix} \begin{bmatrix} v_1 \\ v_2 \\ \theta \end{bmatrix} = \begin{bmatrix} p_1 \\ p_2 \\ M_d \end{bmatrix} \tag{8.36}$$

which has the same form as the matrix equation (8.1). Further, equation (8.36) reduces to equation (8.28) for its two degree of freedom counterpart in which the coordinate θ is suppressed.

8.3 EQUATIONS OF MOTION: LAGRANGE'S FORMULATION

Just as for Newton's method, Lagrange's formulation of the equations of motion requires all of the system's characteristics to be defined: $\boldsymbol{\xi}$, $\mathbf{p}$, $\mathbf{M}$, $\mathbf{K}$, and $\mathbf{C}$. However, unlike Newton's method, Lagrange's formulation does not explicitly require a free body sketch for each lumped mass, nor does it require the elastic restoring forces, but what is required are three scalar energy quantities for the structural system: (1) the kinetic energy K; (2) the potential energy V, which includes the elastic deformation energy and gravitational potential energy; and (3) the virtual work done by all the nonconservative forces acting through their associated virtual generalized displacements $\delta\xi_i$. There are two types of nonconservative forces in the present context: the external, time-varying forces imposed on the structure, and the energy dissipating forces such as viscous drag. For systems with a large number of degrees of freedom, Lagrange's formulation is often preferred because the three mentioned scalar energies are easier to form than the vector restoring forces needed in Newton's method.

System Energies

The first scalar, the system's total kinetic energy K, is assumed to be a function only of the system's generalized coordinates ξ_i and their velocities $\dot{\xi}_i$, or

$$K = K(\xi_1, \xi_2, \dots, \xi_N, \dot{\xi}_1, \dot{\xi}_2, \dots, \dot{\xi}_N) \tag{8.37}$$

Generally in offshore structures, the kinetic energy depends only on velocities of the component masses.

The second scalar, the system's potential energy V, is assumed to depend only on the generalized coordinates, or

$$V = V(\xi_1, \xi_2, \dots, \xi_N) \tag{8.38}$$

The third scalar, the system's virtual work δW, is the sum of the virtual work done on each component mass by each generalized nonconservative force

g_i as its associated mass undergoes a virtual displacement $\delta\xi_i$, or

$$\delta W = \sum_{i=1}^{N} g_i \, \delta\xi_i \tag{8.39}$$

It is emphasized that each virtual or variational displacement $\delta\xi_i$ is a small and *arbitrary* change in the coordinate ξ_i and that this virtual displacement is not to be confused with the actual changes in displacement occurring in structural motion. In the latter case, the notation is $d\xi_i$.

Hamilton's Principle

The scalar quantities K, V, and δW are related through a variational equation which was originally introduced by Hamilton in 1834, and which later became known as Hamilton's Principle in the historical texts (Synge and Griffith, 1959; Whittaker, 1989.) The form of this variational equation and its description given by Clough and Penzien (1993) are particularly appropriate in the present context. This equation and its description are

$$\int_{t_1}^{t_2} \delta(K - V)dt + \int_{t_1}^{t_2} \delta W dt = 0 \tag{8.40}$$

The sum of the time variations of the difference in the kinetic and potential energies and the work done by nonconservative forces over any time interval t_1 to t_2 is zero.

It is now shown how equation (8.40) leads to the equations of motion for structural systems.

Derivation for a Simple System

Lagrange's equations are now derived for a *simple* dynamic system which is defined by all of the following characteristics: a set of N generalized coordinates is assigned, one for each degree of freedom (the system is *scleronomic*); an independent variation can be given to each of the generalized coordinates without violating the system constraints (the system is *holonomic*); the generalized *conservative* forces, such as those due to elastic deformation and those due to gravity, are all derivable from a potential energy function of the form of equation (8.38); the generalized *nonconservative* forces are the externally applied forces and those forces such as viscous fiction that dissipate energy irreversably.

Now compute the first variations δK and δV for equations (8.37) and (8.38). Substitute these results, together with δW of equation (8.39), into the variational equation (8.40), which leads to

$$\int_{t_1}^{t_2} \left(\sum_{i=1}^{N} \frac{\partial K}{\partial \xi_i} \delta\xi_i + \sum_{i=1}^{N} \frac{\partial K}{\partial \dot{\xi}_i} \delta\dot{\xi}_i \right) dt - \int_{t_1}^{t_2} \left(\sum_{i=1}^{N} \frac{\partial V}{\partial \xi_i} \delta\xi_i + \sum_{i=1}^{N} g_i \delta\xi_i \right) dt = 0 \tag{8.41}$$

For the second sum from the left, interchange the sum with integration and then integrate by parts, or

$$\int_{t_1}^{t_2} \frac{\partial K}{\partial \dot{\xi}_i} \delta \dot{\xi}_i \, dt = \int_{t_1}^{t_2} \frac{\partial K}{\partial \dot{\xi}_i} \frac{d}{dt} (\delta \xi_i) \, dt$$

$$= \int \frac{\partial K}{\partial \dot{\xi}_i} d(\delta \xi_i) = \left[\frac{\partial K}{\partial \dot{\xi}_i} \delta \xi_i \right]_{t_1}^{t_2} - \int_{t_1}^{t_2} \frac{d}{dt} \left(\frac{\partial K}{\partial \dot{\xi}_i} \right) \delta \xi_i \, dt \qquad (8.42)$$

Since the variation of each coordinate is independent of time, it follows that $\delta \xi_i(t_1) = \delta \xi_i(t_2) = 0$. Thus, the term without the integral in the last equation vanishes. With this result and with a rearrangement of terms, equation (8.41) becomes

$$\int_{t_1}^{t_2} \left\{ \sum_{i=1}^{N} \left[-\frac{d}{dt} \frac{\partial K}{\partial \dot{\xi}_i} + \frac{\partial K}{\partial \xi_i} - \frac{\partial V}{\partial \xi_i} + g_i \right] \delta \xi_i \right\} dt = 0 \qquad (8.43)$$

Now choose a particular coordinate $i = n$ for which $\delta \xi_n \neq 0$, but for which $\delta \xi_i = 0$ for all remaining $N - 1$ values of i. Again, this can be done since the coordinates are independent. It then follows that the sum in equation (8.43) disappears and the single square bracket that remains in the integrand contains terms all with the subscript n. To satisfy this equation, that square-bracketed term must vanish. Since n is arbitrary, this results holds for all values of n. After a rearrangement of terms in this square bracket, and a change in index from n to i, the result is the set of N Lagrange equations of motion for a simple system, or

$$\frac{d}{dt} \frac{\partial K}{\partial \dot{\xi}_i} - \frac{\partial K}{\partial \xi_i} + \frac{\partial V}{\partial \xi_i} = g_i, \qquad i = 1, 2, \cdots, N \qquad (8.44)$$

For ease of reference for applications, the three scalar energies are summarized in their index and matrix forms. For small motion, the kinetic energy has the following quadratic form:

$$K = \frac{1}{2} \sum_{j=1}^{N} \sum_{i=1}^{N} m_{ij} \dot{\xi}_i \dot{\xi}_j = \frac{1}{2} \dot{\boldsymbol{\xi}}^T \mathbf{M} \dot{\boldsymbol{\xi}} \qquad (8.45)$$

Note that for a diagonal mass matrix, the elements of $\mathbf{M}$ are $m_{ij} = 0$ for $i \neq j$ and $m_{ii} = m_i$ for $i = j$.

For small motion, the potential energy can be expressed as

$$V = \frac{1}{2} \sum_{j=1}^{N} \sum_{i=1}^{N} k_{ij} \xi_i \xi_j + V_g = \frac{1}{2} \boldsymbol{\xi}^T \mathbf{K} \boldsymbol{\xi} + V_g \qquad (8.46)$$

in which the double sum is a quadratic form, and V_g can also be so expressed. The term V_g represents the sum of all gravitational potential energy changes of the system's mass from the system's equilibrium state. This term V_g can be safely omitted for lumped mass frameworks in which each mass has a negligible rise and fall along the gravity vector. (The gravity vector points toward the mass center of the earth). However, V_g can be important in determining the dynamic stability of a relatively rigid gravity platform rocking on an elastic foundation, as will be demonstrated in Chapter 9.

The virtual work in the present context is

$$\delta W = \sum_{i=1}^{N} g_i \, \delta\xi_i = \sum_{i=1}^{N} (p_i - q_{Di})\delta\xi_i = \delta\boldsymbol{\xi}^T \mathbf{g} = \delta\boldsymbol{\xi}^T(\mathbf{p} - \mathbf{q}_D) \qquad (8.47)$$

in which $\mathbf{p}$ is the vector representing the externally applied loads and $\mathbf{q}_D$ is the vector for viscous damping, given by equation (8.22).

Example Problem 8.7. Lagrange's method is now used to derive the equations of motion for the two degree of freedom model of the jacket template structure shown in Figure 8.1. The two generalized coordinates are $(\xi_1, \xi_2) = (v_1, v_2)$, and the corresponding Lagrange equations from (8.44) are

$$\frac{d}{dt}\frac{\partial K}{\partial \dot{v}_1} - \frac{\partial K}{\partial v_1} + \frac{\partial V}{\partial v_1} = g_1 \qquad (8.48a)$$

$$\frac{d}{dt}\frac{\partial K}{\partial \dot{v}_2} - \frac{\partial K}{\partial v_2} + \frac{\partial V}{\partial v_2} = g_2 \qquad (8.48b)$$

The kinetic energy is evaluated from equation (8.45):

$$K = \frac{1}{2}\dot{\boldsymbol{\xi}}^T \mathbf{M}\, \boldsymbol{\xi} = \frac{1}{2}\begin{bmatrix} \dot{v}_1 & \dot{v}_2 \end{bmatrix}\begin{bmatrix} m_1 & 0 \\ 0 & m_2 \end{bmatrix}\begin{bmatrix} \dot{v}_1 \\ \dot{v}_2 \end{bmatrix} = \frac{1}{2}m_1 v_1^2 + \frac{1}{2}m_2 v_2^2 \qquad (8.49)$$

The potential energy based on the conservative forces is evaluated from equation (8.45), assuming that V_g is small relative to the elastic energy. Since $\mathbf{K} = \mathbf{K}^T$, then $k_{21} = k_{12}$ is used.

$$V = \frac{1}{2}\boldsymbol{\xi}^T \mathbf{K}\, \boldsymbol{\xi} = \frac{1}{2}\begin{bmatrix} v_1 & v_2 \end{bmatrix}\begin{bmatrix} k_{11} & k_{12} \\ k_{12} & k_{22} \end{bmatrix}\begin{bmatrix} v_1 \\ v_2 \end{bmatrix}$$

$$= \frac{1}{2}k_{11}v_1^2 + \frac{1}{2}k_{22}v_2^2 + k_{12}v_1 v_2 \qquad (8.50)$$

The virtual work for the nonconservative forces is evaluated from equation (8.47):

$$\delta W = g_1\delta v_1 + g_2\delta v_2 = (p_1 - q_{D1})\delta v_1 + (p_2 - q_{D2})\delta v_2 \qquad (8.51)$$

With the values of q_{D1} and q_{D2} from equation (8.20), the generalized forces, which are the coefficients of the virtual displacements, are as follows:

$$g_1 = p_1 - c_{11}\dot{v}_1 - c_{12}\dot{v}_2; \qquad g_2 = p_2 - c_{21}\dot{v}_1 - c_{22}\dot{v}_2 \qquad (8.52)$$

The terms in equations (8.48) can now be evaluated as follows:

$$\frac{\partial K}{\partial \dot{v}_1} = m_1\dot{v}_1; \qquad \frac{d}{dt}\frac{\partial K}{\partial \dot{v}_1} = m_1\ddot{v}_1; \qquad \frac{\partial K}{\partial v_1} = 0$$

$$\frac{\partial V}{\partial v_1} = k_{11}v_1 + k_{12}v_2$$

$$\frac{\partial K}{\partial \dot{v}_2} = m_2\dot{v}_2; \qquad \frac{d}{dt}\frac{\partial K}{\partial \dot{v}_2} = m_2\ddot{v}_2; \qquad \frac{\partial K}{\partial v_2} = 0$$

$$\frac{\partial V}{\partial v_2} = k_{22}v_2 + k_{21}v_1$$

With equations (8.52) and these last calculated results, the equations of motion (8.48) become

$$m_1\ddot{v}_1 + k_{11}v_1 + k_{12}v_2 = p_1 - c_{11}\dot{v}_1 - c_{12}\dot{v}_2 \qquad (8.53a)$$

$$m_2\ddot{v}_2 + k_{22}v_2 + k_{21}v_1 = p_2 - c_{22}\dot{v}_2 - c_{21}\dot{v}_1 \qquad (8.53b)$$

When the viscous damping forces in the above equations are rearranged to the left sides, then it is observed that equations (8.53a) and (8.53b) are identical to the results obtained using Newton's Method, or equation (8.27) for $i = 1$ and $i = 2$, respectively.

Comments

At this point, it is appropriate to add several comments and recommendations concerning the equations of structural motion formulated in this chapter.

First, it is highly recommended that the stiffness matrix **K** be computed using a well-tested, commercially available structural software package, SAP 2000, for instance. In analyzing the supporting framework for tall offshore structures in relatively deep water and with very massive decks and deck equipment, it is recommended that the chosen software account for the prestress of the structural elements, in particular, the axial compression of the vertical beam elements. Recall that, for the jackup rig analyzed in Chapter 5, the leg's flexural stiffness was reduced by the dead weight compressive load of the deck. However, if the static Euler buckling load for the tower is much smaller than the deck load and its equipment, then such prestress with its accompanying reduction in the flexural stiffness of the framework can be ignored.

Second, it is not difficult to show that the Lagrange equations (8.44) can be equivalent to the matrix form of (8.1), provided that the dynamic system is *simple*, that the motion is small, and that K and V are expressed in quadratic form. To show this equivalence, let K, V, and δW have the forms of equations (8.45)-(8.47), except let $V_g = 0$. When the terms of equations (8.44) are evaluated using equations (8.37)-(8.39), then the result is

$$\mathbf{M}\ddot{\boldsymbol{\xi}} + \mathbf{K}\boldsymbol{\xi} = \mathbf{g} \qquad (8.54)$$

Since the nonconservative force vector is $\mathbf{g} = \mathbf{p} - \mathbf{q}_D$, in which $\mathbf{q}_D = \mathbf{C}\dot{\boldsymbol{\xi}}$ from equation (8.22), then the last equation becomes identical to the linear form, equation (8.1), or

$$\mathbf{M}\ddot{\boldsymbol{\xi}} + \mathbf{C}\dot{\boldsymbol{\xi}} + \mathbf{K}\boldsymbol{\xi} = \mathbf{p} \qquad (8.55)$$

The last comment is that, once the equations of motions are formulated, with all time-varying environmental loads and the constant coefficients identified, then those equations can be solved directly to determine the time-varying responses $\xi_i = \xi_i(t)$, $i = 1, 2, \ldots, N$. To do this, the analyst has a wide choice of a computer software packages, including PSI-Plot (1999) and Mathematica® (1999). For such computations, $2N$ initial conditions must be specified, or

$$\boldsymbol{\xi}(0) = [\xi_1(0), \xi_2(0), \ldots, \xi_N(0)]^T \qquad (8.56a)$$

$$\dot{\boldsymbol{\xi}}(0) = [\dot{\xi}_1(0), \dot{\xi}_2(0), \ldots, \dot{\xi}_N(0)]^T \qquad (8.56b)$$

In general, steady state solutions $\xi_i(t)$ with light damping are sought, and those responses occur in the numerical solutions if the run time t is sufficiently long. In such solutions, the damping eventually eliminates the initial transient responses so that the choice of initial conditions expressed by equations (8.56) is of no consequence. Recall that for a single degree of freedom system, light structural damping was based on the parameter $\zeta = c_1/2\sqrt{k_1 m}$ in which ζ was in the measured range of 0.05 to 0.1. See equations (5.58) and (5.64). Using this latter information, rough estimates of $c_{ij} = c_{ji}$ can be made to obtain the steady state numerical solutions. A more exact way to relate ζ to c_{ij} is discussed later in this chapter.

Although numerical solutions can generally be computed by what is sometimes called the *brute force* method, it is quite appropriate and often more physically meaningful to obtain closed form solutions to the equations of motion by the classical normal mode method. The remainder of this chapter is devoted to this latter method: the computation of the system's characteristic frequencies in free vibration ω_n and their corresponding modal vectors $\mathbf{x}_n$, and the superposition of these modal vectors to obtain the structure's shape in terms of the steady state response vectors $\xi_i = \xi_i(t)$.

8.4   FREE, UNDAMPED MOTION

The values of ω_n and $\mathbf{x}_n$ are computed from the condition of free, undamped structural motion. That is, $\mathbf{C} = \mathbf{p} = \mathbf{0}$, for which the governing equation (8.1) becomes

$$\mathbf{M}\ddot{\boldsymbol{\xi}} + \mathbf{K}\boldsymbol{\xi} = \mathbf{0} \tag{8.57}$$

Frequencies

Assume a harmonic solution to equation (8.57) in the form

$$\boldsymbol{\xi} = \hat{\boldsymbol{\xi}}\, e^{j\omega t} \tag{8.58}$$

in which $\hat{\boldsymbol{\xi}}$ is the time-independent amplitude vector and ω is the frequency parameter. When $\boldsymbol{\xi}$ and $\ddot{\boldsymbol{\xi}}$ from the last equation are substituted into equation (8.57), the result is

$$(-\omega^2 \mathbf{M}\hat{\boldsymbol{\xi}} + \mathbf{K}\hat{\boldsymbol{\xi}})e^{j\omega t} = \mathbf{0} \tag{8.59}$$

Since $e^{j\omega t}$ is arbitrary, then the bracket term in the last equation is zero, which can be expressed as

$$(\mathbf{K} - \omega^2 \mathbf{M})\hat{\boldsymbol{\xi}} = \mathbf{0} \tag{8.60}$$

This last result represents a set of N linear, simultaneous, algebraic, homogeneous equations in the vector components $\hat{\xi}_n$, $n = 1, 2, \ldots, N$. By Cramer's rule, nontrivial solutions for $\hat{\boldsymbol{\xi}}$ exist only if the determinant of the bracket term vanishes, or

$$\det(\mathbf{K} - \omega^2 \mathbf{M}) = 0 \tag{8.61}$$

When this determinant is expanded, the result is an Nth order polynominal in ω^2, for which the N consecutive roots or eigenvalues are positive numbers, designated as $\omega_1^2, \omega_2^2, \ldots, \omega_n^2, \ldots, \omega_N^2$. The numbers obtained from the positive square root of each eigenvalue are the free vibration frequencies of the structural system. For convenience, these frequencies are arranged from the smallest to the largest in order of the ascending subscripts: $\omega_1, \omega_2, \ldots, \omega_n, \ldots, \omega_N$. For $N \geq 3$, the only practical way to determine these frequencies is to employ a computer package, Mathematica® (1999), for instance.

Modal Vectors and Normalization

For the nth frequency ω_n there is a corresponding modal vector $\hat{\boldsymbol{\xi}}_n$ which can be computed from equation (8.60), rewritten as

$$(\mathbf{K} - \omega_n^2 \mathbf{M})\hat{\boldsymbol{\xi}}_n = \mathbf{0} \tag{8.62}$$

The modal vector in component form is

$$\hat{\boldsymbol{\xi}}_n = [\hat{\xi}_{1n},\ \hat{\xi}_{2n},\ \ldots, \hat{\xi}_{Nn}]^T = [1,\ \hat{\xi}_{2n}, \ldots, \hat{\xi}_{Nn}]^T \tag{8.63}$$

where, as is customary, the first component of the modal vector is assigned a value of one. Thus, equation (8.62) can be displayed in component form as

$$
\begin{bmatrix}
k_{11} - \omega_n^2 m_1 & k_{12} & \cdots & k_{1n} \\
k_{21} & k_{22} - \omega_n^2 m_2 & \cdots & k_{2n} \\
\vdots & \vdots & \ddots & \vdots \\
k_{n1} & k_{n2} & \cdots & k_{NN} - \omega_N^2 m_N
\end{bmatrix}
\begin{bmatrix}
1 \\
\hat{\xi}_{2n} \\
\vdots \\
\hat{\xi}_{Nn}
\end{bmatrix}
=
\begin{bmatrix}
0 \\
0 \\
\vdots \\
0
\end{bmatrix}
\tag{8.64}
$$

If any *one* of the algebraic equations in this last display is omitted, then the remaining $N - 1$ linear equations can be solved for the unknown components $\hat{\xi}_{2n}, \hat{\xi}_{3n}, \ldots, \hat{\xi}_{Nn}$. For instance, if $N = 3$, the first two equations of this display can be written as

$$
\begin{bmatrix}
k_{12} & k_{13} \\
k_{22} - \omega_n^2 m_2 & k_{23}
\end{bmatrix}
\begin{bmatrix}
\hat{\xi}_{2n} \\
\hat{\xi}_{3n}
\end{bmatrix}
= -
\begin{bmatrix}
k_{11} - \omega_n^2 m_1 \\
k_{21}
\end{bmatrix}
\tag{8.65}
$$

It is noted that, since each frequency ω_n is distinct, and $\hat{\xi}_{1n} = 1$, then there is a unique modal vector $\hat{\boldsymbol{\xi}}_n$ for each frequency.

After the N modal vectors are calculated in this way, those vectors are normalized with respect to the mass matrix to form the new modal vectors $\mathbf{x}_n$. That is

$$
\mathbf{x}_n = \frac{1}{e_n} \hat{\boldsymbol{\xi}}_n
\tag{8.66}
$$

where e_n is a set of constants computed from the following equation:

$$
\hat{\boldsymbol{\xi}}_n^T \mathbf{M} \hat{\boldsymbol{\xi}}_n = e_n^2
\tag{8.67}
$$

Here, e_n is always a positive real number, which follows since $\mathbf{M}$ is positive definite and symmetric.

Orthogonality of the Modal Vectors

To achieve the uncoupling of the equations of motion (8.1), it is first necessary to show that the modal vectors $\mathbf{x}_n$ are mutually orthogonal with respect to both $\mathbf{M}$ and $\mathbf{K}$. This orthogonality is defined as follows:

$$
\mathbf{x}_n \mathbf{M} \mathbf{x}_r = \delta_{nr}
\tag{8.68}
$$

$$
\mathbf{x}_n \mathbf{K} \mathbf{x}_r = \omega_n^2 \delta_{nr}
\tag{8.69}
$$

in which $\delta_{nr} = 1$ for $r = n$ and $\delta_{nr} = 0$ for $r \neq n$.

Consider the two cases for $r = n$. For $\mathbf{M}$: when equations (8.66) and (8.67) are combined, it follows directly that

$$
\mathbf{x}_n \mathbf{M} \mathbf{x}_n = 1
\tag{8.70}
$$

For $\mathbf{K}$: when equations (8.66) and (8.62) are combined, then

$$\mathbf{K}\,\mathbf{x}_n = \omega_n^2 \mathbf{M}\,\mathbf{x}_n \qquad (8.71)$$

When this last result is premultiplied by $\mathbf{x}_n^T$ and equation (8.70) is used, then

$$\mathbf{x}_n^T \mathbf{K}\,\mathbf{x}_n = \omega_n^2 \, \mathbf{x}_n^T \mathbf{M}\,\mathbf{x}_n = \omega_n^2 \qquad (8.72)$$

Consider the two cases for $r \neq n$. For $\mathbf{M}$: let $n = r$ in equation (8.71) and take the transpose of both sides to obtain

$$\mathbf{x}_r^T \mathbf{K}^T = \omega_r^2 \mathbf{x}_r^T \mathbf{M}^T \qquad (8.73)$$

Since both $\mathbf{K}$ and $\mathbf{M}$ are symmetric, then

$$\mathbf{x}_r^T \mathbf{K} = \omega_r^2 \mathbf{x}_r^T \mathbf{M} \qquad (8.74)$$

Postmultiply this last result by $\mathbf{x}_n$:

$$\mathbf{x}_r^T \mathbf{K}\,\mathbf{x}_n = \omega_r^2 \mathbf{x}_r^T \mathbf{M}\,\mathbf{x}_n \qquad (8.75)$$

Premultiply equation (8.71) by $\mathbf{x}_r^T$, which gives

$$\mathbf{x}_r^T \mathbf{K}\,\mathbf{x}_n = \omega_n^2 \mathbf{x}_r^T \mathbf{M}\,\mathbf{x}_n \qquad (8.76)$$

Subtract equation (8.75) from the last result to give

$$(\omega_n^2 - \omega_r^2)\mathbf{x}_r^T \mathbf{M}\,\mathbf{x}_n = 0 \qquad (8.77)$$

Since the frequencies are distinct ($\omega_n \neq \omega_r$ for $n \neq r$), then

$$\mathbf{x}_r^T \mathbf{M}\,\mathbf{x}_n = 0, \qquad \text{for } n \neq r \qquad (8.78)$$

Thus, this last result, together with equation (8.70), completes the proof of the orthogonality statement (8.68). For $\mathbf{K}$: the right side of equation (8.76) is ω_n^2 for $n = r$ by equation (8.70), but is equal to zero for $n \neq r$ by equation (8.78). This completes the proof of the orthogonality statement (8.69).

8.5 FORCED, DAMPED MOTION

Derived in this section are steady state solutions to the structural equations of motion (8.1), which include both forcing vector $\mathbf{p}$ and system damping. These solutions utilize the free, undamped modal vectors $\mathbf{x}_n$ and their associated undamped frequencies ω_n, both derived in the last section. These solutions depend on two important matrix forms: the modal shape matrix $\mathbf{X}$ and the modal damping matrix $\mathbf{C}$. Following the definition of these forms, a transformation is introduced that uncouples the equations of motion, allowing for the normal mode solutions to be displayed in closed form. The method of solution is then summarized. Numerical examples and applications illustrating the methodology are deferred to Chapter 9.

The Mode Shape Matrix, X

The modal shape matrix is defined as the assembly of the normalized modal vectors $\mathbf{x}_n$, written in the following alternate forms:

$$\mathbf{X} = [\mathbf{x}_1, \mathbf{x}_2, \dots \mathbf{x}_n, \dots \mathbf{x}_N] \tag{8.79}$$

$$= \begin{bmatrix} x_{11} & x_{12} & \cdots & x_{1n} & \cdots & x_{1N} \\ x_{21} & x_{22} & \cdots & x_{2n} & \cdots & x_{2N} \\ \vdots & \vdots & \vdots & \vdots & \vdots & \vdots \\ x_{N1} & x_{N2} & \cdots & x_{Nn} & \cdots & x_{NN} \end{bmatrix} \tag{8.80}$$

Note that the nth normalized modal vector forms the nth column of $\mathbf{X}$, where

$$\mathbf{x}_n = [x_{1n}, x_{2n}, \dots, x_{Nn}]^T \tag{8.81}$$

With this definition of $\mathbf{X}$, the respective orthogonality conditions of equations (8.68) and (8.69) can be restated as

$$\mathbf{X}^T \mathbf{M} \mathbf{X} = \mathbf{I} \tag{8.82}$$

$$\mathbf{X}^T \mathbf{K} \mathbf{X} = \operatorname{diag}(\omega_n) \tag{8.83}$$

in which $\mathbf{I}$ is the $N \times N$ identity matrix, with 1 for all diagonal elements and zero for all off-diagonal elements. Recall that the notation diag(*function of n*) represents an $N \times N$ matrix whose argument defines the n consecutive diagonal elements $(n = 1, 2, \dots, N)$, with zero for all off-diagonal terms.

Modal Damping Matrix, C

Modal damping is assumed in the form first proposed by Lord Rayleigh (1945) and previously given by equation (8.23). That is

$$\mathbf{C} = a_1 \mathbf{K} + a_2 \mathbf{M} \tag{8.84}$$

where a_1 and a_2, the Rayleigh constants, are fixed for a given dynamic system. When the latter equation is premultiplied by $\mathbf{X}^T$ and postmultiplied by $\mathbf{X}$, then

$$\mathbf{X}^T \mathbf{C} \mathbf{X} = a_1 \mathbf{X}^T \mathbf{K} \mathbf{X} + a_2 \mathbf{X}^T \mathbf{M} \mathbf{X} \tag{8.85}$$

With the orthogonal properties of equations (8.82) and (8.83), the last result becomes

$$\mathbf{X}^T \mathbf{C} \mathbf{X} = \operatorname{diag}(a_1 \omega_n^2 + a_2) \tag{8.86}$$

The nth modal damping factor ζ_n is defined in terms of the nth diagonal term in equation (8.86), where

$$\mathbf{X}^T \mathbf{C} \mathbf{X} = \text{diag}(2\zeta_n \omega_n) = \text{diag}(a_1 \omega_n^2 + a_2) \tag{8.87}$$

After equating the nth diagonal terms above, it follows that

$$\zeta_n = \frac{\omega_n}{2} a_1 + \frac{1}{2\omega_n} a_2 \tag{8.88}$$

This last result shows that, for an N degree of freedom system for which the N frequencies $\omega_1, \omega_2, \ldots, \omega_N$ are known, then the two Rayleigh constants a_1 and a_2 are uniquely determined if any *two* values of ζ_n are specified. For instance, if ζ_k and ζ_m are specified, then equation (8.88) yields the following two simultaneous equations from which a_1 and a_2 can be calculated:

$$\zeta_k = \frac{\omega_k}{2} a_1 + \frac{1}{2\omega_k} a_2 \, ; \qquad \zeta_m = \frac{\omega_m}{2} a_1 + \frac{1}{2\omega_m} a_2 \tag{8.89}$$

Then the remaining $N - 2$ values of ζ_n for all $n \neq m$ can then be calculated from equation (8.88). In many applications, the first few modes will dominate the motion, and in such cases it is reasonable to choose ζ_1 and ζ_2 (or $k = 1$ and $m = 2$) as the arbitrary numerical damping factors. Note that, as ω_n (and thus n) becomes large, $\zeta_n \to \omega_n a_1 / 2$ and for all practical purposes the damping increases linearly with frequency. This is consistent with experimental observations in which the higher modes have diminished amplitudes and are difficult to detect because of the increased damping at high frequencies.

Uncoupling the Equations of Motion

Consider the governing equations (8.1) in which the external loading vector $\mathbf{p}$ is an arbitrary function of time at each nodal point, or

$$\mathbf{M}\ddot{\boldsymbol{\xi}} + \mathbf{C}\dot{\boldsymbol{\xi}} + \mathbf{K}\boldsymbol{\xi} = \mathbf{p(t)} \tag{8.90}$$

Define the solution vector for equation (8.90) by the following modal coordinate transformation:

$$\boldsymbol{\xi} = \mathbf{X}\mathbf{y} \quad \text{or} \quad \xi_n = \sum_{k=1}^{N} x_{nk} \, y_k \tag{8.91}$$

Here, $\mathbf{X}$ is the time-invariant matrix of the normalized modal vectors defined by equations (8.80) and the vector $\mathbf{y} = \mathbf{y}(t)$ is to be determined. When this transformation and its appropriate time derivatives are substituted into equation (8.90), then

$$\mathbf{M}\mathbf{X}\ddot{\mathbf{y}} + \mathbf{C}\mathbf{X}\dot{\mathbf{y}} + \mathbf{K}\mathbf{X}\mathbf{y} = \mathbf{p}(t) \tag{8.92}$$

Premultiply this last result by $\mathbf{X}^T$:

$$\mathbf{X}^T\mathbf{M}\mathbf{X}\ddot{\mathbf{y}} + \mathbf{X}^T\mathbf{C}\mathbf{X}\dot{\mathbf{y}} + \mathbf{X}^T\mathbf{K}\mathbf{X}\mathbf{y} = \mathbf{X}^T\mathbf{p}(t) \tag{8.93}$$

In this last equation, it is noted that the coefficients of $\ddot{\mathbf{y}}$, $\dot{\mathbf{y}}$, and $\mathbf{y}$ are defined by equations (8.82), (8.87), and (8.83), respectively, which leads to

$$\ddot{\mathbf{y}} + \text{diag}(2\zeta_n\omega_n)\dot{\mathbf{y}} + \text{diag}(\omega_n^2)\mathbf{y} = \mathbf{X}^T\mathbf{p}(t) \tag{8.94}$$

These transformed equations (8.94) are observed to be uncoupled. The equation for the nth mode is thus

$$\ddot{y}_n + 2\zeta_n\omega_n\dot{y} + \omega_n^2 y_n = \mathbf{x}_n^T\mathbf{p}(t) \tag{8.95}$$

Note that the right side of the last equation is a scalar quantity.

Steady State Solutions

It is recognized that the scalar equation (8.95) is in the form of the single degree of freedom model, equation (5.58). The steady state Duhamel integral solution, equation (5.56), was derived for the latter model in Chapter 5. For the present case, this integral solution can be derived in the same way. The result is

$$y_n = \int_0^t \frac{\mathbf{x}_n^T\mathbf{p}(\tau)}{\omega_{dn}} e^{-\zeta_n\omega_n(t-\tau)} \sin[\omega_{dn}(t-\tau)]\, d\tau \tag{8.96}$$

where the damped frequency for the nth mode is

$$\omega_{dn} = \omega_n\sqrt{1 - \zeta_n^2} \tag{8.97}$$

The vector $\mathbf{y}$ expressed in terms of its components computed from equation (8.96) is

$$\mathbf{y} = [y_1, y_2, \ldots, y_N]^T \tag{8.98}$$

From the transformation equation (8.91), the steady state solution for the modal coordinates in component form is thus

$$\begin{bmatrix} \xi_1 \\ \xi_2 \\ \vdots \\ \xi_N \end{bmatrix} = \begin{bmatrix} x_{11} & x_{12} & \cdots & x_{1N} \\ x_{21} & x_{22} & \cdots & x_{2N} \\ \vdots & \vdots & \vdots & \vdots \\ x_{N1} & x_{N2} & \cdots & x_{NN} \end{bmatrix} \begin{bmatrix} y_1 \\ y_2 \\ \vdots \\ y_N \end{bmatrix} \tag{8.99}$$

8.6 SUMMARY OF THE NORMAL MODE METHOD

Summarized below is a ten-step procedure for obtaining the steady state modal response solutions to the governing equation (8.1):

1. After formulating the N-degree of freedom model of the structure, determine numerical values for the elements of the $\mathbf{M}$ and $\mathbf{K}$ matrices.

2. Evaluate the loading vector $\mathbf{p}(t)$.

3. With $\mathbf{K}$ and $\mathbf{M}$ and the characteristic determinant equation (8.61), calculate the consecutive frequencies $\omega_1, \omega_2, \dots, \omega_N$, and order them from the smallest to the largest.

4. Compute the components of the non-normalized modal vectors $\hat{\boldsymbol{\xi}}_n$ using $N - 1$ of the simultaneous equations (8.64). Note that $\hat{\xi}_n = 1$, $n = 1, 2, \dots, N$.

5. Compute the N scalars e_n using equation (8.67); and then determine the normalized modal vectors $\mathbf{x}_n$ using equation (8.66).

6. Assemble the vectors $\mathbf{x}_n$ to form the $\mathbf{X}$ matrix defined by equations (8.79) and (8.80).

7. Fix two modal damping ratios ζ_k and ζ_m. For instance, pick $k = 1$ and $m = 2$, and $\zeta_1 = \zeta_2 = 0.05$; compute the Rayleigh coefficients a_1 and a_2 from equations (8.89); and compute the remaining $N - 2$ values of ζ_n from equation (8.88).

8. Using the results of ω_n from step 3 and ζ_n from step 7, compute the N damped frequencies ω_{dn} from equation (8.97).

9. Compute $y_n(t)$ from equation (8.98) using numerical integration. That is, generate a table of $y_n(t)$ vs. t for each value of n.

10. Compute the modal coordinates $\xi_n = \xi_n(t)$ from equation (8.99). The peak values of the coordinate components are useful in design.

It is advisable to implement this ten-step solution in a general computer program. Numerical examples will be illustrated in the next chapter.

PROBLEMS

8.1.   Consider a two degree of freedom lumped mass model of the jackup rig shown in Fig. 2.17 and described in *Example Problem 2.8*. Choose the two independent coordinates as v and θ, the displacement and rotation of the deck, respectively. Let p_d and M_d be the arbitrary loadings corresponding to v and θ. For the deck, the mass moment of inertia about its mass center is J_d. Neglect all changes in elevation of the deck. Include damping.

(a) Write down the coordinate vector, the loading vector, and the damping matrix.

(b) Define in words and symbols the two nonzero elements of the mass matrix.

(c) Each leg of the three-legged structure has a bending stiffness EI and a length ℓ. Use classical beam theory to compute the elements of the 2×2 stiffness matrix.

(d) Show a free body sketch for each of the two masses.

(e) Use Newton's method to formulate the two equations of motion in terms of v and θ. Include linear viscous damping in the form of equation (8.20).

8.2.   For the structure described in Problem 8.1, formulate the expressions for the kinetic energy, the potential energy, and the virtual work of the nonconservative forces. With these three scalar quantities, use Lagrange's method to derive the two equations of motion for the structure.

8.3 Suppose that the two frequencies in free vibration for the structural model of Problem 8.1 are computed as ω_1 and ω_2 and that the modal damping factors are estimated from experimental data as $\zeta_1 = 0.05$ and $\zeta_2 = 0.07$. With these damping factors, compute the values of the Rayleigh constants a_1 and a_2 in terms of the two frequencies.

8.4 Suppose that the jacket template structure shown in Figure 8.3a is modeled as a four degree of freedom damped system with the independent coordinates $(v_1, v_2, \theta_1, \theta_2)$. Here, the first two coordinates are defined in Figure 8.3b; the angles θ_1 and θ_2 are the rotations of the deck and the mass m_2; and J_1 and J_2 are the mass moments of inertia for the deck and for m_2, respectively. There are three loads only: the respective horizontal wave loading p_1 and p_2 at v_1 and v_2, and a wave slamming moment M_d on the deck.

(a) Write down the loading vector and the damping matrix.

(b) Identify the elements of the mass matrix. Include the virtual mass where appropriate.

(c) Sketch the four diagrams, analogous to the three diagrams of Figure 8.4, that define the elements of the stiffness matrix.

(d) For each of the two lumped masses, construct a free body sketch, analogous to the sketches in Figure 8.3c.

(e) Use Newton's method to derive the four equations of motion for this structural model.

8.5 For the structure described in Problem 8.4, formulate the expressions for the kinetic energy, the potential energy, and the virtual work of the nonconservative forces. With these three scalar quantities, use Lagrange's method to derive the four equations of motion for the structure.

REFERENCES

Caughey, T. K., and O'Kelly, M. E. J., Classical Normal Modes in Damped Linear Dynamic Systems, *Journal of Applied Mechanics* **32**, 1965.

Chopra, A. K., *Dynamics of Structures: Theory and Applications to Earthquake Engineering*, second ed., Prentice Hall, Upper Saddle River, NJ, 2001.

Clough, R.W., and Penzien, J., *Dynamics of Structures*, second ed., McGraw-Hill, New York, 1993.

Mansour, A. E., and Millman, D. N., Dynamic Random Analysis of Fixed Offshore Platforms, OTC-2049, *Proceedings of the Offshore Technology Conference*, 1974.

Mathematica®, version 4, Wolfram Media, Inc., Champaign, IL, 1999.

PSI-Plot, version 6, Poly Software International, Sandy, UT, 1999.

Rayleigh, Lord, *Theory of Sound* **1**, Dover, New York, 1945.

SAP 2000, *Integrated Structural Analysis and Design Software*, Computers and Structures, Inc., Berkeley, CA, 1997.

Synge, J. L., and Griffith, B. A., *Principles of Mechanics*, third ed., McGraw-Hill, New York, 1959.

Utku, S., *Dynamics of Offshore Structures*, J. F. Wilson, editor, Chapters 8 and 9, Wiley, New York, 1984.

Whittaker, E. T., *A Treatise on the Dynamics of Particles and Rigid Bodies*, fourth ed., Cambridge University Press, Cambridge, UK, 1937. Reprinted 1989.

9

Applications of Multi-Degree of Freedom Analysis

James F. Wilson

This chapter has two main purposes. The first is to apply the time domain theory of linear dynamic analysis formulated in Chapters 7 and 8 to selected offshore structures. For instance, response time histories are calculated for a jacket template platform with deterministic wave and earthquake loadings. The second purpose is to develop in the frequency domain the response statistics for linear structures subject to stationary, random wave excitation. This statistical analysis represents an extension of the theory developed in Chapter 7, now applied to multi-degree of freedom systems.

For the illustrative problems, the mathematical models are simple: they are two degree of freedom systems, chosen more for the purpose of fixing the basic ideas of dynamic analysis in the mind of the reader than for the purpose of design. In actual practice, packaged computer codes utilizing the same essential ideas would be employed in the final design stages where the structure would be represented by perhaps several hundred degrees of freedom. However, back of the envelope calculations for a structure modeled with only a few carefully chosen dynamic coordinates have their place: they give the analyst a feeling for the problem, they are economical to perform, and the results predict the essential characteristics of motion for the same structure modeled as a higher order system. For instance, for the simplified model of the gravity platform, the fundamental frequency and the conclusions about the structural stability are about the same as the results derived from the elaborate counterpart models with many more degrees of freedom.

The emphasis of this chapter is on computational procedures and the interpretation of numerical results for structural responses. As in Chapter 8, the linear equations representing plane structural motion in this chapter are given by the following matrix form:

$$\mathbf{M}\ddot{\boldsymbol{\xi}} + \mathbf{C}\dot{\boldsymbol{\xi}} + \mathbf{K}\boldsymbol{\xi} = \mathbf{p}(t) \tag{9.1}$$

9.1   A FIXED LEG PLATFORM: TIME DOMAIN RESPONSES

Mathematical Model

A fixed leg platform or jacket template structure, discussed in some detail in *Example Problems 8.1-8.5*, is shown again in Figure 9.1. As in these previous examples, the stalk model of this structure is used, as defined in Figure 9.2a, a two degree of freedom model with the lumped virtual masses m_1 and m_2, located by the two independent horizontal displacement coordinates (ξ_1, ξ_2). The corresponding equations of motion for each mass, derived previously as equations (8.28), are

$$\begin{bmatrix} m_1 & 0 \\ 0 & m_2 \end{bmatrix} \begin{bmatrix} \ddot{\xi}_1 \\ \ddot{\xi}_2 \end{bmatrix} + \begin{bmatrix} c_{11} & c_{12} \\ c_{21} & c_{22} \end{bmatrix} \begin{bmatrix} \dot{\xi}_1 \\ \dot{\xi}_2 \end{bmatrix}$$

$$+ \begin{bmatrix} k_{11} & k_{12} \\ k_{21} & k_{22} \end{bmatrix} \begin{bmatrix} \xi_1 \\ \xi_2 \end{bmatrix} = \begin{bmatrix} p_1 \\ p_2 \end{bmatrix} \tag{9.2}$$

Listed in Table 9.1 are the numerical values for the characteristics of the structure and of the incident harmonic water wave. The analysis of the structure's dynamic response begins with a calculation of the undamped natural frequencies and mode shapes.

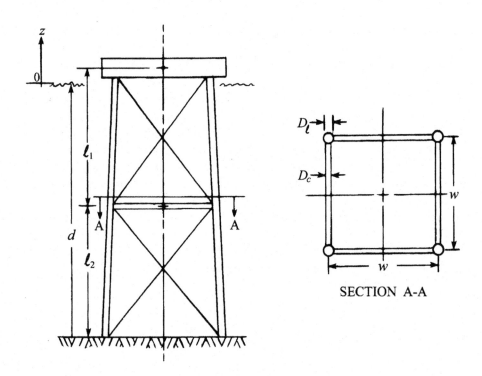

Figure 9.1   A fixed leg platform.

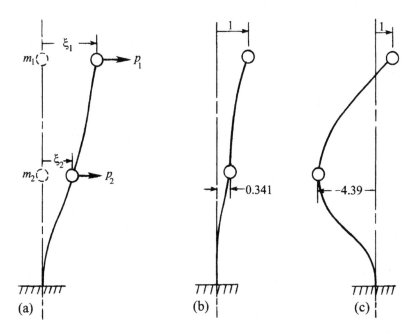

Figure 9.2 Stalk model of the fixed leg platform: (a) definition of coordinates; (b) components of the modal vector $\hat{\xi}_1$; (c) components of the modal vector $\hat{\xi}_2$.

Table 9.1 Characteristics of a Fixed Platform and an Incident Wave

Platform Parameters	Wave Loading Parameters
$m_1 = 4.69 \times 10^6$ kg; $m_2 = 3.13 \times 10^6$ kg	$H = 11.6$ m
$k_{11} = 7.35 \times 10^7$ N/m; $k_{22} = 3.59 \times 10^8$ N/m	$T = 15.4$ s
$k_{12} = k_{21} = -1.15 \times 10^8$ N/m	$k = 0.0201$ m^{-1}
$\ell_1 = \ell_2 = 38$ m; $d = 61$ m	$\lambda = 312$ m
$\zeta_1 = \zeta_2 = 0.05$	$\omega = 0.408$ rad/s
$D_\ell = 5.5$ m; $D_c = 4.3$ m	$\rho = 1031$ kg/m^3
$N_\ell = 4$; $N_c = 2$	$C_M = 2$
$w = 30$ m	

Frequencies

The two undamped structural frequencies for free vibration are computed by solving the characteristic determinant, equation (8.61), for the two positive roots ω. The governing equation is

$$\det(\mathbf{K} - \omega^2 \mathbf{M}) = 0 \tag{9.3}$$

With the numerical values above, this determinant becomes

$$\begin{vmatrix} 73.5 - 4.69\omega^2 & -115 \\ -115 & 359 - 3.13\omega^2 \end{vmatrix} \times 10^6 = 0 \tag{9.4}$$

which reduces to

$$\omega^4 - 130.4\omega^2 + 900.9 = 0 \qquad (9.5)$$

Using the quadratic formula, the two positive roots ω^2 to this equation are calculated as: $\omega_1^2 = 7.322$ rad^2/s^2 and $\omega_2^2 = 123.04$ rad^2/s^2, from which the two frequencies are deduced as

$$\omega_1 = 2.706 \text{ rad/s (or 0.431 Hz)}; \qquad \omega_2 = 11.09 \text{ rad/s (or 1.77 Hz)} \qquad (9.6)$$

Modal Vectors and Normalization

For the nth frequency ω_n there is a modal vector $\hat{\boldsymbol{\xi}}_n$ which can be computed from equation (8.62), or

$$(\mathbf{K} - \omega_n^2 \mathbf{M})\hat{\boldsymbol{\xi}}_n = \mathbf{0} \qquad (9.7)$$

In this case, $\hat{\boldsymbol{\xi}}_n = [1 \quad \hat{\xi}_{2n}]^T$ for $n = 1, 2$ and the components $\hat{\xi}_{2n}$ can be computed from the special case of equation (8.63), or

$$\begin{bmatrix} k_{11} - \omega_n^2 m_1 & k_{12} \\ k_{21} & k_{22} - \omega_n^2 m_2 \end{bmatrix} \begin{bmatrix} 1 \\ \hat{\xi}_{2n} \end{bmatrix} = \begin{bmatrix} 0 \\ 0 \end{bmatrix} \qquad (9.8)$$

The first of these two equations, when solved for $\hat{\xi}_{2n}$, gives

$$\hat{\xi}_{2n} = \frac{1}{k_{12}}(\omega_n^2 m_1 - k_{11}) \qquad (9.9)$$

The reader can verify that the same numerical results for $\hat{\xi}_{2n}$ can be obtained from the second of equations (9.8) as from equation (9.9). For the numerical values of this problem, then $\hat{\xi}_{21} = 0.341$ and $\hat{\xi}_{22} = -4.38$, which correspond to the frequencies of $\omega_1 = 2.706$ and $\omega_2 = 11.09$ rad/s, respectively. These modal vectors, which have no units, are thus

$$\hat{\boldsymbol{\xi}}_1 = [\hat{\xi}_{11} \quad \hat{\xi}_{21}]^T = [1 \quad 0.341]^T \qquad (9.10a)$$

$$\hat{\boldsymbol{\xi}}_2 = [\hat{\xi}_{12} \quad \hat{\xi}_{22}]^T = [1 \quad -4.39]^T \qquad (9.10b)$$

Shown in Figures 9.2b and 9.2c are sketches of $\hat{\boldsymbol{\xi}}_1$ and $\hat{\boldsymbol{\xi}}_2$, respectively. Note that since the components $\hat{\xi}_{11}$ and $\hat{\xi}_{12}$ were arbitrarily chosen as unity, a comparison of magnitude between these two vectors is not meaningful.

The normalized modal vectors $\mathbf{x}_n$ are computed using equations (8.66) and (8.67). That is,

$$\mathbf{x}_n = \frac{1}{e_n}\hat{\boldsymbol{\xi}}_n \qquad (9.11)$$

in which the normalizing constants e_n are computed from

$$e_n^2 = \hat{\boldsymbol{\xi}}_n^T \mathbf{M} \hat{\boldsymbol{\xi}}_n \tag{9.12}$$

For this problem, the two normalizing constants are as follows:

$$e_1^2 = \begin{bmatrix} 1 & 0.341 \end{bmatrix} \begin{bmatrix} 4.69 & 0 \\ 0 & 3.13 \end{bmatrix} \begin{bmatrix} 1 \\ 0.341 \end{bmatrix} \times 10^6 = 5.054 \times 10^6 \text{ kg}$$

$$e_1 = 2248 \text{ kg}^{1/2}$$

$$e_2^2 = \begin{bmatrix} 1 & -4.39 \end{bmatrix} \begin{bmatrix} 4.69 & 0 \\ 0 & 3.13 \end{bmatrix} \begin{bmatrix} 1 \\ -4.39 \end{bmatrix} \times 10^6 = 65.01 \times 10^6 \text{ kg}$$

$$e_2 = 8063 \text{ kg}^{1/2}$$

With these results and equation (9.11), the two normalized modal vectors are calculated as

$$\mathbf{x}_1 = \frac{1}{e_1} \begin{bmatrix} 1 & 0.341 \end{bmatrix}^T = \begin{bmatrix} 4.45 \\ 1.52 \end{bmatrix} \times 10^{-4} \text{ kg}^{-1/2} \tag{9.13a}$$

$$\mathbf{x}_2 = \frac{1}{e_2} \begin{bmatrix} 1 & -4.39 \end{bmatrix}^T = \begin{bmatrix} 1.24 \\ -5.44 \end{bmatrix} \times 10^{-4} \text{ kg}^{-1/2} \tag{9.13b}$$

The modal shape matrix $\mathbf{X}$, defined previously by equations (8.79) and (8.80) as the assembly of the modal vectors $\mathbf{x}_n$, is thus

$$\mathbf{X} = \begin{bmatrix} x_{11} & x_{12} \\ x_{21} & x_{22} \end{bmatrix} = \begin{bmatrix} 4.45 & 1.24 \\ 1.52 & -5.44 \end{bmatrix} \times 10^{-4} kg^{-1/2} \tag{9.14}$$

Response to a Harmonic Wave

Consider the steady state response of the structure in Figure 9.1 to a plane, harmonic storm wave that has a recurrence interval of 100 years. This wave, based on studies of severe storms in the Gulf of Mexico (Mansour and Millman, 1974), has a significant wave height of $H = 11.6$ m and a dominant wave period of $T = 15.4$ s. The wave frequency is thus $\omega = 2\pi/T = 0.408$ rad/s.

To determine the appropriate wave theory needed for the structural loading, first compute the two wave parameters, which are the abscissa and ordinate of Figure 3.10. These are

$$\frac{d}{T^2} = \frac{61 \text{ m} \times 3.28 \text{ ft/m}}{15.4^2 \text{ s}^2} = 0.844 \text{ ft/sec}^2$$

$$\frac{H}{T^2} = \frac{11.6 \text{ m} \times 3.28 \text{ ft/m}}{15.4^2 \text{ s}^2} = 0.0489 \text{ ft/sec}^2$$

These two parameters place this wave in the region of the Stoke's second order theory, in the intermediate water depth range. As discussed in Chapter 3 in

Example Problem 3.2, the wave length λ can be expressed in terms of the water depth d, the wave period T, and the height H. In this case, the dispersion relation has the following form:

$$\lambda = T\sqrt{\frac{g\lambda}{2\pi}\tanh\frac{2\pi d}{\lambda}} = 15.4\sqrt{\frac{9.81}{2\pi}\lambda\tanh\frac{2\pi \cdot 61}{\lambda}} \qquad (9.15)$$

The latter equation, implicit in λ, was solved using Mathematica$^{\circledR}$ (1999) with the subroutine RootFind, with the result that $\lambda = 312$ m. From this, the wave number becomes $k = 2\pi/\lambda = 0.0201$ m^{-1}. These two wave parameters are listed in Table 9.1.

In computing the structural loading associated with this wave, the following assumptions are made: (1) the motion of the structure is much smaller than the motion of the wave, so that Morison's equation (2.14) applies; (2) the flow is predominately in the inertia regime so that the structural loading term of Morison's equation that involves C_M dominates the fluid drag term that involves C_D; (3) the four vertical legs ($N_\ell = 4$) plus the two horizontal cross braces ($N_c = 2$), which are normal to the flow at node point 2, account for most of the structural wave loading; (4) because of their relatively small diameter compared to the legs, the wave loading of the cross bracings is mainly fluid drag, a loading that is relatively small compared to the inertia loading on the other six members to which C_M applies; (5) since the wave length $\lambda = 312$ m is much larger than the distance $w = 30$ m between the vertical legs in the direction of wave propagation, the phase of the wave can be neglected, or $x = 0$ in the expression for the wave acceleration $\dot{u}$. With these assumptions, $\dot{u} = \dot{u}(z,t)$ of Table 3.2 has the form

$$\dot{u} = -\frac{2\pi^2 H}{T^2}\frac{\cosh k(z+d)}{\sinh kd}\sin \omega t - \frac{3\pi^3 H^2}{T^2\lambda}\frac{\cosh 2k(z+d)}{\sinh^4 kd}\sin 2\omega t \qquad (9.16)$$

The loading per unit length of the four vertical legs, and of the two cross braces normal to the wave direction are, respectively

$$\bar{q}_\ell(z,t) = N_\ell C_M\frac{\pi}{4}\rho D_\ell^2\dot{u}(z,t) \qquad (9.17a)$$

$$\bar{q}_c(z,t) = N_c C_M\frac{\pi}{4}\rho D_c^2\dot{u}(z,t), \qquad \text{at} \quad z = -(d - \ell_2) \qquad (9.17b)$$

The respective total loads lumped at nodes 1 and 2 are computed by integrating these loadings over the appropriate structural members. As a conservative measure, all of the wave loading on the four legs from the sea surface to node 2 is lumped at node 1 located at the deck level. Also, all of the wave loading on the legs extending from node 2 to the sea floor is lumped at node 2. Note that the two cross braces normal to the flow are located node 2 also. With these assumptions, the nodal loads can be expressed as follows:

$$p_1(t) = \int_{-(d-\ell_2)}^{0} \bar{q}_\ell(z,t)dz \qquad (9.18a)$$

$$p_2(t) = \int_{-d}^{-(d-\ell_2)} \bar{q}_\ell(z,t)dz + w\bar{q}_c[-(d-\ell_2),t]dz \qquad (9.18b)$$

When $\dot{u}$ of equation (9.16) is substituted into equations (9.17), and those results are integrated according to equations (9.18), the nodal loads become

$$p_1(t) = b_1 \sin \omega t + b_3 \sin 2\omega t \qquad (9.19a)$$

$$p_2(t) = b_2 \sin \omega t + b_4 \sin 2\omega t \qquad (9.19b)$$

in which the coefficients of the harmonic terms are

$$a_1 = N_\ell C_M \frac{\pi}{4}\rho D_\ell^2; \qquad a_2 = -\frac{2\pi^2 H}{T^2 \sinh kd}$$

$$a_3 = -\frac{3\pi^3 H^2}{T^2 \lambda \sinh^4 kd}; \qquad a_4 = N_c C_M \frac{\pi}{4}\rho D_c^2$$

$$b_1 = \frac{1}{k}a_1 a_2(\sinh kd - \sinh k\ell_2); \qquad b_2 = \frac{1}{k}a_1 a_2 \sinh k\ell_2 + wa_2 a_4 \cosh k\ell_2$$

$$b_3 = \frac{1}{2k}a_1 a_3(\sinh 2kd - \sinh 2k\ell_2); \qquad b_4 = \frac{1}{2k}a_1 a_3 \sinh 2k\ell_2 + wa_3 a_4 \cosh 2k\ell_2$$
$$(9.20)$$

These coefficients, when evaluated using the system parameters of Table 9.1, lead to the following explicit results for the nodal loads, in units of newtons:

$$p_1(t) = -4.334 \times 10^6 \sin 0.408t - 0.500 \times 10^6 \sin 0.816t \text{ N} \qquad (9.21a)$$

$$p_2(t) = -6.534 \times 10^6 \sin 0.408t - 0.432 \times 10^6 \sin 0.816t \text{ N} \qquad (9.21b)$$

With this loading, together with the normalized vectors $\mathbf{x}_n$, the steady state solutions to the governing equations of motion (9.2) can be computed using equations (8.96)-(8.99) and the procedure outlined in Section 8.6. The two scalar products in the integral solution (8.96) are computed using the vectors of equations (9.13) and (9.21), or

$$\mathbf{x}_1^T \mathbf{p}(\tau) = [4.45 \quad 1.52] \times 10^{-4} \begin{bmatrix} b_1 \sin \omega\tau + b_3 \sin 2\omega\tau \\ b_2 \sin \omega\tau + b_4 \sin 2\omega\tau \end{bmatrix}$$

$$= -2922 \sin \omega\tau - 288.4 \sin 2\omega\tau \quad \text{kg}^{-1/2}\text{N} \qquad (9.22a)$$

$$\mathbf{x}_2^T \mathbf{p}(\tau) = [1.24 \quad -5.44] \times 10^{-4} \begin{bmatrix} b_1 \sin \omega\tau + b_3 \sin 2\omega\tau \\ b_2 \sin \omega\tau + b_4 \sin 2\omega\tau \end{bmatrix}$$

$$= 3017 \sin \omega\tau + 173.2 \sin 2\omega\tau \ \mathrm{kg}^{-1/2}\mathrm{N} \qquad (9.22b)$$

The undamped and damped frequencies are

$$\omega_1 = 2.7060 \ \mathrm{rad/s}; \quad \omega_{d1} = 2.7026 \ \mathrm{rad/s} \qquad (9.23a)$$

$$\omega_2 = 11.090 \ \mathrm{rad/s}; \quad \omega_{d1} = 11.076 \ \mathrm{rad/s} \qquad (9.23b)$$

With these results, $y_1 = y_1(t)$ and $y_2 = y_2(t)$ were computed by numerical integration of equation (8.96) at each of the following times $t = 0.2, 0.4, 0.6, \ldots, 30.0$ s. For each of these times, the structural displacements ξ_1 and ξ_2 were calculated using the transformation of equation (8.99), or equivalently equation (8.91). That is

$$\xi_1(t) = x_{11}y_1 + x_{12}y_2 = 4.45 \times 10^{-4}y_1 + 1.24 \times 10^{-4}y_2 \ \mathrm{m} \qquad (9.24a)$$

$$\xi_2(t) = x_{21}y_1 + x_{22}y_2 = 1.52 \times 10^{-4}y_1 - 5.44 \times 10^{-4}y_2 \ \mathrm{m} \qquad (9.24b)$$

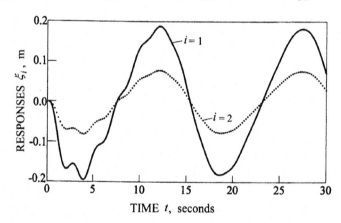

Figure 9.3   Responses for the fixed leg platform to a harmonic wave.

Shown in Figure 9.3 are numerical results for these two structural steady state displacements over a time of 30 s, or for approximately two cycles of the wave loading. The absolute values of the first peaks are:

$$|\xi_1|_{\max} = 0.1950 \ \mathrm{m}; \qquad |\xi_2|_{\max} = 0.0810 \ \mathrm{m} \qquad (9.25)$$

and the respective subsequent peaks change very little from these values. Also, after the first cycle the responses become more smooth, a result of the light damping.

This same problem was also solved using linear wave theory instead of Stoke's second order wave theory. For linear wave theory, the wave loading is given by equations (9.19) with $b_3 = b_4 = 0$, and with b_1 and b_2 defined by equations

(9.20) as before. Computations showed that the response curves were nearly identical to those of Figure 9.3, and that the absolute values of the first peaks were:

$$|\xi_1|_{\max} = 0.1937 \text{ m}; \qquad |\xi_2|_{\max} = 0.0805 \text{ m} \qquad (9.26)$$

For this problem, then, linear wave theory is adequate for preliminary dynamic design.

Response to Earthquake Excitation

The two-mass model of the offshore platform is shown in Figure 9.4a, now subject to a horizontal ground displacement $v_g = v_g(t)$ that simulates one type of earthquake excitation. As shown in Figure 9.4b, the coordinates ξ_1 and ξ_2 now represent the displacements of m_1 and m_2 *relative* to the rigid base of the structure, and it is these displacements that give rise to the elastic restoring force and the damping force. In the absence of other external excitations, the equations of motion are derived from the general equations (9.1) by replacing the acceleration vector $\ddot{\boldsymbol{\xi}}$ for each lumped mass by the *new* absolute acceleration vector $(\ddot{\boldsymbol{\xi}} + \mathbf{1}\ddot{v}_g)$, in which the unit vector in this two-mass example is

$$\mathbf{1} = \begin{bmatrix} 1 & 1 \end{bmatrix}^T \qquad (9.27)$$

With the indicated substitutions, equation (9.1) now becomes

$$\mathbf{M}\ddot{\boldsymbol{\xi}} + \mathbf{C}\dot{\boldsymbol{\xi}} + \mathbf{K}\boldsymbol{\xi} = \mathbf{p}(t) = \mathbf{M}\mathbf{1}\ddot{v}_g \qquad (9.28)$$

In this last result, the negative sign on the right was omitted since the sign of $\boldsymbol{\xi}$ is of no consequence in this problem. Further, this last result also applies to stalk models in plane motion with horizontal, rigid base excitation in which the number of degrees of freedom $N > 2$, provided that the unit vector $\mathbf{1}$ has the same dimension as N. For stalk models that include soil-structural interactions, see Clough and Penzien (1993), Chapter 27.

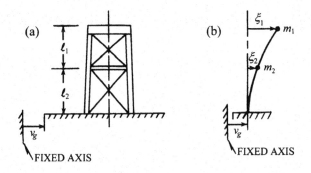

Figure 9.4   Earthquake excitation of the fixed leg platform.

The steady state solution to equation (9.28) is given by $\boldsymbol{\xi} = \mathbf{X}\mathbf{y}$, equation (8.91). The solution components y_n are given by equation (8.96), which for earthquake excitation becomes

$$y_n = \mathbf{x}_n^T \mathbf{M1} \int_0^t \frac{\ddot{v}_g(\tau)}{\omega_{dn}} e^{-\zeta_n \omega_n (t-\tau)} \sin[\omega_{dn}(t-\tau)]\, d\tau \qquad (9.29)$$

The peak values of the $\boldsymbol{\xi}$ vector, or $\boldsymbol{\xi}_{\max}$, are sought for a given experimental earthquake time history $\ddot{v}_g$. This is done by first forming the product $\omega_n y_n$ and then computing the maximum value of the above integral for each frequency ω_n. That is, compute the pseudovelocity of the nth mode, defined by

$$S_{n\xi} = \max \left(\int_0^t \frac{\ddot{v}_g(\tau)}{\sqrt{1-\zeta_n^2}} e^{-\zeta_n \omega_n (t-\tau)} \sin[\omega_{dn}(t-\tau)]\, d\tau \right) \qquad (9.30)$$

Form the pseudodisplacement components $S_{n\xi}/\omega_n$, now defined as the diagonal elements of the 2×2 matrix: $\mathrm{diag}(S_{n\xi}/\omega_n)$. In these terms and with equation (8.91), the solution can be written as

$$\boldsymbol{\xi}_{\max} = \mathbf{X}\,\mathrm{diag}(S_{n\xi}/\omega_n)\mathbf{X}^T \mathbf{M1} \qquad (9.31)$$

As a numerical example, let the structure of Figure 9.4 have the characteristics given in Table 9.1. The undamped frequencies ω_n and the modal shape matrix $\mathbf{X}$ are given by equations (9.6) and (9.14), respectively. Choose the El Centro earthquake as the design condition since the pseudovelocity $S_{n\xi}$ has been computed for this case as a function of damping and structural or modal period $T_0 = T_n$ (see Figure 5.8). Compute $T_0 = T_n$ for each frequency as

$$T_1 = \frac{2\pi}{\omega_1} = \frac{2\pi}{2.706} = 2.32 \text{ s}; \qquad T_2 = \frac{2\pi}{11.09} = 0.567 \text{ s} \qquad (9.32)$$

Let $\xi_1 = \xi_2 = 0.05$ and use Figure 5.8 to obtain the respective pseudovelocities for these two periods:

$$S_{1\xi} = 25 \text{ in./sec} = 0.635 \text{ m/s}; \qquad S_{2\xi} = 30 \text{ in./sec} = 0.762 \text{ m/s} \qquad (9.33)$$

The peak responses are then computed from equation (9.31), or

$$\begin{bmatrix} \xi_{1\,\max} \\ \xi_{2\,\max} \end{bmatrix} = \begin{bmatrix} 4.45 & 1.24 \\ 1.52 & -5.44 \end{bmatrix} \times 10^{-4} \text{ kg}^{-1/2} \begin{bmatrix} 0.235 & 0 \\ 0 & 0.0687 \end{bmatrix} \text{ m} \times$$

$$\begin{bmatrix} 4.45 & 1.52 \\ 1.24 & -5.44 \end{bmatrix} \times 10^{-4} \text{ kg}^{-1/2} \begin{bmatrix} 4.69 \\ 3.13 \end{bmatrix} \times 10^6 \text{ kg} = \begin{bmatrix} 0.258 \\ 0.133 \end{bmatrix} \text{ m} \qquad (9.34)$$

Thus, the displacement of the deck at mass m_1 is 0.285 m, and the displacement at the 38 m height at mass m_2 is 0.133 m, both relative to the bottom of the legs. Using these results, the maximum horizontal shear loads at m_1 and m_2 are deduced as

$$\mathbf{f}_{max} = \mathbf{K}\boldsymbol{\xi}_{max} = \begin{bmatrix} 0.735 & \text{-1.15} \\ \text{-1.15} & 3.59 \end{bmatrix} \begin{bmatrix} 0.258 \\ 0.133 \end{bmatrix} \times 10^7 = \begin{bmatrix} 3.69 \\ 18.3 \end{bmatrix} \times 10^5 \text{ N}$$

(9.35)

The sum of these two horizontal shear loads is 2.20×10^6 N, which is an upper bound of the shear load shared by all four legs at the base of the structure. Further, an upper bound on the base overturning moment due to these shear loads, also shared by all four legs of lengths $\ell_1 = \ell_2 = 38$ m, is given by

$$M_{max} = 3.69 \times 10^5(\ell_1 + \ell_2) + 18.3 \times 10^5 \ell_2 = 9.74 \times 10^7 \text{ N} \cdot \text{m}$$

(9.36)

The responses computed in this numerical example are upper bound values since they are based on $S_{1\xi}$ and $S_{2\xi}$ derived from the maximum of the Duhamel integral, irrespective of the time of occurrence. Thus, the maximum shear loads were assumed to be in phase. Nevertheless, this type of calculation serves as an economical, approximate check on results derived from the more involved models where the modal displacements are matched in time.

9.2 A MONOPOD GRAVITY PLATFORM: FREE VIBRATION AND STABILITY

Mathematical Model

A monopod gravity platform on a flexible soil foundation is modeled as shown in Figure 9.5. The most important assumptions are that the structure is a rigid body with two degrees of freedom system in which the respective coordinates for horizontal base sliding and for rotation in the plane are denoted as $\xi_1 = v$ and $\xi_2 = \theta$. Typical numerical parameters describing the structure and the soil foundation are listed in Table 9.2. For structural sliding and rocking, the soil's stiffness and damping behavior is modeled after equations (2.76)-(2.79). The purposes of this section are to use this model to set up the equations of motion, to compute the structure's undamped frequencies, and to discuss briefly the general criterion for structure's dynamic stability.

Although neither the applied structural loads nor the damping are needed to compute the structural rocking and sliding motion in free vibration, those quantities are included for the sake of completeness in the following derivation of the equations of sructural motion. The fluid loadings shown in Figure 9.5b are: $F = F(t)$ which represents the net, time-dependent horizontal load due to current, wind and waves acting at an equivalent height h_0; and $M_{pc} = M_{pc}(t)$ which is the net time-dependent moment about the base point 0 due to the time-varying pressure imbalance across the *top* of the caisson. In this mathematical model, it is assumed that the viscous damping forces of the soil foundation, represented by the constants c_1 for structural sliding and c_θ for structural rotation, are much larger than the viscous damping effects of the surrounding water, so that the latter damping is ignored.

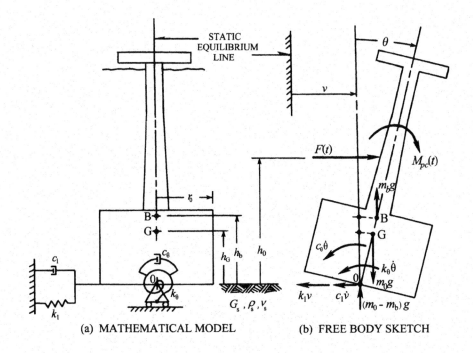

Figure 9.5   Two degree of freedom model of a monotower gravity platform on an elastic foundation.

Table 9.2   System Parameters and Results for the Monotower

Soil modulus	$G_s = 10$ MPa
Soil density; Poisson's ratio	$\rho_s = 2000$ kg/m^3; $\nu_s = 0.33$
Soil stiffness, sliding	$k_1 = 2.16 \times 10^9 \,(1 - 0.0318\omega_0)$ N/m
Soil stiffness, rocking	$k_\theta = 3.63 \times 10^{12}(1 - 0.137\omega_0)$ N·m
Structural mass, virtual	$m = 6.15 \times 10^8$ kg
Structural mass, actual	$m_0 = 3.56 \times 10^8$ kg
Structural mass, buoyant	$m_b = 2.59 \times 10^8$ kg
Structural inertia, virtual	$J_G = 8.80 \times 10^{11}$ kg·m^2
Structural inertia ratio	$J_0/J_G = 1.66$
Center of actual mass	$h_G = 30.7$ m
Center of buoyant mass	$h_b = 31.7$ m
Caisson radius	$r_0 = 45$ m
Computed frequencies	$\omega_1 = 1.24$ rad/s; $\omega_2 = 2.73$ rad/s

Lagrange's method as discussed in Chapter 8 is now used to derive the equations of motion of this monotower. For this two degree of freedom model, the two equations of motion in the form of equation (8.44) are written in terms of the structure's independent coordinates $\xi_1 = v$ and $\xi_2 = \theta$, or

$$\frac{d}{dt}\frac{\partial K}{\partial \dot{v}} - \frac{\partial K}{\partial v} + \frac{\partial V}{\partial v} = g_v \tag{9.37}$$

$$\frac{d}{dt}\frac{\partial K}{\partial \dot{\theta}} - \frac{\partial K}{\partial \theta} + \frac{\partial V}{\partial \theta} = g_\theta \tag{9.38}$$

The kinetic energy K, the potential energy V, and the nonconservative virtual work δW of the system's nonconservative generalized forces g_v and g_θ are, respectively,

$$K = \frac{1}{2}m(\dot{v} + h_G\dot{\theta})^2 + \frac{1}{2}J_G\dot{\theta}^2 \tag{9.39}$$

$$V = \frac{1}{2}k_1 v^2 + \frac{1}{2}k_\theta \theta^2 - m_0 g h_G(1 - \cos\theta) + m_b g h_b(1 - \cos\theta) \tag{9.40}$$

$$\delta W = g_v \delta v + g_\theta \delta\theta = [-c_1\dot{v} + F]\delta v + [-c_\theta\dot{\theta} + h_0 F + M_{pc}]\delta\theta \tag{9.41}$$

When the last three equations are used with equations (9.37), and then with (9.38), the respective equations for plane motion become

$$m\ddot{v} + mh_G\ddot{\theta} + c_1\dot{v} + k_1 v = F \tag{9.42}$$

$$mh_G\ddot{v} + (J_G + mh_G^2)\ddot{\theta} + c_\theta\dot{\theta} + (k_\theta - m_0 g h_G + m_b g h_b)\theta = h_0 F + M_{pc} \tag{9.43}$$

The above two equations can be written in standard form in the following way: Multiply equation (9.42) by $(-h_G)$ and add this result to equation (9.43) to give

$$J_G\ddot{\theta} + c_\theta\dot{\theta} - h_G c_1\dot{v} + (k_\theta - m_0 g h_G + m_b g h_b)\theta - h_G k_1 v = M_{pc} + (h_0 - h_G)F \tag{9.44}$$

Then substitute $\ddot{\theta}$ from this result into equation (9.42). Thus

$$m\ddot{v} + \frac{J_0}{J_G}c_1\dot{v} - \frac{mh_G}{J_G}c_\theta\dot{\theta} - \frac{mh_G}{J_G}[k_\theta - m_0 g h_G + m_b g h_b]\theta + \frac{J_0}{J_G}k_1 v$$

$$= \left[1 - \frac{mh_G}{J_G}(h_0 - h_G)\right]F - \frac{mh_G}{J_G}M_{pc} \tag{9.45}$$

In the last two equations, which are now in standard form, the parallel axis theorem was used, where

$$J_0 = J_G + mh_G^2 \tag{9.46}$$

Equations (9.44) and (9.45) are rewritten in matrix form as

$$
\begin{bmatrix} m & 0 \\ 0 & J_G \end{bmatrix} \begin{bmatrix} \ddot{v} \\ \ddot{\theta} \end{bmatrix} + \begin{bmatrix} \frac{J_0}{J_G}c_1 & -\frac{mh_G}{J_G}c_\theta \\ -h_G c_1 & c_\theta \end{bmatrix} \begin{bmatrix} \dot{v} \\ \dot{\theta} \end{bmatrix}
$$

$$
+ \begin{bmatrix} \frac{J_0}{J_G}k_1 & -\frac{mh_G}{J_G}(k_\theta - m_0 g h_G + m_b g h_b) \\ -h_G k_1 & (k_\theta - m_0 g h_G + m_b g h_b) \end{bmatrix} \begin{bmatrix} v \\ \theta \end{bmatrix}
$$

$$
= \begin{bmatrix} \left[1 - \frac{mh_G}{J_G}(h_0 - h_G)\right] F - \frac{mh_G}{J_G} M_{pc} \\ M_{pc} + (h_0 + h_G)F \end{bmatrix} \tag{9.47}
$$

Frequencies and Mode Shapes

Suppose that the monotower is at rest in still water. Then, displaced by a sudden wind gust that subsequently subsides, the tower undergoes free vibration. The tower's motion will then be partly translational and partly rotational, with an exchange of energy between these modes. This should be expected since the equations of motion (9.47) are coupled. The two characteristic free vibration frequencies and their corresponding mode shapes are now computed by the methods in Chapter 8, Section 8.4. Let $\mathbf{C} = \mathbf{p} = \mathbf{0}$ in equations (9.47) and form the characteristic determinant

$$
\det(\mathbf{K} - \omega^2 \mathbf{M}) = 0 \tag{9.48}
$$

With the numerical parameters of Table 9.2 substituted for $\mathbf{K}$ and $\mathbf{M}$ in the governing equations (9.47), this determinant becomes

$$
\begin{vmatrix} (3.59 - 0.114\omega_0 - 0.615\omega^2) \times 10^9 & (0.0772 + 0.0107\omega_0) \times 10^{12} \\ (-66.3 + 2.11\omega_0) \times 10^9 & (3.60 - 0.497\omega_0 - 0.880\omega^2) \times 10^{12} \end{vmatrix} = 0 \tag{9.49}
$$

Assume that the soil foundation stiffnesses k_1 and k_θ are especially sensitive to the structure's fundamental frequency ω_1, which is the smallest positive root that satisfies this determinant. It is logical then to let $\omega_0 = \omega_1$. With this condition imposed, a trial and success procedure is used to extract the frequencies from equation (9.49). As a first trial, choose $\omega_0 = 1.41$ rad/s, the result obtained for this same structure under pure rocking motion. See *Example Problem 5.2*. The characteristic determinant then reduces to

$$
\omega^4 - 8.867\omega^2 + 11.08 = 0 \tag{9.50}
$$

Using the quadratic formula, the smallest positive root of this polynomial is calculated as $\omega_1 = 1.23$ rad/s. Next, choose $\omega_0 = 1.24$ rad/s, for which the polynomial from equation (9.49), when solved for the lowest two positive roots, yields the convergent results to three significant figures, or

$$
\omega_1 = 1.24 \text{ rad/s } (0.197 \text{ Hz}); \qquad \omega_2 = 2.73 \text{ rad/s } (0.434 \text{ Hz}) \tag{9.51}
$$

It follows that the $\mathbf{K}$ matrix from equation (9.47) is given by

$$\mathbf{K} = \begin{bmatrix} 3.44 \times 10^9 \text{ N/m} & -0.0641 \times 10^{12} \text{ N} \\ -0.0638 \times 10^{12} \text{ N} & 2.99 \times 10^{12} \text{ N·m} \end{bmatrix} \tag{9.52}$$

Because of roundoff errors and possible variations in the experimental values of the soil foundation constants, the last matrix is not symmetric to three significant figures.

Following equations (8.62)-(8.64), there is a corresponding modal vector $\hat{\boldsymbol{\xi}}_n$ for each frequency ω_n, computed from

$$(\mathbf{K} - \omega_n^2 \mathbf{M})\hat{\boldsymbol{\xi}}_n = \mathbf{0}, \qquad n = 1, 2 \tag{9.53}$$

in which the component form of the nonnormalized modal vector is

$$\hat{\boldsymbol{\xi}}_n = \begin{bmatrix} 1 & \hat{\xi}_{2n} \end{bmatrix}^T \tag{9.54}$$

The last two equations, when combined and written in component form, are

$$\begin{bmatrix} k_{11} - \omega_n^2 m & k_{12} \\ k_{21} & k_{22} - \omega_n^2 J_G \end{bmatrix} \begin{bmatrix} 1 \\ \hat{\xi}_{2n} \end{bmatrix} = \begin{bmatrix} 0 \\ 0 \end{bmatrix} \tag{9.55}$$

The first of these two equations is solved for $\hat{\xi}_{2n}$ to give

$$\hat{\xi}_{2n} = \frac{1}{k_{12}}(\omega_n^2 m - k_{11}) \tag{9.56}$$

With this equation, with the numerical values given in equations (9.51), (9.52) and Table 9.2, and with $\hat{\xi}_{11} = \hat{\xi}_{12} = 1$, the two modal vectors are evaluated as

$$\hat{\boldsymbol{\xi}}_1 = \begin{bmatrix} \hat{\xi}_{11} & \hat{\xi}_{21} \end{bmatrix}^T = \begin{bmatrix} 1 & 0.0390 \end{bmatrix}^T \tag{9.57a}$$

$$\hat{\boldsymbol{\xi}}_2 = \begin{bmatrix} \hat{\xi}_{12} & \hat{\xi}_{22} \end{bmatrix}^T = \begin{bmatrix} 1 & -0.0178 \end{bmatrix}^T \tag{9.57b}$$

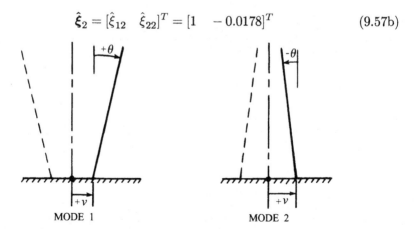

Figure 9.6   Mode shape envelopes for the monopod gravity platform.

The modal vectors of the last two equations are shown in Figure 9.6. For mode 1 corresponding to the lowest frequency, both components of $\hat{\boldsymbol{\xi}}_1$ are positive, which indicates a positive or right displacement for v (the $\hat{\xi}_{11}$ term) and a positive or clockwise rotation for θ (the $\hat{\xi}_{21}$ term), signs which are consistent with the chosen coordinate directions. The broken line is also possible since $\hat{\xi}_{11}$ could have been chosen as (-1) instead of $(+1)$, with the result that both v and θ would then be negative. The motions of mode 1 are in phase since both components of $\hat{\boldsymbol{\xi}}_1$ have the same sign. The motions of mode 2 are out of phase since the signs of the $\hat{\boldsymbol{\xi}}_2$ components are always opposite. In free vibration, the actual motion is a combination of both mode shapes, which depend on the structure's initial displacement and velocity. Once in motion, the mode shapes continually change as potential and kinetic energies are transferred in the foundation restraints.

With the frequencies and mode shapes in hand, the deterministic responses of this monotower to given time histories of loading, $F(t)$ and $M_{pc}(t)$, can be calculated in a straightforward way using the normal mode solutions derived in Chapter 8 and illustrated for the fixed leg platform at the beginning of this chapter.

This particular problem gives some interesting insights into the frequency behavior of multi-degree of freedom systems. For instance, it is observed that the fundamental undamped frequency is depressed by about 12 percent, from 1.41 rad/s for its single degree of freedom counterpart of *Example Problem 5.2*, to $\omega_1 = 1.24$ rad/s for the two degree of freedom analysis. This is characteristic of linear systems: as more flexibility is incorporated by allowing more degrees of freedom, ω_1 decreases. In either model, the effect of foundation damping and viscous damping due to the surrounding water is to depress the undamped frequency.

Nataraja and Kirk (1977) analyzed a similar structure. They calculated a value of 1.02 rad/s for the fundamental frequency of a three-legged gravity platform modeled to include leg flexibility and the same type of soil foundation elasticity (k_1 and k_θ) as used here. Three factors in the Nataraja and Kirk model account for their lower value of ω_1: their structural mass was somewhat higher, their model included soil foundation damping, and their model had added degrees of freedom because of leg flexibility.

Dynamic Stability

There is a vast literature on criteria and methods for determining the dynamic stability of coupled, linear systems such as the gravity platform. The classical works of Liapunov (1907) and Ziegler (1956) are especially noteworthy. For present purposes, however, the criteria are relatively simple. The dynamic stability of a linear, undamped structure in free vibration about its static equilibrium position can be tested by investigating the nature of the roots ω_i^2 as calculated from the characteristic determinant, equation (9.3). If every ω_i^2 is real and positive, then the characteristic frequencies given by $\omega_i = \sqrt{\omega_i^2}$ are real and positive, and the system undergoes stable, bounded, harmonic oscillations.

That is, the displacements ξ_i behave as

$$\xi_i \approx \hat{\xi}_i \, e^{j\omega_i t} \approx \hat{\xi}_i \sin \omega_i t \tag{9.58}$$

However, if any value of ω_i^2 is negative or complex, the system will be dynamically unstable. For instance, if $\omega_1^2 = -\alpha^2$ where α is real and positive, then $\omega_1 = \pm j\alpha$. With $\omega_1 = -j\alpha$, equation (9.58) becomes

$$\xi_1 \approx \hat{\xi}_1 e^{\alpha t} \tag{9.59}$$

which is unbounded as time increases. If ω_i^2 is complex, one of the roots produces this same unboundedness. Computer codes that extract roots of a determinant, such as subroutine *Eigenvalues* of Mathematica® (1999), are generally available to the engineer, so it is a relatively straightforward task to check the dynamic stability of a structure.

With respect to a gravity platform on an elastic soil foundation, dynamic instability leading to toppling would occur for critical combinations of its mass, its center of buoyancy, its center of gravity, and its foundation stiffness. For instance, for sufficiently high values of platform mass and h_G, it is visualized that a moderately weak soil foundation could offer an insufficient restoring moment to resist both the angular momentum of the structure and the overturning moment due to its deadweight. Then the structure would topple.

In conclusion, it is noted that if the criterion for dynamic stability is observed (that all values of ω_i^2 are real and positive), then the static criteria discussed in Chapter 1, applied to gravity platforms, will also be satisfied. It is now well recognized that a dynamic stability analysis includes the results obtained from a static stability analysis; but the converse is not true.

9.3 STRUCTURAL RESPONSE STATISTICS FOR WAVE LOADING

As discussed in Chapters 6 and 7, the wave data available to the analyst and designer of offshore structures are most often in the form of a surface wave height spectrum, $S_\eta(\omega)$. In Chapter 6 a method was presented for representing this spectrum as harmonic waves forms. These forms, when converted to structural forces through suitable transfer functions $G(\omega)$, are then used to calculate the time history of structural response. An alternative approach was discussed in Chapter 7 where the spectral density of the response and its variance were calculated directly from $S_\eta(\omega)$ and $G(\omega)$ for the single degree of freedom case. This latter approach is now employed for the case of a linear structure with N degrees of freedom. Summarized in the following ten steps are the critical assumptions and the methodology leading to an analytical form for the spectral density defined as $S(\xi_k, \omega)$ and the variance $\sigma^2(\xi_k)$, calculated for each independent physical coordinate ξ_k. The load transfer functions in this case are defined by the vector $\mathbf{G}(p, \omega)$.

1. Set up the mathematical model and derive the equations of motion for the structure in the form of equation (9.1). Identify the coefficient matrices $\mathbf{M}$ and $\mathbf{K}$.

2. Calculate each component $p_k(t)$ of the loading vector $\mathbf{p}(t)$. Select linear wave theory. Use Morison's equation and diffraction theory where appropriate to determine the components of loading.

3. Calculate each component function $G(p_k, \omega)$ of the load transfer vector $\mathbf{G}(p, \omega)$.

4. Calculate the undamped frequencies ω_k and the normalized mode shape matrix $\mathbf{X}$. The consecutive columns of $\mathbf{X}$ are modal vectors $\mathbf{x}_1, \mathbf{x}_2, \ldots, \mathbf{x}_N$. The kth modal vector has components $x_{1k}, x_{2k}, \ldots, x_{Nk}$.

5. Define the generalized loading vector in modal coordinates as $\bar{\mathbf{q}}(t) = \bar{\mathbf{q}}$. Then calculate the corresponding transfer functions $G(\bar{q}_k, \omega)$ for each component $\bar{q}_k$ of $\bar{\mathbf{q}}$ where

$$\bar{q}_k = \mathbf{x}_k^T \mathbf{p}(t) \tag{9.60}$$

Thus the generalized load transfer function for the kth mode is

$$G(q_k, \omega) = \mathbf{x}_k^T \mathbf{G}(p, \omega) \tag{9.61}$$

Note that $\mathbf{G}(p, \omega)$ was calculated in step 3.

6. With equations (8.95) and (9.60), the uncoupled equations of motion become

$$\ddot{y}_k + 2\zeta_k \omega_k \dot{y} + \omega_k^2 y_k = \bar{q}_k \tag{9.62}$$

Calculate $H_k(\omega)$, the harmonic response function for the kth mode, by substituting the following quantities into the last equation:

$$\bar{q}_k(t) = e^{j\omega t}; \qquad y_k(t) = H_k(\omega) e^{j\omega t} \tag{9.63}$$

The results, including the modulus, are

$$H_k(\omega) = (\omega_k^2 - \omega^2 + 2j\zeta_k \omega_k \omega)^{-1} \tag{9.64}$$

$$|H_k(\omega)| = [(\omega_k^2 - \omega^2)^2 + (2\zeta_k \omega_k \omega)^2]^{-1/2} \tag{9.65}$$

7. Assume that $\mathbf{p}(t)$ is a stationary ergodic process. It follows that $\bar{\mathbf{q}}(t)$ will be stationary and ergodic also since the components of $\mathbf{p}(t)$ and $\bar{\mathbf{q}}(t)$ are related by the linear transformation (9.60). Using the analysis in Chapter 7 for the single degree of freedom system in the form of equation (9.62), it follows that the spectral density of y_k is

$$S(y_k, \omega) = |H_k(\omega)|^2 S(\bar{q}_k, \omega) \tag{9.66}$$

Here $S(\bar{q}_k, \omega)$ is the spectral density of the generalized force component $\bar{q}_k$.

8. Assume linear wave theory. From *Example Problem 7.2*, deduce the following analogous relationship between the spectral density of the kth load, $S(p_k, \omega)$, the wave height spectrum $S_\eta(\omega)$, and the transfer function $G(p_k, \omega)$:

$$S(p_k, \omega) = |G(p_k, \omega)|^2 S_\eta(\omega) \tag{9.67}$$

Since the system is linear, it follows that

$$S(\bar{q}_k, \omega) = |G(\bar{q}_k, \omega)|^2 \, S_\eta(\omega) \tag{9.68}$$

which, with equation (9.61), becomes

$$S(\bar{q}_k, \omega) = \left|\mathbf{x}_k^T \mathbf{G}(p, \omega)\right|^2 S_\eta(\omega) \tag{9.69}$$

With this last result, equation (9.66) is then

$$S(y_k, \omega) = |H_k(\omega)|^2 \cdot \left|\mathbf{x}_k^T \mathbf{G}(p, \omega)\right|^2 \cdot S_\eta(\omega) \tag{9.70}$$

in which all components on the right side of this last equation are known.

9. Deduce the relationship between $S(y_k, \omega)$ of this last result and the spectral density in terms of the physical coordinates, $S(\xi_k, \omega)$. To do this, use the definition of the autocorrelation function given by equation (7.19). For the kth coordinate ξ_k, this function is

$$R(\xi_k, \tau) = E[\xi_k(t)\xi_k(t + \tau)] \tag{9.71}$$

With this definition and the component form of the modal coordinate transformation given by

$$\xi_k = \sum_{n=1}^{n=N} x_{kn} y_n \tag{9.72}$$

It follows that

$$R(\xi_k, \tau) = E\left[\sum_{n=1}^{N}\sum_{m=1}^{N} x_{kn} x_{km} y_n(t) y_m(t + \tau)\right] \tag{9.73}$$

In the sums of this last result, there are N autocorrelation functions of the form

$$R_n(\tau) = E[y_n(t)y_n(t + \tau)], \qquad n = m \tag{9.74}$$

and $N(N-1)/2$ cross-correlation functions

$$R_{nm}(\tau) = E[y_n(t)y_m(t + \tau)], \qquad n \neq m \tag{9.75}$$

Assume now that modal coupling is negligible, meaning that each $y_n(t)$ is a statistically independent process. Thus, the cross-correlation functions of the last equation vanish. It follows from equations (9.73) and (9.74) that

$$R(\xi_k, \tau) = \sum_{n=1}^{N} x_{kn}^2 R_n(\tau) \tag{9.76}$$

The following two results are based on the definition from equation (7.20):

$$S(\xi_k, \omega) = \frac{1}{2\pi} \int_{-\infty}^{\infty} R(\xi_k, \tau) e^{-j\omega\tau} d\tau \tag{9.77}$$

$$S(y_n, \omega) = \frac{1}{2\pi} \int_{-\infty}^{\infty} R_n(\tau) e^{-j\omega\tau} d\tau \tag{9.78}$$

Substitute $R(\xi_k, \tau)$ of equation (9.76) in (9.77). Then interchange the order of integration and summation, assuming a well-behaved argument. Use equation (9.78) in this result to obtain

$$S(\xi_k, \omega) = \sum_{n=1}^{N} x_{kn}^2 S(y_n, \omega) \tag{9.79}$$

Now substitute equation (9.65) into (9.70), change the index from k to n, and substitute this rewritten form of $S(y_n, \omega)$ into equation (9.79). The result is

$$S(\xi_k, \omega) = S_\eta(\omega) \cdot \sum_{n=1}^{N} \frac{x_{kn}^2 \left| \mathbf{x}_n^T \mathbf{G}(p, \omega) \right|^2}{(\omega_n^2 - \omega^2)^2 + (2\zeta_n \omega_n \omega)^2} \tag{9.80}$$

10. Calculate the variance for each physical coordinate ξ_k. Using the definition given by equation (7.22) and $S(\xi_k, \omega)$ of the last result, compute

$$\sigma^2(\xi_k) = 2 \int_0^{\infty} S(\xi_k, \omega) d\omega \tag{9.81}$$

With the wave height spectrum $S_\eta(\omega)$ in the Pierson-Moskowitz or JON-SWAP form, and the limits of integration in equation (9.81) replaced by 0.16 and 1.4 rad/s, the variance of each coordinate can be obtained by numerical integration. Assuming that $\mathbf{p}(t)$ is Gaussian with zero mean, then the responses will also be Gaussian with zero mean, since the system is linear. Thus, the rms value of the kth physical coordinate is given by the square root of equation (9.81). For practical purposes, the extreme limits of ξ_k are $\pm 3\sigma(\xi_k)$, $k = 1, 2, \ldots, N$. For these extreme values, if the static stresses and deflections of the members are within the allowable limits, then the structure is assumed *safe*, without considering other loadings and fatigue failure. For design purposes, it is generally acceptable to superimpose the effects of the static or steady drag loadings due to winds and currents.

9.4   A FIXED LEG PLATFORM: STATISTICAL RESPONSES

The statistical responses derived in the last section are now computed for the two degree of freedom jacket template platform modeled in Figures 9.1 and 9.2. The numerical parameters of this platform are summarized in Table 9.1. The results of the first three steps of this calculation were obtained in Section 9.1 and are summarized as follows in terms of these numerical parameters:

1. The mathematical model is defined by the following two equations of motion. The form of damping is defined later.

$$\begin{bmatrix} 4.69 \times 10^6 & 0 \\ 0 & 3.13 \times 10^6 \end{bmatrix} \begin{bmatrix} \ddot{\xi}_1 \\ \ddot{\xi}_2 \end{bmatrix} + \begin{bmatrix} c_{11} & c_{12} \\ c_{21} & c_{22} \end{bmatrix} \begin{bmatrix} \dot{\xi}_1 \\ \dot{\xi}_2 \end{bmatrix}$$

$$+ \begin{bmatrix} 7.35 \times 10^7 & -1.15 \times 10^8 \\ -1.15 \times 10^8 & 3.59 \times 10^8 \end{bmatrix} \begin{bmatrix} \xi_1 \\ \xi_2 \end{bmatrix} = \begin{bmatrix} p_1 \\ p_2 \end{bmatrix} \qquad (9.82)$$

2. Assume that linear wave is sufficiently accurate for this analysis. The wave loading vector is thus given by equations (9.19) and (9.20) with $b_3 = b_4 = 0$, or

$$\begin{bmatrix} p_1 \\ p_2 \end{bmatrix} = -\alpha \frac{\omega^2 H}{k \sinh 61k} \begin{bmatrix} \sinh 61k - \sinh 38k \\ \sinh 38k + \beta k \cosh 38k \end{bmatrix} \sin \omega t \qquad (9.83)$$

Here, the numerical values of the constants α and β are

$$\alpha = N_\ell C_M \frac{\pi}{8} \rho D_\ell^2 = 97\,980 \text{ kg/m}; \quad \beta = w \frac{N_c}{N_\ell} \left(\frac{D_c}{D_\ell}\right)^2 = 9.169 \text{ m} \qquad (9.84)$$

From each load component, the corresponding component of the transfer function is deduced from the definition that the real part of $\mathbf{G}(p, \omega)$ is equal to $\mathbf{p}(t)/H$, or

$$\begin{bmatrix} G(p_1, \omega) \\ G(p_2, \omega) \end{bmatrix} = j\alpha \frac{\omega^2}{k \sinh 61k} \begin{bmatrix} \sinh 61k - \sinh 38k \\ \sinh 38k + \beta k \cosh 38k \end{bmatrix} e^{j\omega t} \qquad (9.85)$$

3. The two undamped frequencies and the normalized modal matrix, given by the respective equations (9.6) and (9.14), are

$$\omega_1 = 2.706 \text{ rad/s}; \quad \omega_2 = 11.09 \text{ rad/s} \qquad (9.86)$$

$$\mathbf{X} = \begin{bmatrix} x_{11} & x_{12} \\ x_{21} & x_{22} \end{bmatrix} = \begin{bmatrix} 4.45 & 1.24 \\ 1.52 & -5.44 \end{bmatrix} \times 10^{-4} \text{ kg}^{-1/2} \qquad (9.87)$$

With the results of these three steps, the statistical responses are then calculated directly from steps 9 and 10, or equations (9.80) and (9.81) of Section 9.3. These calculations do require a wave height spectrum, which is now chosen as the Pierson-Moskowitz spectrum of equation (7.59), with a corresponding significant wave height of $H_s = 15$ m. That is, for $g = 9.81$ m/s^2,

$$S_\eta(\omega) = \frac{0.780}{\omega^5} e^{-0.0138/\omega^4} \text{ m}^2 \cdot \text{s/rad} \qquad (9.88)$$

Using the normalized modal vectors of $\mathbf{X}$, equation (9.87), the components of the transformed transfer function in equation (9.80) are computed as follows:

$$\mathbf{x}_1^T \mathbf{G}(p, \omega) = [4.45 \quad 1.52] \times 10^{-4} \begin{bmatrix} G(p_1, \omega) \\ G(p_2, \omega) \end{bmatrix}$$

$$= [4.45 G(p_1, \omega) + 1.52 G(p_2, \omega)] \times 10^{-4} \text{ kg}^{1/2}\text{s}^{-2} \qquad (9.89)$$

$$\mathbf{x}_2^T \mathbf{G}(p, \omega) = [1.24 \quad -5.44] \times 10^{-4} \begin{bmatrix} G(p_1, \omega) \\ G(p_2, \omega) \end{bmatrix}$$

$$= [1.24G(p_1, \omega) - 5.44G(p_2, \omega)] \times 10^{-4} \text{ kg}^{1/2}\text{s}^{-2} \tag{9.90}$$

The squares of the moduli of these last two results are, respectively,

$$\left| \mathbf{x}_1^T \mathbf{G}(p, \omega) \right|^2 = (4.45P + 1.52Q)^2 \tag{9.91}$$

$$\left| \mathbf{x}_2^T \mathbf{G}(p, \omega) \right|^2 = (1.24P - 5.44Q)^2 \tag{9.92}$$

in which the quantities P and Q are defined by

$$P = \frac{\alpha\omega^2 \times 10^{-4}}{k \sinh 61k} (\sinh 61k - \sinh 38k) \tag{9.93}$$

$$Q = \frac{\alpha\omega^2 \times 10^{-4}}{k \sinh 61k} (\sinh 38k + \beta k \cosh 38k) \tag{9.94}$$

Note that α and β are constants given by equations (9.84). Also, the wave number k is related to the wave frequency ω by dispersion equation (3.16), which in this example is

$$\omega^2 = 9.81k \tanh 61k \tag{9.95}$$

With equations (9.86)-(9.88) and (9.91)-(9.94), the spectral densities of the horizonal deflections based on equation (9.80) then have the following forms:

$$S(\xi_1, \omega) = S_\eta(\omega) \frac{(4.45 \times 10^{-4})^2 (4.45P + 1.52Q)^2}{(2.706^2 - \omega^2)^2 + 0.01(2.706)^2\omega^2}$$

$$+ S_\eta(\omega) \frac{(1.24 \times 10^{-4})(1.24P - 5.44Q)^2}{(11.09^2 - \omega^2)^2 + 0.01(11.09)^2\omega^2} \tag{9.96}$$

$$S(\xi_2, \omega) = S_\eta(\omega) \frac{(1.52 \times 10^{-4})^2 (4.45P + 1.52Q)^2}{(2.706^2 - \omega^2)^2 + 0.01(2.706)^2\omega^2}$$

$$+ S_\eta(\omega) \frac{(-5.44 \times 10^{-4})(1.24P - 5.44Q)^2}{(11.09^2 - \omega^2)^2 + 0.01(11.09)^2\omega^2} \tag{9.97}$$

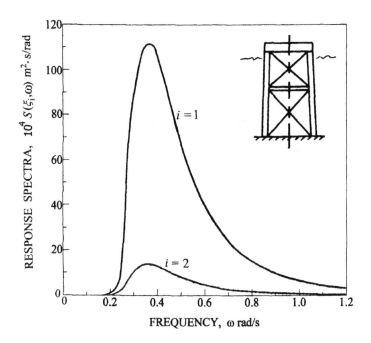

Figure 9.7   Response spectra for the displacement coordinates of the fixed leg
platform.

Numerical results for these response spectra, computed using Mathematica®
(1999), and PSI-Plot (1999), are shown in Figure 9.7. As one would expect, the
higher spectrum is associated with ξ_1 or the deck level at 76 m. These peaks
for $S(\xi_1, \omega)$ and $S(\xi_2, \omega)$ are 111.2×10^{-4} m²·s/rad and 13.6×10^{-4} m²·s/rad,
respectively, where both peaks occur at $\omega = 0.372$ rad/s. The variance of each
response was computed from these results using equation (9.81), where the limits
of integration $(0, \infty)$ were replaced by $(0.16, 1.4)$ rad/s. Those results and their
rms values are as follows:

$$\sigma^2(\xi_1) = 69.62 \times 10^{-4} \text{ m}^2; \quad \sigma(\xi_1) = 0.0834 \text{ m} \qquad (9.98)$$

$$\sigma^2(\xi_2) = 8.564 \times 10^{-4} \text{ m}^2; \quad \sigma(\xi_2) = 0.0293 \text{ m} \qquad (9.99)$$

Assuming a Gaussian process, there is only a 0.026 percent chance that each
response exceeds the $\pm 3\sigma$ limits, or that ξ_1 exceeds 0.250 m and that ξ_2 exceeds
0.0878 m. Note that these latter two displacements of 0.250 m and 0.0878 m
are close to the respective values of $\xi_1 = 0.1937$ m and $\xi_2 = 0.0805$ m derived
in Section 9.1 as the peak responses to a single incident harmonic wave with a
significant wave height of 15 m.

PROBLEMS

9.1   For a water depth of $d = 300$ m, plot the value of wave number k as a function of wave frequency ω where $0.1 < \omega < 2$ rad/s. Use the relationship $\omega^2 = gk \tanh kd$. For what range of frequency is k given by ω^2/g to within 5 percent?

9.2   Calculate the damping matrix $\mathbf{C}$ for the fixed leg platform of Section 9.1. Assume $\zeta_1 = \zeta_2 = 0.07$. Indicate the units of $\mathbf{C}$.

9.3   Derive in detail the result of equation (9.28), the mathematical model for an N degree of freedom stalk structure subjected to a horizontal base excitation. Why is the negative sign on the right of equation (9.28) ignored in the response solution of equation (9.30)?

9.4   For the gravity platform modeled in Section 9.2, calculate $\mathbf{X}$ and show the units for each component vector $\mathbf{x}_i$.

9.5   Suppose that the gravity monotower of Section 9.2 is subjected to the horizontal El Centro earthquake. Calculate the maximum values of v and θ, the peak absolute value of horizontal displacement for the deck, the peak shear force at the base, and the peak overturning moment. Assume that the modal damping for each mode is 0.05.

9.6   Show that if one of the frequencies ω_i of the characteristic determinant of an undamped linear system is a complex number, then the system's behavior is divergent.

9.7   For an N degree of freedom linear structure, write a computer program to calculate the coordinate displacement spectra, the corresponding variances, and the rms values. Use the results of equations (9.80) and (9.81) where the following quantities are specified as input: $S_\eta(\omega)$, $\mathbf{X}$, $\mathbf{G}(p, \omega)$, ζ_k, and ω_k, $k = 1, 2, \ldots, N$. As a numerical example, check the results for $S(\xi_k, \omega)$ and $\sigma(\zeta_k)$ obtained for the linear structure in Section 9.4, for $k = 1, 2$.

REFERENCES

Clough, R. W., and Penzien, J., *Dynamics of Structures*, second ed., McGraw-Hill, New York, 1993.

Liapunov, A. M., Probleme general de la stabilite du mouvement, translated into French by E. Davaux, *Annales de Toulouse* (2), **9**, 1907. Reprinted by Princeton University Press, Princeton, NJ, 1949.

Mansour, A. E., and Millman, D. N., Dynamic Random Analysis of Fixed Offshore Platforms, OTC-2049, *Proceedings of the Offshore Technology Conference*, 1974.

Mathematica®, version 4, Wolfram Media, Inc., Champaign, IL, 1999.

Nataraja, R., and Kirk, C. L., Dynamic Response of a Gravity Platform Under Random Wave Forces, OTC-2904, *Proceedings of the Offshore Technology Conference*, 1977.

PSI-Plot, version 6, Poly Software International, Sandy, UT, 1999.

Ziegler, H., On the Concept of Elastic Stability, *Advances in Applied Mechanics, Vol. 4*, H. L. Dryden and T. von Karman, editors, Academic Press, New York, 1956.

10

Continuous Systems

James F. Wilson

The dynamic models of offshore structures discussed so far have involved only a finite number of independent coordinates and ordinary differential equations of motion. In the single degree of freedom systems, one coordinate was chosen to describe the dominant structural vibration mode in the plane. In the multi-degree of freedom systems, examples included the rigid gravity platform with coordinates to describe sliding and rocking motion and jacket template platforms with coordinates to describe the motion of discrete masses lumped at node points.

For line components such as rather long beams, pipelines, and cables, alternative continuous system models may provide more precise and sometimes more economical descriptions of component motion. Since a partial differential equation is used to characterize the motion of a continuous line component, solutions are generally more involved mathematically than for a corresponding lumped system. However, if the continuous models are chosen judiciously, closed form expressions can be derived for the characteristic frequencies and mode shapes of line components, which then lead to upper and lower bounds on their dynamic responses.

Two classes of continuous line components are analyzed in this chapter. The first component is designated as a beam for which bending stiffness and longitudinal tension are incorporated in the model. Examples include: pipelines for dredging manganese nodules from the sea floor as depicted in Figure 1.10a; pipelines for ocean thermal energy conversion as shown in Figure 1.10b; gathering lines and risers; and the long structural bracing members of the various offshore platforms. The second structural component considered here is the cable which resists tension but whose bending stiffness is negligible. Examples include the steel and synthetic fiber ropes and steel chains used to stay buoys, floating platforms, compliant towers, and ships.

Excitation of line components comes about in several ways. For instance, a flexible cylinder in a steady current may undergo adverse transverse motion and can be destroyed by the periodic shedding of vortices. This can occur if one of the lower natural frequencies of the cylinder is coincident with that of the vortices. There is also direct transverse excitation due to waves, and there is end excitation of line components attached to ships and towers that move with the waves. Two types of end motion for a mooring line, transverse and longitudinal (or parametric) excitation, are depicted in Figure 10.1.

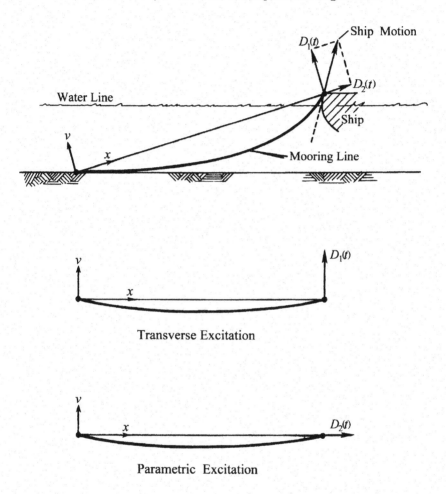

Figure 10.1 Types of end excitation for a mooring line.

The purpose of this chapter is to describe the dynamic behavior of these continuous line components in a plane. As in the analysis of the discrete, finite degree of freedom systems, this analysis involves setting up appropriate mathematical models, solving for free vibration frequencies, determining mode shapes, and then solving for forced motion. Both deterministic and statistical response models are discussed. The chapter concludes with a practical example – numerical response calculations for an ocean thermal energy conversion (OTEC) pipeline attached to barge subjected to random wave excitation.

10.1 MODELING BEAMS AND CABLES

Governing Equations

The beam or cable model is shown in Figure 10.2. This line component has a virtual mass per unit length of $\bar{m}$, a length ℓ, and a fixed longitudinal axis x intersecting the end points. The transverse dynamic displacement from its equilibrium position at $v = 0$ is $v = v(x, t)$ which is assumed to be small enough so that its slope $\theta = \partial v / \partial x$ is always much less than one. A further assumption, which is also consistent with classical beam theory, is that transverse planes at equilibrium, or $v = 0$, remain planes during motion, when $v \neq 0$. The infinitesimal element of length dx shows the bending moment M, the transverse shear load V, and the tension load P, all of which are subject to small changes across the element. The transverse excitation force per unit length is $\bar{q} = \bar{q}(x, t)$, and the linear damping force per unit length is $\bar{c}\dot{v}$, which correspond to the average element loads $\bar{q}dx$ and $\bar{c}\dot{v}dx$, respectively, acting at the element's center. For a small slope, first-order changes in θ, M, V, and P are appropriate. That is

$$\theta + d\theta \simeq \frac{\partial v}{\partial x} + \frac{\partial^2 v}{\partial x^2}dx \; ; \qquad M + dM \simeq M + \frac{\partial M}{\partial x}dx$$

$$V + dV \simeq V + \frac{\partial V}{\partial x}dx \; ; \qquad P + dP \simeq P + \frac{\partial P}{\partial x}dx \qquad (10.1)$$

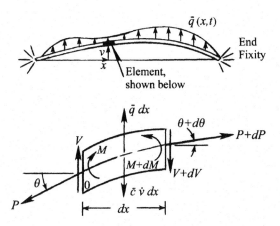

Figure 10.2   Dynamic model of a line element.

Now apply Newton's second law to the element in the direction of v:

$$\sum F_{v \text{ direction}} = \bar{m} \, dx \frac{\partial^2 v}{\partial t^2}$$

$$-P\theta + (P+dP)(\theta + d\theta) + V - (V+dV) + \bar{q}\,dx - \bar{c}\frac{\partial v}{\partial t}dx = \bar{m}\,dx\frac{\partial^2 v}{\partial t^2} \quad (10.2)$$

After combining equations (10.1) and (10.2), expanding, and then dropping the higher order terms involving $(dx)^2$, the result is

$$-\frac{\partial V}{\partial x} + \frac{\partial}{\partial x}\left(P\frac{\partial v}{\partial x}\right) + \bar{q} - \bar{c}\frac{\partial v}{\partial t} = \bar{m}\frac{\partial^2 v}{\partial t^2} \quad (10.3)$$

Next sum moments about point 0, the lower left corner of the element, and equate this sum to zero. With this approximation, rotational inertia, or rotatory inertia as it is sometimes called, is neglected. For the relatively low frequencies encountered in beams and cables of ocean structures (below 100 Hz), the rotational energy is much less than that due to transverse motion, which justifies this assumption. It follows that

$$\sum M_0 \simeq 0$$

$$M - (M + dM) - \bar{q}\,dx\left(\frac{dx}{2}\right) + \bar{c}\dot{v}\,dx\left(\frac{dx}{2}\right) + (V+dV)dx = 0 \quad (10.4)$$

Here, the terms involving $\bar{q}, \bar{c}$, and dV are multiplied by $(dx)^2$, and are thus neglected as higher-order terms. Equation (10.4) thus yields the shear load as

$$V = \frac{\partial M}{\partial x} \quad (10.5)$$

For elastic members, elementary beam theory gives

$$M = EI\frac{\partial^2 v}{\partial x^2} \quad (10.6)$$

where EI is the bending stiffness. Differentiating equations (10.5) and (10.6), and then combining the results, leads to

$$\frac{\partial V}{\partial x} = \frac{\partial^2 M}{\partial x^2} = \frac{\partial^2}{\partial x^2}\left(EI\frac{\partial^2 v}{\partial x^2}\right) \quad (10.7)$$

When the last result is combined with equation (10.3), the final result is obtained, or

$$\frac{\partial^2}{\partial x^2}\left(EI\frac{\partial^2 v}{\partial x^2}\right) - \frac{\partial}{\partial x}\left(P\frac{\partial v}{\partial x}\right) + \bar{c}\frac{\partial v}{\partial t} + \bar{m}\frac{\partial^2 v}{\partial t^2} = \bar{q}(x,t) \quad (10.8)$$

This is a general form of the linear Bernoulli-Euler dynamic beam-cable equation where EI, P, and $\bar{m}$ are arbitrary functions of x and the excitation load $\bar{q}(x,t)$ is arbitrary in both x and t.

For submerged beams or cables, a more realistic form of damping than the linear approximation $\bar{c}(\partial v/\partial t)$ in equation (10.8) is *velocity squared* damping, defined by

$$c' \frac{\partial v}{\partial t} \cdot \left| \frac{\partial v}{\partial t} \right| \tag{10.9}$$

Here c' is an experimental constant which generally depends on the frequency of oscillation of the structural component. In velocity squared damping, the absolute value sign is necessary to ensure that the damping force always opposes the direction of beam motion. Since equation (10.9) renders the Bernoulli-Euler equation nonlinear and intractable for closed form solutions, for simplicity the linear form is assumed for this chapter. However, equation (10.9) has been successfully included in computer codes that solve for the nonlinear responses of submerged beams (Wilson et al., 1982).

Very specific initial and boundary conditions must be specified for solutions to equation (10.8) to be unique. For a cable ($EI = 0$) and for a beam ($EI > 0$), the following two initial conditions are always required:

$$v(x,0) \quad \text{and} \quad \frac{\partial v}{\partial t}(x,0), \quad \text{for} \quad 0 \leq x \leq \ell \tag{10.10}$$

In addition, solutions to cable problems require that the end displacements $v(0,t)$ and $v(\ell, t)$ be specified, and solutions to beam problems require four boundary conditions, two at $x = 0$ and two at $x = \ell$. These conditions will be described presently.

Two special cases of equation (10.8) that are of practical importance in offshore structural systems are modeled as follows:

1. For an undamped, flexible cable ($EI = 0$) subjected to a tension load P which is independent of x, equation (10.8) becomes

$$-P\frac{\partial^2 v}{\partial x^2} + \bar{m}\frac{\partial^2 v}{\partial t^2} = \bar{q}(x,t) \tag{10.11}$$

Here m can vary with the longitudinal dimension, but it is constant for most applications.

2. For an undamped beam of constant stiffness EI, a constant $\bar{m}$, and a tension P which is independent of x, equation (10.8) becomes

$$EI \frac{\partial^4 v}{\partial x^4} - P\frac{\partial^2 v}{\partial x^2} + \bar{m}\frac{\partial^2 v}{\partial t^2} = \bar{q}(x,t) \tag{10.12}$$

The procedures for calculating the free, undamped frequencies and mode shapes for cables and beams described by these last two models are now illustrated.

Cable Frequencies and Mode Shapes

Consider the free, undamped vibrations of a flexible cable of constant $\bar{m}$ and constant tension $P = P_0$. The corresponding equation of motion is deduced from equation (10.11) as

$$P_0 \frac{\partial^2 v}{\partial x^2} - \bar{m} \frac{\partial^2 v}{\partial t^2} = 0 \tag{10.13}$$

Let $X = X(x)$ be the general expression for the mode shape, and assume harmonic motion with ω denoting the frequency parameter, or

$$v = X e^{j\omega t} \tag{10.14}$$

Combine the last two equations to give

$$X'' + \gamma^2 X = 0 \tag{10.15}$$

where (′) denotes the operator d/dx. The frequency parameter is given by

$$\omega = \gamma \sqrt{\frac{P_0}{\bar{m}}} \tag{10.16}$$

where γ has yet to be calculated. With each end of the cable fixed, then

$$v(0,t) = 0 \quad \text{or} \quad X(0) = 0; \qquad v(\ell,t) = 0 \quad \text{or} \quad X(\ell) = 0 \tag{10.17}$$

where the end conditions on X are determined from the end conditions on v through equation (10.14).

The general solution of equation (10.15) in terms of two arbitrary constants D_1 and D_2 is

$$X(x) = D_1 \sin \gamma x + D_2 \cos \gamma x \tag{10.18}$$

When each condition of equation (10.17) is applied to equation (10.18), the two results are

$$D_1 \sin 0 + D_2 \cos 0 = 0 \tag{10.19}$$

$$D_1 \sin \gamma \ell + D_2 \cos \gamma \ell = 0 \tag{10.20}$$

It follows from equation (10.19) that $D_2 = 0$. With this, the last equation implies either the trivial result that $D_1 = 0$ or that no motion exists. The alternative is that $\sin \gamma \ell = 0$, which leads to

$$\gamma = \frac{n\pi}{\ell}, \quad n = 1, 2, \ldots \tag{10.21}$$

With this result and equation (10.16), it follows that the cable frequencies are given by

$$\omega_n = \frac{n\pi}{\ell} \sqrt{\frac{P_0}{\bar{m}}}, \quad n = 1, 2, \ldots \tag{10.22}$$

where the subscript n on the frequency parameter is a reminder that there are multiple frequencies. For each characteristic value γ there is a frequency ω_n and a corresponding mode shape $X = X_n$ given by equations (10.18) and (10.22), where $D_1 = C_n$ and $D_2 = 0$. That is

$$X_n = C_n \sin \frac{n\pi x}{\ell}, \quad n = 1, 2, \ldots \tag{10.23}$$

in which the coefficients C_n are arbitrary. The first two mode shapes are shown in Figure 10.3. It is observed that the fundamental frequency ω_1 corresponds to the half sine wave, and the next highest frequency ω_2 corresponds to a full sine wave.

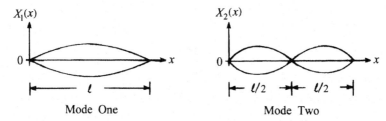

$X_1(x)$ $X_2(x)$

Mode One   Mode Two

Figure 10.3   The first two mode shapes for both a fixed end cable and a simply supported beam.

This mathematical model predicts that ever increasing frequencies are possible as n becomes larger and larger. In reality, damping and cross-coupling effects omitted in the mathematical model reduce these higher frequencies and accompanying mode shapes to insignificance. Thus it is the practice of many engineers to take $n = 20$ as a physically realistic upper limit for the purposes of analysis and design of offshore cables and other continuous components as well.

Beam Frequencies and Mode Shapes

The free undamped vibrations of a uniform beam with a negligible axial tension are described by equation (10.12) for $P_0 = \bar{q} = 0$, or

$$EI \frac{\partial^4 v}{\partial x^4} + \bar{m} \frac{\partial^2 v}{\partial t^2} = 0 \tag{10.24}$$

The characteristic frequencies and mode shapes depend on the support or boundary conditions. For purposes of illustration, the beam's supports are chosen to be at $x = 0$ and $x = \ell$ only. Further, each end is subjected to any one of the following three sets of conditions in which v_e designates the deflection at the end $x = 0$ or at $x = \ell$:

1. Simple support, or zero displacement and zero moment at a pin or roller:

$$v_e = 0; \quad EI \frac{\partial^2 v_e}{\partial x^2} = 0 \tag{10.25}$$

2. Clamped support, or zero displacement and zero slope at the end:

$$v_e = 0; \qquad \frac{\partial v_e}{\partial x} = 0 \qquad\qquad (10.26)$$

3. No support, or zero moment and zero transverse shear at the end:

$$EI\,\frac{\partial^2 v_e}{\partial x^2} = 0; \qquad EI\,\frac{\partial^3 v_e}{\partial x^3} = 0 \qquad\qquad (10.27)$$

Let $X = X(x)$ denote the mode shape and ω the frequency parameter. For harmonic beam vibrations of the form given by equation (10.14) applied to equation (10.24), the characteristic beam equation becomes

$$X'''' - \alpha^2 X = 0 \qquad\qquad (10.28)$$

where the frequency parameter is

$$\omega = \alpha^2 \sqrt{\frac{EI}{\bar{m}}} \qquad\qquad (10.29)$$

and where α has yet to be calculated. The boundary conditions associated with equation (10.28) are recast using equation (10.14). Letting $X(e)$ designate either $X(0)$ or $X(\ell)$, equations (10.25)-(10.27) become, respectively:

$$X(e) = X''(e) = 0, \qquad \text{simple support} \qquad\qquad (10.30)$$

$$X(e) = X'(e) = 0, \qquad \text{clamped end} \qquad\qquad (10.31)$$

$$X''(e) = X'''(e) = 0, \qquad \text{free end} \qquad\qquad (10.32)$$

The general solution to equation (10.28) is given by

$$X(x) = D_1 \sin \alpha x + D_2 \cos \alpha x + D_3 \sinh \alpha x + D_4 \cosh \alpha x \qquad\qquad (10.33)$$

where D_1, D_2, D_3, and D_4 are constants. This solution, together with the appropriate end conditions lead to the calculation of α, the free vibration frequencies, and the mode shapes. This procedure is now illustrated.

Example Problem 10.1. Compute the range of undamped, free vibration frequencies for a submerged, uniform cross brace of a jacket template platform, as shown in Figure 10.4a. Because each end of this brace is welded to a leg, the ends are not simple supports, but because of leg and joint flexibility, these ends are not fully clamped either. Since the true end fixity is somewhere between these extremes, the actual frequencies will lie between those calculated for each extreme end fixity.

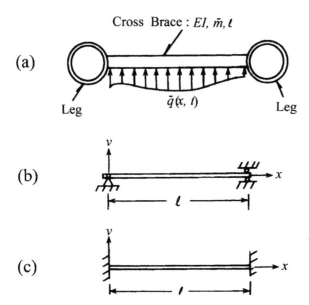

Figure 10.4   (a) Top view of the cross brace; (b) simple support model; (c) clamped
end model.

A lower bound on the frequencies corresponds to the brace model with the least constraints, or with the simple supports as shown in Figure 10.4b. The associated boundary conditions are those of equation (10.30), each applied at each end, or

$$X(0) = X''(0) = X(\ell) = X''(\ell) = 0 \tag{10.34}$$

The consecutive application of these last four conditions to equation (10.33) yields the following four equations:

$$D_2 + D_4 = 0 \tag{10.35a}$$

$$-D_2 + D_4 = 0 \tag{10.35b}$$

$$D_1 \sin \alpha\ell + D_2 \cos \alpha\ell + D_3 \sinh \alpha\ell + D_4 \cosh \alpha\ell = 0 \tag{10.35c}$$

$$-D_1 \sin \alpha\ell - D_2 \cos \alpha\ell + D_3 \sinh \alpha\ell + D_4 \cosh \alpha\ell = 0 \tag{10.35d}$$

When equations (10.35a) and (10.35b) are added, it follows that $D_4 = 0$, from which $D_2 = 0$. The last two equations thus reduce to

$$D_1 \sin \alpha\ell + D_3 \sinh \alpha\ell = 0; \quad -D_1 \sin \alpha\ell + D_3 \sinh \alpha\ell = 0 \tag{10.36}$$

If at least one of the remaining constants D_1 or D_3 is to be nonzero, it follows that

$$\begin{vmatrix} \sin \alpha \ell & \sinh \alpha \ell \\ -\sin \alpha \ell & \sinh \alpha \ell \end{vmatrix} = 0 \tag{10.37}$$

This determinant is expanded to give the following transcendental equation:

$$(\sin \alpha \ell)(\sinh \alpha \ell) = 0 \tag{10.38}$$

Since $\sinh \alpha \ell \neq 0$ for $\alpha \ell > 0$, then it is necessary that $\sin \alpha \ell = 0$ to satisfy equation (10.38), or

$$\alpha = \frac{n\pi}{\ell}, \quad n = 1, 2, \ldots \tag{10.39}$$

With this last result, the beam frequencies from equation (10.29) become

$$\omega_n = \frac{n^2 \pi^2}{\ell^2} \sqrt{\frac{EI}{\bar{m}}}, \quad n = 1, 2, \ldots \tag{10.40}$$

in which ω_n replaces ω.

Since $\sin \alpha \ell = 0$ and thus $D_3 = 0$ from equations (10.36), the only remaining nonzero constant is D_1. For each ω_n there is a corresponding mode shape $X = X_n$ at an arbitrary amplitude $D_1 = C_n$ given by equation (10.33) or

$$X_n = C_n \sin \frac{n\pi x}{\ell}, \quad n = 1, 2, \ldots \tag{10.41}$$

In this case the mode shapes are identical to those for the fixed end cable, shown in Figure 10.3. However, the frequencies of this beam vary as n^2 rather than as n for the cable.

Calculate next the upper bound frequencies, or those corresponding to the brace with clamped ends, Figure 10.4c. The appropriate boundary conditions are those of equation (10.31), each applied at each end, or

$$X(0) = X'(0) = X(\ell) = X'(\ell) = 0 \tag{10.42}$$

The consecutive application of these four conditions to the general solution, equation (10.33), leads to the following four equations:

$$D_2 + D_4 = 0 \tag{10.43a}$$

$$D_1 + D_3 = 0 \tag{10.43b}$$

$$D_1 \sin \alpha \ell + D_2 \cos \alpha \ell + D_3 \sinh \alpha \ell + D_4 \cosh \alpha \ell = 0 \tag{10.43c}$$

$$D_1 \cos \alpha \ell - D_2 \sin \alpha \ell + D_3 \cosh \alpha \ell + D_4 \sinh \alpha \ell = 0 \tag{10.43d}$$

From these last results, the determinant of the coefficients of D_1, D_2, D_3, and D_4 is formed, expanded, and set equal to zero. The simple result is

$$(\cos \alpha \ell)(\cosh \alpha \ell) = 1 \tag{10.44}$$

It is easily verified that $\alpha_1 = 4.730$ is the first nonzero root of equation (10.44) and that this is just 0.37 percent lower than the approximate value given for $n = 1$ given by

$$\alpha_n \ell = (n + 0.5)\pi \tag{10.45}$$

For $n = 2, 3, \ldots$, equation (10.45) yields successive roots of equation (10.44) which are even more accurate than this. Thus the upper bound frequencies for this problem are given by equation (10.29) with (10.45), or

$$\omega_n = (n + 0.5)^2 \frac{\pi^2}{\ell^2} \sqrt{\frac{EI}{\bar{m}}}, \quad n = 1, 2, \ldots \tag{10.46}$$

The corresponding mode shapes for this fixed end beam are found by consecutively eliminating the arbitrary constants D_2, D_3 and D_4 from equations (10.43) and writing the result in terms of D_1 only. For the nth mode, let $D_1 = C_n$, and the result is

$$X_n(x) = C_n[\sin \alpha_n x - \sinh \alpha_n x + \beta_n(\cos \alpha_n x - \cosh \alpha_n x)] \tag{10.47}$$

where α_n is given by equation (10.45) and

$$\beta_n = \frac{\sin \alpha_n \ell - \sinh \alpha_n \ell}{\cosh \alpha_n \ell - \cos \alpha_n \ell} \tag{10.48}$$

The first two of these mode shapes are shown in Figure 10.5. Unlike the mode shapes for the cable and the simply supported beam, these shapes have zero slope at each end.

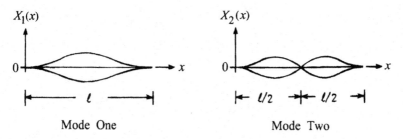

Figure 10.5   The first two mode shapes for a beam clamped at both ends.

In conclusion, the frequencies ω_n for the cross bracing which is uniform, undamped, without end tension, and of virtual mass per unit length of $\bar{m}$, are bounded by equations (10.40) and (10.46), or

$$\frac{n^2 \pi^2}{\ell^2} \sqrt{\frac{EI}{\bar{m}}} < \omega_n < (n + 0.5)^2 \frac{\pi^2}{\ell^2} \sqrt{\frac{EI}{\bar{m}}}, \quad n = 1, 2, \ldots \tag{10.49}$$

This result shows that for $n = 1$, the fundamental frequency for the most constrained case is 2.25 times higher than that for the least constrained case. This factor decreases rapidly as n increases, where the factor is approximately 1.2 for $n = 5$ and 1.1 for $n = 10$.

10.2   CABLE RESPONSES

Transverse motion or motion perpendicular to the longitudinal axis of a submerged cable is caused by three main types of excitation: vortex shedding in a constant current stream leading to cable galloping; transverse end excitation due to the motion of the platform, tower, ship, or buoy to which the cable is attached; and longitudinal or parametric end excitation, also imparted by the cable-supported structure. Experiments on cable responses to coupled vortex and parametric excitation were reported by Trogdon et al. (1976), but such analytical studies are sparse. The subjects of this section are transverse and parametric end excitation, as depicted in Figure 10.1.

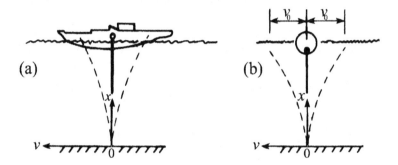

Figure 10.6   Transverse end excitation of a submerged cable: (a) attached to a ship; (b) attached to a buoy.

Transverse End Excitation

Consider a single cable of length ℓ and with a uniform virtual mass per unit length $\bar{m}$. One end of the cable is fixed at $x = 0$ and the other end is subjected to transverse harmonic excitation. This excitation can occur with regular seas for the mooring line shown in Figure 10.1, for the nearly vertical anchor line shown in Figure 10.6a, and for the buoy chain shown in Figure 10.6b. In any case, the average line tension is P_0, and the end conditions are chosen as

$$v(0, t) = 0 \tag{10.50}$$

$$v(\ell, t) = v_0 \cos \bar{\omega} t \tag{10.51}$$

where v_0 is the amplitude of motion and $\bar{\omega}$ is the excitation frequency, both of which depend on the sea surface wave height spectrum and the type of ship or buoy involved. A way to calculate $\bar{\omega}$ for a moored barge is discussed in Section 10.4. Neglect all damping and all transverse loading except at $x = \ell$. The model for transverse cable motion is assumed as equation (10.13), for which a steady state solution is chosen in the same form as for the end excitation imposed at $x = \ell$, or

$$v(x, t) = X \cos \bar{\omega} t \tag{10.52}$$

where $X = X(x)$. Combine equation (10.52) with (10.13) from which

$$X'' + \bar{\gamma}^2 X = 0 \tag{10.53}$$

$$\bar{\gamma}^2 = \frac{\bar{m}}{P_0} \bar{\omega}^2 \tag{10.54}$$

The general solution to equation (10.53) is given by

$$X = D_1 \sin \bar{\gamma} x + D_2 \cos \bar{\gamma} x \tag{10.55}$$

Applying the fixed boundary condition of equation (10.50) to (10.52), then $X(0) = 0$. This same condition applied to equation (10.55) gives $D_2 = 0$. Combining equation (10.52) with (10.55), and that result with the other end condition, equation (10.51) leads to

$$v(\ell, t) = D_1 \sin \bar{\gamma} \ell \cos \bar{\omega} t = v_0 \cos \bar{\omega} t \tag{10.56a}$$

$$D_1 = \frac{v_0}{\sin \bar{\gamma} \ell} \tag{10.56b}$$

From equation (10.54) and the expression for the natural frequencies ω_n given by equation (10.22), it follows that

$$\bar{\gamma} \ell = \frac{\bar{\omega} \ell}{\sqrt{P_0/\bar{m}}} = n\pi \frac{\bar{\omega}}{\omega_n} \tag{10.57}$$

With equations (10.55)-(10.57), the solution to equation (10.52) becomes

$$v(x, t) = \frac{v_0}{\sin (n\pi \bar{\omega}/\omega_n)} \sin \left(\frac{n\pi x \bar{\omega}}{\ell \omega_n} \right) \cos \bar{\omega} t \tag{10.58}$$

This solution shows that the transverse cable displacement at any point $0 < x < \ell$ becomes unbounded if the excitation frequency $\bar{\omega}$ coincides with ω_n, since then the term $\sin n\pi = 0$ in the denominator of equation (10.58). Had light damping been included in the mathematical model from the very beginning, the peak response for $\bar{\omega} = \omega_n$ would have been bounded, but still amplified compared to the imposed end displacement amplitude v_0. As discussed in Problem 10.4 at the end of this chapter, this resonance phenomenon can be observed in a simple laboratory experiment.

Parametric Excitation

Longitudinal or parametric excitation of a relatively taut cable is depicted in Figure 10.1b. This type of excitation may also occur in the vertical lines or chains shown in Figures 10.6 when the ship or buoy undergoes heave motion in regular waves. In such cases the average cable tension is P_0, the amplitude of the harmonically fluctuating force is $P_1 < P_0$, and the excitation frequency is $\bar{\omega}$. Parametric excitation is thus defined as

$$P = P_0 + P_1 \cos \bar{\omega} t \qquad (10.59)$$

The mathematical model for cable motion is chosen as equation (10.11) in which the bending stiffness, damping, and all transverse loadings are neglected. With equation (10.59), the equation for transverse motion is thus

$$-(P_0 + P_1 \cos \bar{\omega} t)\frac{\partial^2 v}{\partial x^2} + \bar{m}\frac{\partial^2 v}{\partial t^2} = 0 \qquad (10.60)$$

To study the effects of only parametric excitation on transverse motion, all transverse motion is suppressed at each end of the cable, or

$$v(0,t) = v(\ell,t) = 0 \qquad (10.61)$$

The following solution to equation (10.60) is assumed, a solution that already satisfies the two boundary conditions just stated.

$$v(x,t) = \sum_{n=1}^{\infty} y_n(t)\sin\frac{n\pi x}{\ell} \qquad (10.62)$$

Here $y(t)$ denotes the generalized coordinates, $n = 1, 2, \ldots$. Combining equation (10.62) with (10.60), it follows that

$$\sum_{n=1}^{\infty}\left[(P_0 + P_1 \cos\bar{\omega}t)\left(\frac{n\pi}{\ell}\right)^2 y_n(t) + \bar{m}\frac{d^2 y_n}{dt^2}\right]\sin\frac{n\pi x}{\ell} = 0 \qquad (10.63)$$

Since the sine term of the last equation is not zero for all values of x, then its coefficient must vanish, or

$$\bar{m}\frac{d^2 y_n(t)}{dt^2} + \left(\frac{n\pi}{\ell}\right)^2 (P_0 + P_1 \cos\bar{\omega}t)\, y_n(t) = 0 \qquad (10.64)$$

This last equation can be transformed to a standard form using the following four parameters:

$$\tau = \bar{\omega}t; \qquad \bar{\alpha}_n = \frac{\omega_n^2}{\bar{\omega}^2}; \qquad \bar{\beta}_n = \frac{P_1}{P_0}\frac{\omega_n^2}{\bar{\omega}^2} \qquad (10.65)$$

$$\omega_n = \frac{n\pi}{\ell}\sqrt{\frac{P_0}{\bar{m}}}, \qquad n = 1, 2, \ldots \qquad (10.66)$$

The result is the famous Mathieu equation, which is

$$\frac{d^2 y_n(\tau)}{d\tau^2} + (\bar{\alpha}_n + \bar{\beta}_n \cos \tau) y_n(\tau) = 0 \tag{10.67}$$

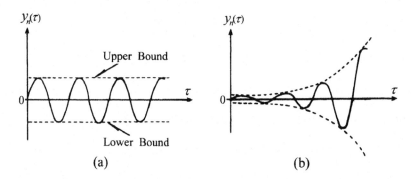

Figure 10.7   Dynamic system behavior: (a) stable or bounded response; (b) unstable or unbounded response.

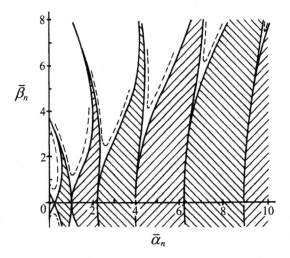

Figure 10.8   Haines-Strett stability plot (after Lubkin and Stokes, 1943).

It is observed from equation (10.62) that the behavior of solutions $y_n(\tau)$ to this equation produce the same behavior for the transverse displacement response $v(x,t)$. Thus, if $y_n(\tau)$ is stable, then the response is bounded in time as shown in Figure 10.7a. If the solutions $y_n(\tau)$ are unstable, then $v(x,t)$ is unstable and exhibits the divergent response as shown in Figure 10.7b. Lubkin and Stokes (1943) made extensive analytical studies of equation (10.67) to determine which combinations of system parameters $\bar{\alpha}_n$ and $\bar{\beta}_n$ yield stable and unstable solutions $y_n(\tau)$. Those results are shown in Figure 10.8, which is some-times called the Heines-Strett diagram. If the parameter set $(\bar{a}_n, \bar{\beta}_n)$ lies in the cross hatched regions, $y_n(\tau)$ is stable, but if this parameter set is elsewhere, the responses are unstable. Further studies have shown that the addition of light damping increases the parameter range for stability where the cross hatched regions are extended to the broken lines of Figure 10.8.

Example Problem 10.2. Shown in Figure 10.9 is a landing ship-tank (LST), spread-moored in the Gulf of Mexico. For this ship, O'Brien and Muga (1964) took simultaneous measures of the spectra for the surface wave height and the tension forces in the restraining lines. The measured wave height spectra are shown in Figure 10.10, and the measured longitudinal force response spectra for port bow chain No. 2 are shown in Figure 10.11. The problem is to explain the reason for the double peaks in these bow chain force response spectra. Are those peaks due to transverse or parametric excitation, or do they arise from some other reason?

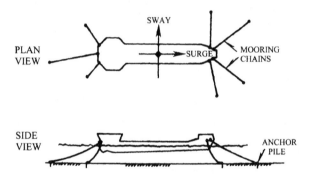

Figure 10.9 A spread mooring configuration for an LST in the Gulf of Mexico.

The data for the overall mooring restraint system and the ship's surge and sway frequencies were given in *Example Problems 2.10* and *5.5*. Pertinent data on bow chain No. 2 are as follows. The actual unit mass of this chain is increased by 3 percent to give its virtual unit mass.

$$\bar{m} = (1.03)\bar{m}_0 = 1.03(35 \text{ lb/ft})(32.2 \text{ ft/sec}^2)^{-1} = 1.12 \text{ lb-sec}^2/\text{ft}^2$$
$$\ell = 465 \text{ ft, length;} \qquad P_0 = 30,000 \text{ lb;} \qquad P_1 = 0.5P_0$$

The first step is to calculate the natural frequencies of the bow chain. From equation (10.22) and the preceding data, it follows that

$$\omega_n = \frac{n\pi}{\ell}\sqrt{\frac{P_0}{\bar{m}}} = \frac{n\pi}{465}\sqrt{\frac{30,000}{1.12}} = 1.11n \text{ rad/sec,} \quad n = 1, 2, \cdots$$

For resonance due to transverse excitation, one would expect peak bow chain responses near ω_n or near 1.11, 2.22, ... rad/sec. Since Figure 10.11 shows no significant peaks at these frequencies, it is concluded that there is no significant bow chain resonance response.

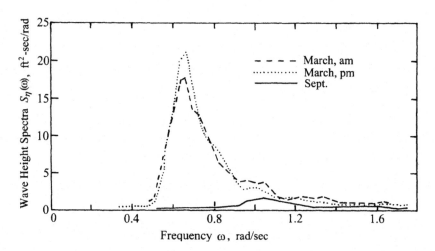

Figure 10.10   Water wave amplitude spectra for a spread-moored LST (after
O'Brien and Muga, 1964).

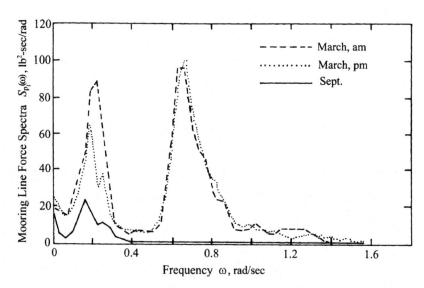

Figure 10.11   Force amplitude spectra for port bow chain No. 2 of an LST (after
O'Brien and Muga, 1964).

To investigate possible bow chain resonance due to parametric excitation,
choose a driving frequency $\bar{\omega} = 0.65$ rad/sec, or the frequency at the highest
concentration of energy for the wave. With the data and equations (10.65), the
characteristic parameters are

$$\bar{\alpha}_n = \omega_n^2/\bar{\omega}^2 = (1.11n)^2/(0.65)^2 = 2.92n^2$$
$$\bar{\beta}_n = (P_1/P_0)\bar{\alpha}_n = 0.5(2.92n^2) = 1.46n^2 \quad \text{(maximum)}$$

For $n = 1$, then $(\bar{\alpha}_1; \bar{\beta}_1) = (2.92; 1.46)$. This coordinate set lies in the cross hatched region of Figure 10.8, and thus the transverse chain displacements (and tension forces) are bounded. This same conclusion is reached for $(\bar{\alpha}_n; \bar{\beta}_n)$ based on $n = 2, 3, \ldots$. However, if $\bar{\omega} = 1.1$ rad/sec, which is at the high end of the driving frequency for this mooring chain, then $\alpha_n \simeq n^2$ and $\bar{\beta}_n \simeq 0.5n^2$. For $n = 1$, then $(\bar{\alpha}_1; \bar{\beta}_1) = (1; 0.5)$, for which Figure 10.8 shows unstable motion in the absence of damping, but stable motion in the presence of small, realistic damping. For $n = 2, 3, \ldots$, the motion is stable even without damping.

The conclusion of the preceeding calculations is that the peak responses in Figure 10.11 are not due to either transverse or parametric resonance of this particular bow chain. How, then, can these two major response peaks recorded in the month of March be reconciled with the single peak of wave excitation in Figure 10.10? The answer can be found by studying the motion of the whole ship.

Referring again to *Example Problems 2.10* and *5.5*, the ship motion $v = v(t)$ in either surge or sway can be modeled by

$$m\ddot{v} + c_1\dot{v} + k_1 v + k_3 v^3 = p_0 \sin \bar{\omega} t \tag{10.68}$$

where the coefficients on the left side are given by equation (2.72) or (2.73), where p_0 is the wave force in line with v, and where $\bar{\omega} \simeq 0.65$ rad/sec. The results of Chapter 5 showed that peak responses of such a nonlinear system under harmonic excitation occur not only at the excitation frequency $\bar{\omega} = 0.65$ rad/sec but also for $\bar{\omega}/3$, which is near ω_0. Since $\omega_0 = 0.144$ rad/sec for sway (*Example Problem 5.5*), then the peak at $\omega \simeq 0.2$ rad/sec can be explained as a subharmonic ship response of order one-third. The existence of the one-third subharmonic for equation (10.68) is shown in Section 5.5, where the amplitude parameter is shown in Figure 5.11. Thus, the lower frequency (or longer period) ship response reflected in mooring chain No. 2 arises from the group behavior of all the mooring lines and comes about because of the nonlinear restoring force constant k_3 of equation (10.68), during sway motion of the ship. Whether drift currents at a frequency of about 0.22 rad/sec existed during these sea tests and also contributed to the lower resonance peak of Figure 10.11 is not known.

10.3 BEAM RESPONSES

As for cables, submerged beams and pipelines are subjected to three main types of excitation: transverse loading due to vortex shedding, transverse end motion, and parametric excitation. Considered first in this section are deterministic responses of uniform beams with common types of end supports and with an arbitrary transverse load per unit length $\bar{q} = \bar{q}(x, t)$. Following this is a stability study of a simply supported beam subjected to parametric excitation. This section concludes with a calculation of the statistical responses of beams to stationary, ergodic excitation. Modal analysis is employed throughout and closed form solutions are sought.

Transverse Excitation

The undamped motion of a uniform Bernoulli-Euler beam with negligible longitudinal tension $(P = 0)$ is given by equation (10.12), or

$$EI \frac{\partial^4 v}{\partial x^4} + \bar{m} \frac{\partial^2 v}{\partial t^2} = \bar{q}(x, t) \tag{10.69}$$

Of practical interest is a beam or pipeline of length ℓ supported at the ends only, according to one of the six sets of boundary or end conditions defined in Table 10.1.

Table 10.1   Types of End Conditions for a Beam

$C - C$	Both ends clamped
$C - SS$	Clamped and simply supported
$SS - SS$	Both ends simply supported
$C - F$	Clamped and free
$SS - F$	Simply supported and free
$F - F$	Both ends free

It is not difficult to visualize that the first and third end conditions of Table 10.1 represent the limits of constraints for the cross beams of offshore structures. The second end condition is discussed in Problem 10.2 at the end of this chapter. The fourth and fifth conditions can represent cases of support for pipeline segments during deployment, and the last condition can model an unconstrained, floating pipeline such as proposed to transport water along the Pacific coast of the United States.

The free vibration mode shapes $X_n = X_n(x)$ and their corresponding frequencies ω_n are employed in the following modal analysis of forced beam vibrations. The quantities X_n and ω_n are calculated for one of the six sets of boundary conditions listed above by using the solution to equation (10.28), which is given by equation (10.33), together with the appropriate transformed boundary conditions chosen from equations (10.30)-(10.32). The modal analysis also requires the following two conditions of orthogonality for X_n:

$$\int_0^\ell X_m X_n \, dx = 0, \quad \text{for } m \neq n \tag{10.70}$$

$$\int_0^\ell X_m X_n \, dx \neq 0, \quad \text{for } m = n \tag{10.71}$$

The proof of orthogonality for X_n corresponding to each of the six boundary conditions of Table 10.1 is shown as follows. Consider two solutions of equation (10.28) as $X = X_m$ and $X = X_n$, with corresponding values of $\alpha = \alpha_m$ and $\alpha = \alpha_n$. For free vibrations

$$X_m'''' - \alpha_m^4 X_m = 0 \tag{10.72}$$

$$X_n'''' - \alpha_n^4 X_n = 0 \tag{10.73}$$

where the frequency parameters are

$$\alpha_i^4 = \frac{\omega_i^2 \bar{m}}{EI}, \quad i = m, n \tag{10.74}$$

Multiply equations (10.72) and (10.73) by X_n and X_m, respectively, subtract the two resulting equations, and integrate the results over the interval $(0, \ell)$. Thus

$$(\alpha_n^4 - \alpha_m^4) \int_0^\ell X_m X_n \, dx = \int_0^\ell (X_m X_n'''' - X_n X_m'''') dx \tag{10.75}$$

Integrate the right side of the last equation by parts four times. The first integration, for instance, gives the result

$$\int_0^\ell (X_m X_n'''' - X_n X_m'''') dx = [X_m X_n''' - X_n X_m''']_{x=\ell}^{x=0} + \int_0^\ell (X_m' X_n''' - X_n' X_m''') dx \tag{10.76}$$

Integrate the integral on the right side of equation (10.76) by parts. Repeat this procedure twice more to give

$$(\alpha_n^4 - \alpha_m^4) \int_0^\ell X_m X_n \, dx = 2Q + \int_0^\ell (X_n X_m'''' - X_m X_n'''') dx \tag{10.77}$$

where

$$Q = [X_m X_n''' - X_n X_m''' - X_m' X_n'' - X_n' X_m'']_{x=\ell}^{x=0} \tag{10.78}$$

With equations (10.72) and (10.73), the integral on the right hand side of equation (10.77) becomes

$$\int_0^\ell (X_n X_m'''' - X_m X_n'''') dx = (\alpha_m^4 - \alpha_n^4) \int_0^\ell X_m X_n \, dx \tag{10.79}$$

Combining equation (10.79) with (10.77), the result is

$$(\alpha_n^4 - \alpha_m^4) \int_0^\ell X_m X_n \, dx = Q \tag{10.80}$$

If $m = n$, then $\alpha_n = \alpha_m$ and Q given by the last equation is identically zero regardless of the value of the integral. However, if $m \neq n$, the system frequency parameters α_m and α_n are distinct and different, so that if $Q = 0$ the integral of equation (10.80) must vanish. This is indeed the case for the six sets of end conditions of Table 10.1. Written in terms of $X(= X_m$ or $X_n)$, these conditions are, respectively,

$$
\begin{aligned}
C-C \quad & X(0) = X'(0) = X(\ell) = X'(\ell) = 0 \\
C-SS \quad & X(0) = X'(0) = X(\ell) = X''(\ell) = 0 \\
SS-SS \quad & X(0) = X''(0) = X(\ell) = X''(\ell) = 0 \\
C-F \quad & X(0) = X'(0) = X''(\ell) = X'''(\ell) = 0 \\
SS-F \quad & X(0) = X''(0) = X''(\ell) = X'''(\ell) = 0 \\
F-F \quad & X''(0) = X'''(0) = X''(\ell) = X'''(\ell) = 0
\end{aligned}
\tag{10.81}
$$

It is observed from equation (10.78) that each of these six conditions leads to $Q = 0$. Thus from equation (10.80) the validity of the orthogonality of normal modes as stated by equations (10.70) and (10.71) is now apparent.

Return now to the forced vibration problem defined by equation (10.69). Assume a particular solution to that equation as a product of the normal modes $X_n(x)$ and a generalized coordinate $y_n(t)$ which is to be determined. That is, let

$$
v(x, t) = \sum_{n=1}^{\infty} X_n(x)\, y_n(t)
\tag{10.82}
$$

When this last equation is substituted into equation (10.69), the result is

$$
\sum_{n=1}^{\infty} [EI\, X_n''''\, y_n(t) + \bar{m} X_n \ddot{y}_n(t)] = \bar{q}(x, t)
\tag{10.83}
$$

Multiplying each term in this last equation by X_m and then integrating the result over $(0, \ell)$ gives

$$
EI \int_0^\infty \sum_{n=1}^{\infty} X_m X_n'''' y_n(t) dx + \bar{m} \int_0^\ell \sum_{n=1}^{\infty} X_m X_n \ddot{y}_n(t) dx = \int_0^\ell X_m\, \bar{q}(x, t) dx
\tag{10.84}
$$

Assume that the series of the last result is uniformly convergent over the interval $(0, \ell)$, which allows for the order of integration and summation to be interchanged. This leads to

$$
\sum_{n=1}^{\infty} \int_0^\infty [EI\, X_m X_n'''' y_n(t) dx + \bar{m}\, X_m X_n\, \ddot{y}_n(t)] dx = \int_0^\ell X_m\, \bar{q}(x, t) dx
\tag{10.85}
$$

The following identity is obtained by multiplying equation (10.73) by X_m and integrating the result over the interval $(0, \ell)$. With equation (10.74), the result is

$$
EI \int_0^\ell X_m X_n'''' \, dx = \omega_n^2\, \bar{m} \int_0^\ell X_m X_n \, dx
\tag{10.86}
$$

Combining the last two equations gives

$$
\sum_{n=1}^{\infty} [\omega_n^2\, y_n(t) + \ddot{y}_n(t)] \int_0^\ell X_m X_n \, dx = \frac{1}{\bar{m}} \int_0^\ell X_m\, \bar{q}(x, t) dx
\tag{10.87}
$$

With the orthogonality conditions of equations (10.70) and (10.71), each term in the sum of equation (10.87) vanishes except for $m = n$, in which case equation (10.87) reduces to

$$\ddot{y}_n(t) + \omega_n^2\, y_n(t) = p_n(t) \tag{10.88}$$

in which

$$p_n(t) = \frac{1}{\bar{m}C_0} \int_0^\ell X_n\, \bar{q}(x,t)dx \tag{10.89}$$

$$C_0 = \int_0^\ell X_n^2\, dx \tag{10.90}$$

Assume that the beam is initially at rest, or that $v(x,0) = \partial v(x,t)/\partial t = 0$. Using equation (10.82), the initial rest position implies that $y_n(0) = \dot{y}_n(0) = 0$. The solution to equation (10.88) is in the form of the Duhamel integral for $\zeta = 0$, or

$$y_n(t) = \frac{1}{\omega_n} \int_0^t p_n(\tau) \sin \omega_n(t - \tau)d\tau \tag{10.91}$$

This is verified by comparing equation (5.58) to equation (10.88) and then by comparing their respective solutions given by equations (5.76) and (10.91).

These results are summarized. The transverse response $v(x,t)$ for arbitrary transverse unit loading $\bar{q}(x,t)$ of an undamped Bernoulli-Euler beam, subjected to one of the six constraints of Table 10.1, is given by equation (10.82). Here X_n and ω_n are calculated as outlined in Section 10.1, and $y_n(t)$ is calculated from equations (10.89)-(10.91).

Parametric Excitation

Consider the dynamic behavior of a vertical dredge pipe attached at the top to a barge as shown in Figure 10.12a. The barge undergoes harmonic heave motion in regular waves. This pipeline is modeled as the beam of Figure 10.12b where the ends are simply supported to avoid adverse bending stresses at these points of fixity. The average longitudinal load on the pipeline is assumed in the form

$$P = P_0 + P_1 \cos \bar{\omega}t \tag{10.92}$$

where P_0 is approximated as the sum of the ballast weight (in water) applied at the lower end $(x = 0)$ and one-half of the pipeline's weight (in water). Note that in reality the hinge at $x = \ell$ carries all of the pipeline's weight, whereas the hinge at $x = 0$ carries none of the pipeline's weight. The approximation given for P_0 becomes more accurate for increasingly high ratios of ballast weight to pipeline weight.

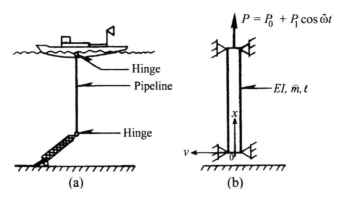

Figure 10.12 (a) Barge-dredge pipe system; (b) model of the dredge pipe.

Implied also in this mathematical model is that the fundamental longitudinal pipeline frequency ω_ℓ is always significantly higher than the barge heave frequency $\bar{\omega}$. Thus if $\omega_\ell \gg \bar{\omega}$, the longitudinal load at the top hinge is essentially the same as at the bottom hinge at any instant of time. For an elastic pipeline without clamped ends, its longitudinal frequency is given by

$$\omega_\ell = \frac{\pi}{\ell}\sqrt{\frac{E}{\rho_p}} \tag{10.93}$$

where ρ_p and E are the mass density and Young's modulus for the pipeline of length ℓ (Timoshenko and Goodier, 1951). Applying equation (10.93) to a steel pipeline 3000 ft in length, we find $\omega_\ell = 17.7$ rad/sec, which is more than ten times the highest excitation frequency expected to be imparted by the barge through wave action. (Motion of a typical barge in waves is discussed in Section 10.4.) Although $\omega_\ell \gg \bar{\omega}$ holds true for this steel pipeline, the inequality may fail for pipelines manufactured of polymeric materials such as polyethylene. This is because the ratios E/ρ_p for polymeric materials are generally much smaller than for steel.

With this mathematical model, the governing equation of motion is then equation (10.12) with the parametric excitation load of equation (10.92), or

$$EI\frac{\partial^4 v}{\partial x^4} - (P_0 + P_1\cos\bar{\omega}t)\frac{\partial^2 v}{\partial x^2} + \bar{m}\frac{\partial^2 v}{\partial t^2} = 0 \tag{10.94}$$

The four homogeneous boundary conditions corresponding to simple end supports are

$$v(0,t) = \frac{\partial^2 v(0,t)}{\partial x^2} = v(\ell,t) = \frac{\partial^2 v(\ell,t)}{\partial x^2} = 0 \tag{10.95}$$

It is easily verified that all four of these latter conditions satisfy the following form chosen as a solution to equation (10.94):

$$v(x,t) = \sum_{n=1}^{\infty} y_n(t)\sin\frac{n\pi x}{\ell} \tag{10.96}$$

Using the same procedure as for parametric excitation of a cable, results analogous to equations (10.63) and (10.64) are obtained when equation (10.96) is combined with (10.94). The following parameters are then applied to the resulting second order ordinary differential equation:

$$\tau = \bar{\omega} t \qquad (10.97)$$

$$\bar{\alpha}_n = \frac{\omega_n^2}{\bar{\omega}^2} + \frac{P_0}{\bar{m}\bar{\omega}^2}\left(\frac{n\pi}{\ell}\right)^2 ; \qquad \bar{\beta}_n = \frac{P_1}{\bar{m}\bar{\omega}^2}\left(\frac{n\pi}{\ell}\right)^2 \qquad (10.98)$$

$$\omega_n = \frac{n^2\pi^2}{\ell^2}\sqrt{\frac{EI}{\bar{m}}}, \qquad n = 1, 2, \cdots \qquad (10.99)$$

It is noted that the last result was derived in *Example Problem 10.1*, equation (10.40), which expresses the free transverse vibration frequencies of a simple beam. The result is again the Mathieu equation (10.67), but with differently defined coefficients, or

$$\frac{d^2 y_n(\tau)}{d\tau^2} + (\bar{\alpha}_n + \bar{\beta}_n \cos\tau)\, y_n(\tau) = 0 \qquad (10.100)$$

Thus, for a given set of system parameters $(\bar{\alpha}_n; \bar{\beta}_n)$ specified by equation (10.98), the stability of $y_n(t)$ and therefore the stability of the pipeline is determined from the Haines-Strett plot of Figure 10.8.

Example Problem 10.3. Investigate the dynamic stability of a steel pipeline under parametric excitation by a barge with a heave frequency $\bar{\omega}$. The pipeline is modeled in Figure 10.12b. Assume that $P_1 \leq P_0/2$ and that n has an upper limit of five.

Consider first the condition for which the parameters of equations (10.98) become

$$\bar{\alpha}_n \simeq \frac{P_0}{\bar{m}\bar{\omega}^2}\left(\frac{n\pi}{\ell}\right)^2 ; \qquad \bar{\beta}_n \leq \frac{\bar{\alpha}_n}{2} \qquad (10.101)$$

From Figure 10.8 it is observed that pipeline instability occurs *near* $\bar{\alpha}_n = 0.25$ and 1.0. With equations (10.101), the barge heave frequencies at which such instabilities occur are deduced as

$$\bar{\omega} \simeq \frac{n\pi}{\ell}\left(\frac{P_0}{\bar{m}\bar{\alpha}_n}\right)^{1/2} ; \qquad \bar{\alpha}_n \simeq 0.25 \;\; \text{or} \;\; 1.0 \qquad (10.102)$$

in which $\bar{\omega} \gg \omega_n$. If, however, $\bar{\omega} \simeq \omega_n$ and $0 \leq P_1 \leq P_0/2$, then

$$\bar{\alpha}_n = 1 + \frac{P_0}{\bar{m}\bar{\omega}^2}\left(\frac{n\pi}{\ell}\right)^2 = 1 + D_0 ; \qquad \bar{\beta}_n = \frac{P_1}{P_0}D_0 \leq \frac{D_0}{2} \qquad (10.103)$$

In this case, the coordinate pairs always lie below the dashed lines on the Haines-Strett plot. Without damping, instability would occur near $\bar{\alpha}_n = 2.2$, 4.0, 6.2, and so on. However, since natural damping always exists, this pipeline is stable for $\bar{\omega} \simeq \omega_n$ under parametric excitation.

Statistical Wave Excitation

Most of the background analysis needed to calculate the statistical dynamic responses of beams to stationary, ergodic wave excitation has already been developed. Outlined now are the basic assumptions, methodology, and results of this statistical analysis. In this section, the reader is encouraged to carry through the details of the mathematical manipulations. Further expositions on this topic are given by Clough and Penzien (1993) and Gould and Abu-Sitta (1980).

The mathematical model chosen for analysis is the uniform Bernoulli-Euler beam without longitudinal loading ($P = 0$) but with light, linear viscous damping. From equation (10.8) this model is

$$EI \frac{\partial^4 v}{\partial x^4} + \bar{c} \frac{\partial v}{\partial t} + \bar{m} \frac{\partial^2 v}{\partial t^2} = \bar{q}(x, t) \tag{10.104}$$

where the loading per unit length $\bar{q} = \bar{q}(x, t)$ is stationary, ergodic and Gaussian, defined by the spectral density $S_{\bar{q}}(\omega)$.

Equation (10.104) is reduced to a familiar form by expanding both the displacement solution $v = v(x, t)$ and the loading $\bar{q} = \bar{q}(x, t)$ in terms of the undamped normal modes $X_n = X_n(x)$ and the generalized modal coordinates $y_n = y_n(t)$ and $\bar{q}_n = \bar{q}_n(t)$. That is, let

$$v(x, t) = \sum_{n=1}^{\infty} X_n \, y_n \tag{10.105}$$

$$\bar{q}(x, t) = \sum_{n=1}^{\infty} X_n \, \bar{q}_n \tag{10.106}$$

The procedure now is analogous to that used previously to calculate beam responses with deterministic loading where equation (10.88) was derived starting with equation (10.82) and the undamped beam model given by equation (10.69). That is, substitute equations (10.105) and (10.106) into equation (10.104); multiply each term by $X_m = X_m(x)$; from equation (10.73) let $X_n'''' = \alpha_n^4 X_n$; integrate each term of the resulting equation over the interval $(0, \ell)$; interchange the order of integration and summation under the assumption that the series are uniformly convergent; and apply the orthogonality condition of (10.70) and (10.71) to the result, which is valid for any of the six sets of beam support conditions of (10.81). Thus all terms in the series vanish except for $m = n$, which leads to

$$\bar{m} \ddot{y}_n + \bar{c} \dot{y}_n + EI \alpha_n^4 y_n = \bar{q}_n(t) \tag{10.107}$$

Define

$$\frac{\bar{c}}{\bar{m}} = 2\zeta_n \omega_n \tag{10.108}$$

With the last equation and the natural frequencies ω_n given by equation (10.74), equation (10.107) becomes

$$\ddot{y}_n + 2\zeta_n \omega_n \dot{y}_n + \omega_n^2 y_n = \frac{1}{\bar{m}} \bar{q}_n(t) \tag{10.109}$$

It is observed that equation (10.109) is in the form of equation (9.62) for a finite degree of freedom system. Assume now that the nth modal response $y_n(t)$ is statistically independent of the mth modal response $y_m(t)$, the same assumption that previously led to the spectral density of the response given by equation (9.70). By analogy the response spectral density $S_v(\omega)$ for the beam is given by superimposing modal responses according to equation (10.105) and equating its cross-spectral density terms to zero. That is,

$$S_v(\omega) = \sum_{n=1}^{\infty} \frac{X_n}{(\bar{m}\omega_n^2)^2} |H_n(\omega)|^2 S_{\bar{q}n}(\omega) \tag{10.110}$$

Here $H_n(\omega)$ is the harmonic response function for the nth mode, derived from equation (10.109) by letting

$$y_n = \frac{\bar{q}_0}{\bar{m}\omega_n^2} H_n(\omega) e^{j\omega t} \tag{10.111}$$

$$\bar{q}_n = \bar{q}_0 e^{j\omega t} \tag{10.112}$$

where $\bar{q}_0$ is an arbitrary constant. The results are

$$H_n(\omega) = \left[1 - \frac{\omega^2}{\omega_n^2} + 2\zeta_n \frac{\omega}{\omega_n} j \right]^{-1} \tag{10.113}$$

$$|H_n(\omega)|^2 = \left[\left(1 - \frac{\omega^2}{\omega_n^2} \right)^2 - \left(2\zeta_n \frac{\omega}{\omega_n} \right)^2 \right]^{-1} \tag{10.114}$$

The last ingredient needed in equation (10.110) is $S_{\bar{q}n}(\omega)$, the spectral density of the generalized force. To derive this term, first observe that the beam loading $\bar{q}(x,t)$ is related to the surface wave height $\eta(t)$ by

$$\bar{q}(x,t) = |G(\omega)| \eta(t) \tag{10.115}$$

where the transfer function $G(\omega)$ is calculated for the chosen wave theory and flow regime as discussed in Chapter 4. Rewrite equation (10.115) by replacing $\bar{q} = \bar{q}(x,t)$ with its expanded form given by equation (10.106). Now multiply the resulting equation by $X_m = X_m(x)$, and integrate each term over the interval $(0, \ell)$. Applying the orthogonality conditions (10.70) and (10.71), the result is

$$\bar{q}_n(t) = \frac{1}{C_0} \int_0^\ell X_n \, \bar{q}(x, t) dx \qquad (10.116)$$

$$C_0 = \int_0^\ell X_n^2 \, dx \qquad (10.117)$$

With equation (10.115), this generalized load becomes

$$\bar{q}_n(t) = \frac{\eta(t)}{C_0} \int_0^\ell X_n \, |G(\omega)| \, dx \qquad (10.118)$$

Note that for a horizontal beam normal to simple, incident plane waves, $G(\omega)$ does not depend on the beam's longitudinal coordinate x. However, for a *vertical* beam such as a pile subjected to these same waves, then $G(\omega)$ does depend on the beam's longitudinal coordinate x, and two adjustments in nomenclature for the wave theory appearing in Chapter 3 are in order. The first adjustment is to replace the depth coordinate z in the wave theory by $z = -x$. Thus, the origin of the beam coordinate is at the still water line and is positive downward. The second adjustment is to let $x = 0$ in the wave theory, which puts the origin of the wave at the location of the vertical beam.

Assume now that each $\bar{q}_n(t)$ is a statistically independent process. Then using the right side of equation (10.118), write in full the autocorrelation function of $\bar{q}_n$, designated as $R_{\bar{q}n}(\tau)$. Then obtain the spectral density $S_{\bar{q}n}(\omega)$ by taking the Fourier transform of $R_{\bar{q}n}(\tau)$. This leads to

$$S_{\bar{q}n}(\omega) = \frac{1}{C_0^2} \left[\int_0^\ell X_n \, |G(\omega)| \, dx \right]^2 S_\eta(\omega) \qquad (10.119)$$

Now combine equations (10.114) and (10.119) with equation (10.110) to obtain the time average of the response spectrum, or

$$S_v(\omega) = S_\eta(\omega) \sum_{n=1}^{\infty} \frac{X_n^2 \left[\int_0^\ell X_n \, |G(\omega)| \, dx \right]^2}{(\bar{m}\omega_n^2 C_0)^2 \left[\left(1 - \frac{\omega^2}{\omega_n^2}\right)^2 - \left(2\zeta_n \frac{\omega}{\omega_n}\right)^2 \right]} \qquad (10.120)$$

where C_0 is given by equation (10.117) and modal damping has the form of equation (10.108). The time average for the variance of this displacement is given by

$$\sigma_v^2 = 2 \int_0^\infty S_v(\omega) d\omega \qquad (10.121)$$

and the *rms* value of the displacement is the square root of σ_v^2 since $v(x, t)$ was assumed to have a zero mean. A *space* average response for this variance for x in the interval $(0, \ell)$ is defined as

$$\bar{\sigma}_v^2 = \frac{1}{\ell} \int_0^\ell \sigma_v^2 \, dx \qquad (10.122)$$

The results of this section are summarized. Begin by identifying the beam's properties $El, \bar{m}, \ell$, and its constraint condition of (10.81). Compute the natural frequencies and corresponding mode shapes for free vibration as outlined in Section 10.1. For a selected wave theory and flow regime, calculate $G(\omega)$ as discussed in Chapter 4. Then select the wave height spectrum $S_\eta(\omega)$ and assume the modal damping factors ζ_n subject to the constraint of equation (8.88). Evaluate numerically the statistical responses given by equations (10.120)-(10.122). Generally, the upper limit for n can be taken as 20, and the limits of integration $(0, \infty)$ of equation (10.121) are (0.05, 1.5) rad/sec for the commonly used wave height spectrum.

10.4 DEPLOYMENT OF AN OTEC PIPELINE

A vertical pipe of about 1000 m in length and 10 to 20 m in diameter is required in typical ocean thermal energy conversion (OTEC) units. Such a cold-water pipe (CWP) is used to raise the cooler water from the ocean depths to the warmer surface water where the resulting fluid temperature difference of 10 to 20 deg C is sufficient to produce net power through heat exchange. Motion analyses of several CWP systems under the excitation of ocean currents, waves, and the deployment barge to which the upper end is attached have been compared and summarized by Barr and Johnson (1979), Hove and Grote (1980), and Scotti and Galef (1980). Other relevant articles are by McGuiness et al. (1979), Griffin and Mortaloni (1980), Green et al. (1980), and Whitney and Chung (1981). In the 1990s, there was ongoing research in OTEC systems by the national laboratories in India.

The following analysis, based on the work of Wilson et al. (1982), addresses the problem of dynamic stability for a uniform, continuous, vertical CWP which allows for an arbitrary elastic rotational restraint at the barge end. The barge heave and sway motions are included as the end excitation parameters, and Morison's equation is employed to account for the wave drag and inertial forces along the pipe length. The wave environment is simulated by discretizing the Pierson-Moskowitz wave spectrum. Using the Crank-Nicolson implicit finite difference method to solve the equations of CWP motion, the dynamic deflection and moment distributions are calculated along a typical CWP as it is lengthened by adding vertical segments at the barge end.

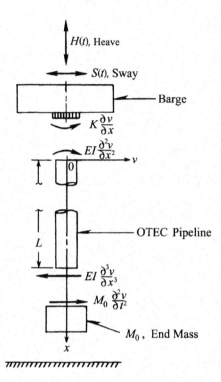

Figure 10.13   Mathematical model of the OTEC pipeline-barge system.

Mathematical Model

The mathematical model of the OTEC pipeline attached to a barge is shown in Figure 10.13. To the extent that classical beam theory is valid, the transverse displacement $v = v(x,t)$ for the pipeline can be approximated by solutions of the Bernoulli-Euler form given by equation (10.8). That is

$$EI\frac{\partial^4 v}{\partial x^4} - \frac{\partial}{\partial x}\left(P\frac{\partial v}{\partial x}\right) + \bar{m}\frac{\partial^2 v}{\partial t^2} = \bar{q}(x,t) \qquad (10.123)$$

where damping is neglected. Both EI, the bending stiffness, and $\bar{m}$, the virtual mass per unit length of the pipe including its contents, are constant. However, $P = P(x,t)$, the longitudinal tension, and $\bar{q} = \bar{q}(x,t)$, the transverse wave loading per unit length, depend on both x and time t. Motion is restricted to the plane in which $\bar{q}$ is a maximum, which is in the direction of the waves.

Consider the boundary conditions. The transverse motion at the top is assumed to be that of the barge in sway, or

$$v(0,t) = S(t) \qquad (10.124)$$

An elastic rotational restraint characterized by the constant K is provided at the barge end. Such a restraint should facilitate the attachment of vertical pipe segments during deployment. This end condition is

$$EI\frac{\partial^2 v(0,t)}{\partial x^2} = K\frac{\partial v(0,t)}{\partial x} \tag{10.125}$$

The transverse motion at the bottom is minimized by the addition of a ballast mass M_0 which is pinned at $x = L$. Thus

$$EI\frac{\partial^2 v(L,t)}{\partial x^2} = 0; \qquad EI\frac{\partial^3 v(L,t)}{\partial x^3} = -M_0\frac{\partial^2 v(L,t)}{\partial t^2} \tag{10.126}$$

which express, respectively, the condition of zero end moment and the compatibility of the transverse shear force at the pin with the motion of the ballast mass.

At location x the average tension load on the cross section depends on two dead weight terms and on the barge heave motion, $H(t)$. That is,

$$P(x,t) = M_1 g + (L-x)\bar{m}_1 g + (L-x)\bar{m}\ddot{H}(t) + M_0\ddot{H}(t) \tag{10.127}$$

where $M_1 g$ is the ballast weight in water, $\bar{m}_1 g$ is the weight per unit length of the pipe and its contents in water, and $\ddot{H}(t)$ is the barge heave acceleration. It is asumed that $\ddot{H}(t)$ is approximately the same at the top and bottom of the CWP since the natural period of longitudinal pipe oscillations (T_1) is significantly smaller than the shortest excitation period (T_2). For a steel pipe 1000 m long, $T_1 = 0.3$ s is smaller by about an order of magnitude than the shortest realistic excitation period of $T_2 = 3$ s.

Pipe Excitations by Barge and Waves

Assume that the motions of the barge from which the pipeline is deployed are completely determined by the ocean waves. That is, the barge motion is not affected by the pipe motion, which is a reasonable assumption since the barge mass and the wave forces on it are both much larger than the pipe mass and the pipe's wave forces. Choose the modified Pierson-Moskowitz spectral density function for the sea surface elevation $\eta(t)$, or

$$S_\eta(\omega) = A\omega^{-5}e^{-B/\omega^4} \tag{10.128}$$

where A and B are constants and the wave frequency has the range $\omega_1 \le \omega \le \omega_N$. Partition the spectrum into N components for which the nth component is $S_\eta(\omega_n)$. As discussed by Borgman (1969), the sea surface wave amplitude at each central frequency $\omega_n = n\Delta\omega$ is given by

$$\eta(\omega_n) = [2S_\eta(\omega_n)\Delta\omega]^{1/2}, \quad n = 1, 2, \dots, N \tag{10.129}$$

where the bandwidth $\Delta\omega$ is sufficiently small.

The heave motion of the barge is thus

$$H(t) = \sum_{n=1}^{N} R(\omega_n) \cdot \eta(\omega_n) \cos(\omega_n t + \phi_n) \tag{10.130}$$

where $(\phi_1, \phi_2, \ldots, \phi_N)$ is a set of random phase variables distributed uniformly over the interval $(0, 2\pi)$. Approximate the empirical weighting function as

$$R(\omega_n) = 1, \qquad \text{for} \ \ 0 < \omega_n \leq \omega_b$$

$$R(\omega_n) = 1 - \frac{\omega_n - \omega_b}{\omega_N - \omega_b}, \qquad \text{for} \ \ \omega_b \leq \omega_n \leq \omega_N$$

$$R(\omega_n) = 0, \qquad \text{for} \ \ \omega_n > \omega_N \tag{10.131}$$

Here ω_b is related to the barge length B_0 by the empirical result given by Kim et al. (1971): $\omega_b = c/B_0^{1/2}$, where c is a constant. With $\ddot{H}(t)$ obtained from equation (10.130), the description of $P(x, t)$ is complete.

The characterization of barge sway motion is somewhat more subtle than that for heave motion. In the usual type of pipe-laying barges considered here, second order sway motions or slow drift oscillations dominate the first order sway motions. Thus it is appropriate to ignore the first order sway motions, or the motions having the same frequencies as the incident waves, with amplitudes proportional to the first power of the wave amplitudes. In this analysis the amplitudes of sway oscillations are assumed to be directly proportional to the incident wave amplitudes. As is usual for slow drift motion, the associated periods are based on the envelopes of these wave amplitude time histories. With these assumptions, the sway motion of the barge sway is given by

$$S(t) = \eta(\omega_p) \cdot \sin\left(e_1 \omega_p t + \phi_p\right) + \sum_{p+1}^{N} e_2 \, \eta(\omega_n) \cdot \sin\left(e_1 \omega_n t + \phi_n\right) \tag{10.132}$$

Here the dominant amplitude $\eta(\omega_p)$ at $\omega = \omega_p$ is that of the highest energy incident waves, and e_1 and e_2 are constants. For practical barge dimensions and mooring lines of wire or chain, a dominant sway period of 170 s is estimated. This corresponds to a dominant frequency of $0.093\omega_p$ rad/s, where $e_1 = 0.093$ and $\omega_p = 0.4$ rad/s. A realistic value for the fractional reduction of the higher frequency waves, as these translate into barge sway, is $e_2 = 0.05$.

The direct pipe excitation by waves is based on a modified form of Morison's equation (4.1), or

$$\bar{q} = 0.25\pi C_M \rho D^2 \dot{u} - 0.25\pi C_A \rho D^2 \ddot{v} + 0.5 C_D \rho D (u - \dot{v}) |u - \dot{v}| \tag{10.133}$$

Here the coefficients C_M, C_D and C_A represent inertia, drag, and added mass; ρ is the mass density of water; D is the pipe diameter; and u is the horizontal water particle velocity. Vortex shedding is neglected. By the superposition of N simple, deepwater waves, $\dot{u}$ from Table 3.1 with $x = 0$ becomes

$$\dot{u} = \sum_{n=1}^{N} \omega_n^2 \, \eta(\omega_n) \cdot \frac{\exp k_n (z + d)}{\sinh k_n d} \sin(\omega_n t + \phi_n) \tag{10.134}$$

where $4\pi^2/T^2 = \omega_n^2$, $A = \eta(\omega_n)$, $\cosh k_n(z+d) \simeq \exp k_n(z+d)$, d is the water depth, g is the acceleration due to gravity, and $k_n = \omega_n^2/g$.

The foregoing formulation of Morison's equation was derived from laboratory tests of groups of circular cylinders in oscillatory flow. These tests showed that whenever the ratio of cylinder diameter D to water wave length λ exceeded 0.2, the Kuelegan-Carpenter number is small, and the drag-related viscous term becomes unimportant. In such cases, Morison's equation (10.133) is no longer correct since the flow field is modified by the presence of the cylinder. In such cases, a diffraction analysis based on potential flow theory is used.

In the present case of the OTEC pipe, this ratio D/λ approaches 0.2 for the shorter wavelengths in the wave excitation spectrum. Therefore, although a diffraction analysis is not required, it is clear that the inertia term dominates the excitation history. Separate calculations not included here indicate that the drag-related term accounts for only about 10 percent of the maximum induced loads. This observation was not particularly surprising since it is in agreement with the results of other investigators, notably Hogben (1976) and Hogben and Standing (1975). Therefore an attractive and obvious alternative, which is the one employed here, is to neglect the drag term in equation (10.133) and at the same time increase the remaining inertia term by 10 percent. The effect of this approximation on pipe stability is negligible.

The excitations needed for solutions $v(x,t)$ of equation (10.123) are defined by equations (10.127)-(10.134). The solutions lead to the bending moment responses given by

$$M(x,t) = EI\frac{\partial^2 v}{\partial x^2} \tag{10.135}$$

From these moments, the critical bending stresses are calculated using elementary theory.

For convenience, all 25 system parameters for this OTEC pipeline system are summarized in Table 10.2, together with typical numerical values used for exploratory solutions. In a given ocean location, the values of d, g and ρ remain essentially constant. The remaining 22 parameters fall in the following four categories:

Six wave height spectrum parameters: $(A, B, N, \omega_1, \omega_n, \omega_p)$

Five barge sway motion parameters: $(B_0, c, e_1, e_2, \omega_b)$

Nine pipe, restraint, and ballast parameters: $(D, E, EI, K, L, M_0, M_1, \bar{m}, \bar{m}_1)$

Two wave-pipe interaction parameters: (C_M, C_D)

Table 10.2   Typical Parameters for an OTEC System

Symbol	Meaning	Numerical Value
A	Wave spectrum amplitude	0.780 m$^2 \cdot$ (s/rad)^6
B	Wave spectrum constant	0.0311 (rad/s)^4
B_0	Barge length	122 m
c	Barge frequency parameter	5.55 m$^{1/2} \cdot$rad/s
C_A	Added mass coefficient	1.0
C_M	Inertia coefficient	2.0
d	Water depth	1067 m
D	Outside pipe diameter	9.144 m
e_1	Sway frequency scale factor	0.093
e_2	Sway amplitude scale factor	0.05
E	Young's modulus for pipe	2.07×10^{11} N/m^2
EI	bending stiffness for pipe	3.94×10^{11} N$\cdot$m^2
g	Acceleration due to gravity	9.81 m/s^2
K	Rotational spring constant	1.13×10^8 N$\cdot$m/rad
L	Pipeline length	61 m to 1037 m
M_0	Ballast mass	4.54×10^4 kg; 2.27×10^5 kg
M_1	Ballast mass in water	$0.6 M_0$
$\bar{m}$	Pipe mass/unit length, incl. contents	6.43×10^4 kg/m
$\bar{m}_1$	Pipe mass/unit length in water	1.25×10^3 kg/m
N	Number of waves in spectrum	15
ρ	Mass density of water	962 kg/m^3
ω_1	Spectrum frequency, lower bound	0.1 rad/s
ω_b	Sway weighting frequency	0.503 rad/s
ω_N	Spectrum frequency, upper bound	1.5 rad/s
ω_p	Frequency at peak of $S_\eta(\omega)$	0.4 rad/s

Numerical Results

Using the numerical values of the 25 system parameters listed in Table 10.2, computer solutions were obtained to equation (10.123), subjected to the boundary conditions of equations (10.124)-(10.126), and the local pipe tension given by equation (10.127). A numerically stable implicit finite difference method was employed, using the Crank-Nicolson approximation (Carnahan et al.,1964). Fifty spatial steps in the interval $0 \leq x \leq L$ and 300 time steps were used in solving the difference equations. The central difference form was used except for the shear-moment equation (10.126), which was expressed in terms of forward and backward differences. In all calculations, zero initial conditions were assumed: the pipe was vertical and at rest at time $t = 0$. The explicit difference equations and a series of test runs performed to validate the computer program for this problem were described in detail by Pandey (1980).

A total excitation time of 600 s was used for each of the 17 pipe lengths, ranging from 61 to 1073 m. This simulation time was sufficient to encompass

143 heave cycles at the dominant heave frequency of $\omega_N = 1.5$ rad/s, and to encompass about 3.6 dominant sway cycles of frequency $e_1\omega_p = 0.0372$ rad/s. Since trial simulation times longer than 600 s produced no further increases in peak responses, the 600 s simulation was considered in the present calculations to give a steady state upper bound on the responses, where the deployment rate was about 1.7 m/s.

Table 10.3   Peak Pipeline Responses During Deployment for a 600 s Simulation Time, for $K = 1.13 \times 10^8$ N·m/rad (Asterisks Represent Unbounded Responses)

Pipeline Length L, m	Deflection $v(L, t_c)$, m		Moment $M_c(x,t)$, 10^5 N·m	
	$M_0 =$ 45 400 kg	$M_0 =$ 227 000 kg	$M_0 =$ 45 400 kg	$M_0 =$ 227 000 kg
61	9.63	7.94	−51 800	−51 800
122	24.4	12.5	13 000	13 000
183	*	55.3	*	−15 900
244	6.27	5.84	3 390	−3 400
305	−6.28	−4.87	−2 290	−2 290
366	9.85	7.71	−1 730	−1 730
427	−5.00	−4.14	−1 420	−1 420
488	13.2	13.6	2 730	2 690
549	*	−82.4	*	1 190
610	*	*	*	*
671	216	108	40 200	20 300.
732	*	*	*	*
793	−108	−45.7	40 300	15 200
854	−139	−98.0	89 100	60 100
915	*	*	*	*
975	*	*	*	*
1037	*	*	*	*

The time histories of transverse pipe deflection and moment were calculated at 50 points along each chosen length L. The peak values of those responses, for two values of ballast mass, are listed in Table 10.3. The peak deflection $v(L, t_c)$ is bounded and occurred at the ballast end ($x = L$) and at a critical time $t = t_c$. However, the peak or critical moment $M_c(x, t)$ usually occurred close to but not exactly at the barge end ($x = 0$). It is noted that, for this 9.144 m diameter steel pipe where the stiffness EI is based on a 6.35-mm wall thickness, the bending moment at which yielding would begin (for most structural steels) would be about 1500×10^5 N·m. These severe pipe moments are probably due mainly to the transverse pipe wave forces, which are strongest near the surface and drop off exponentially with water depth. Thus for an average deployment rate for this pipe of about 1.7 m/s, care must be taken to reduce the bending moment near the barge, perhaps by attaching cable stays to the pipe or by restraining barge motion.

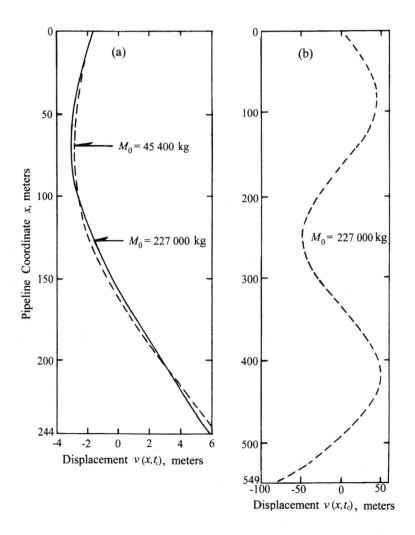

Figure 10.14   Effect of ballast on critical pipeline displacements: (a) $L = 244$ m; (b) $L = 549$ m.

Shown with asterisks in Table 10.3 are the pipeline lengths at which the responses are unbounded. Note that the larger ballast mass (227 000 kg instead of 45 400 kg) increased the system stability for lengths of 183 m and 545 m, but this additional ballast did not alleviate unbounded responses at lengths greater than 610 m.

Shown in Figures 10.14a and 10.14b are typical stable pipe profiles corresponding to the peak transverse displacement $v(L, t_c)$ occurring during the 600 s simulations. Figure 10.14a shows that an increase of a factor of five in ballast mass does little either to straighten the pipe to relieve its peak moment occurring near $x = 75$ m, or to reduce its peak deflection at $x = L$. These two profiles correspond to critical times t_c of 146 s and 264 s for the light and heavy ballast, respectively. In Figure 10.14b, where $L = 549$ m, the profile for the lighter ballast is not shown since those responses are unbounded, as shown in Table 10.3. The heavier ballast did stabilize the system, however, where $t_c = 580$ s. In these studies, an unbounded response was defined whenever the ratio $v(L, t_c)/L$ exceeded unity.

Shown in Table 10.4 are the effects of K, the rotational stiffness restraint at the barge end, on the dynamic pipe response and stability. These results show that, for a ballast mass of $M_0 = 45\,400$ kg, an increase in K by a factor of 100 increased the range of pipe length for which the system remained dynamically stable. For instance, the system was stable at $L = 183$ m for the higher rotational restraint but not for the lower rotational restraint. Also, the higher K reduced the peak deflection in the range up to $L = 183$ m, but beyond this length the stable deflections often exceeded those corresponding to the lower K value.

Table 10.4   Effects of Torsional Restraint on Peak Pipe Responses for a Ballast of $M_0 = 45\,000$ kg and a Simulation Time of 600 s

Pipeline Length L, m	Torsion Constant, 1.13×10^8 N·m/rad		Torsion Constant, 1.13×10^{10} N·m/rad	
	Deflection $v(L, t_c)$, m	Moment $M_c(x, t)$, 10^5 N·m	Deflection $v(L, t_c)$, m	Moment $M_c(x, t)$, 10^5 N·m
61	9.63	−51 800	7.44	−53 490
122	24.4	13 000	−6.01	14 100
183	*	*	6.14	6 290
244	6.27	3 390	−11.40	4 400
305	−6.28	−2 290	17.79	3 180
366	9.85	−1 730	−26.89	−2 470
427	−5.00	−1 420	−10.80	−1 960
488	13.2	2 730	*	*
549	*	*	−13.12	−3 140
610	*	*	−8.28	1 700
671	216	40 200	−20.86	5 530
732	*	*	−71.29	15 800
793	−108	40 300	−329	−94 880
854	−139	89 100	4.54	−2910
915	*	*	*	*
976	*	*	−249	−118 800
1 037	*	*	*	*

It is concluded that this analysis can be used to determine the feasibility of deploying any vertical cold-water pipe from a typical barge in a given sea state. These exploratory studies showed that deployment is a delicate task since there may exist a range of pipe lengths longer than 500 m, for instance, where unbounded deflections and moments occur. Such pipe instabilities may be avoided in some cases by imposing a relatively fast deployment rate. However, these exploratory studies indicate that subsequent anchoring to the sea floor of the end ballast and of selected points along the pipeline using cable stays is needed to maintain dynamic stability if a barge wave excitation, based on a Pierson-Moskowitz wave spectrum, continues even for just ten minutes. However, if the barge rotational restraint is chosen carefully and possibly changed during deployment, stable pipelines can be achieved for longer times.

PROBLEMS

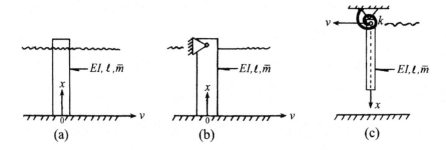

Figure 10.15   Three uniform, submerged structures: (a) a fixed-free pile; (b) a fixed-hinged pile; (c) a pipeline with torsional restraint.

10.1   The submerged, uniform, cylindrical pile shown in Figure 10.15a has full fixity at the sea floor and is unrestrained at the top. Neglecting axial loading due to self-weight, set up the determinant from which the undamped bending frequencies can be calculated. From the transcendental equation derived from this determinant, calculate the lowest three values of $\alpha_n \ell$, where the corresponding frequencies are given by

$$\omega_n = \frac{(\alpha_n \ell)^2}{\ell^2} \sqrt{\frac{EI}{\bar{m}}}$$

10.2   Calculate the lowest three undamped frequencies for a submerged pile pinned to a deck structure, as shown in Figure 10.15b. Express the results in the same form as in Problem 10.1.

10.3   The uniform, submerged pipeline shown in Figure 10.15c is hinged to a barge and is restrained from rotating about that hinge by a linear spring support of constant k. The restraining moment at $x = 0$ is

$$M(0, t) = k \frac{\partial v(0, t)}{\partial x}$$

The lower end is unrestrained. Account for self-weight by assuming a mean value for the pipe tension, or $P \simeq W/2 = $ constant where W is the pipe's weight in water. Set up the determinant from which the transverse bending frequencies can be calculated and from this derive the transcendental equation. Discuss briefly how one could obtain a computer-aided solution for the consecutive roots of this equation. How are these roots related to the frequencies ω_n?

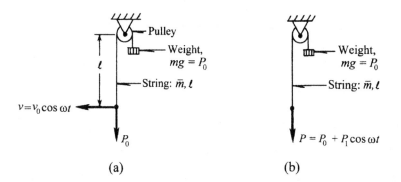

(a)   (b)

Figure 10.16 Experimental setup: (a) Problem 10.4; (b) Problem 10.5.

10.4 Set up an experimental laboratory model such as shown in Figure 10.16a to complement the analysis of resonance frequencies and mode shapes for transverse excitation of a cable. Measure: the length ℓ; the mass per unit length of the string $\bar{m}$; and the applied weight that gives the string tension, P_0. Apply the transverse excitation of amplitude v_0 and frequency $\bar{\omega}$, each of which can be adjusted independently by means of a magnetic shaker power supply and frequency generator. Record consecutive frequencies $\bar{\omega}$ which give rise to observable mode shapes one through eight. Compare these results with the corresponding theoretical frequencies predicted by equation (10.22). Discuss briefly the reasons for any discrepancies between the measured and predicted values of ω_n.

10.5 In a laboratory experiment using a magnetic shaker, impose parametric excitation to the cable or string as shown in Figure 10.16b. Keep the magnitude of the oscillating load, P_1, much less than the constant weight load P_0. As can be observed in Figure 10.8, at small β_n, transverse instability is predicted near certain imposed frequencies $\bar{\omega}$, where

$$\alpha_n \simeq \frac{k^2}{4}, \quad k = 1, 2, 3, \ldots$$

$$\bar{\omega} = \frac{2\omega_n}{k} = 2\omega_n, \omega_n, \frac{2\omega_n}{3}, \frac{\omega_n}{2}, \ldots$$

Experimentally verify these predicted results as follows. Calculate ω_n for the laboratory setup and record them in Table 10.5. Then calculate $\bar{\omega} = 2\omega_n/k$ and compare these with the corresponding experimental values of $\bar{\omega}$ where transverse unstable motion is observed for the stated (n, k) combinations. Discuss briefly reasons for any discrepancies.

Table 10.5   Data Sheet for Problem 10.5

n	ω_n (rad/sec)	ω_n (Hz)	k	Theory $\bar{\omega}$ (Hz)	Measured $\bar{\omega}$ (Hz)
1	$\omega_1 = (\pi/\ell)\sqrt{P_0/\bar{m}} =$		1		
1			2		
1			3		
1			4		
2	$\omega_2 =$		1		
2			2		
2			3		
2			4		
3	$\omega_3 =$		1		
3			2		
3			3		
3			4		

10.6   Based on the findings of Trogdon et al. (1976), design a series of water tunnel experiments in which a submerged, tensioned cable undergoes combined vortex and parametric excitation. Consider how the new hypothesized constant $\bar{K}$, related to the amplitude of the vortex-shedding force and the coupling effect of longitudinal excitation, can be deduced from a set of experimental measurements.

10.7   A stranded steel mooring cable has a density of 0.2 lb/in.^{3}, a diameter of 2.0 in., and a length of 500 ft. If the cable tension has a mean value of 100,000 lb, what current velocities (perpendicular to the cable's longitudinal axis) will give rise to vortex shedding? The kinematic viscosity and density of the seawater at 45°F are 1.8×10^{-5} ft^2/sec and 0.0372 lb/in.^{3}, respectively. Take $\bar{m}$, the virtual mass per unit length of cable, as the sum of its actual mass per unit length and the mass of the seawater that it displaces per unit length.

10.8   A uniform cross beam of an offshore structure, modeled as the simple beam of Figure 10.4b, is subjected to the harmonic wave loading $\bar{q} = \bar{q}_0 \sin \bar{\omega} t$. (a) Calculate the steady state displacement $v(x, t)$ based on Bernoulli-Euler beam theory and the normal mode method . (b) Compute the series expression for the dynamic bending moment at midspan and also the dynamic shear load at the points of fixity. Neglect P, the longitudinal loading.

10.9   Solve Problem 10.8, but replace the simple end supports with clamped ends as shown in Figure 10.4c. Neglect P.

10.10   Consider a uniform cable under constant tension, fixed at each end, and subjected to an arbitrary transverse load per unit length of $\bar{q}(x,t)$. Use the normal mode method to predict the transverse cable displacement $v(x,t)$, assumed in the form of equation (10.82). Follow the same procedure used to obtain the beam responses to the same loading, which leads to the results given by equations (10.88) through (10.91). Show that $X_n(x)$, $n = 1,2,\ldots$ for the cable are orthogonal functions.

10.11   Refer to a text such as Clough and Penzien (1993) or Timoshenko and Goodier (1951) to derive the equation of *longitudinal* vibrations for a uniform, elastic pipeline. Let $u = u(x,t)$ be the longitudinal displacement of a material point at position x on its longitudinal axis. For a pipeline without end constraints, deduce that $\partial u/\partial x = 0$ on each end. Then, following the same method used in Section 10.1 to derive the free transverse vibration frequencies for a fixed end cable, to calculate the free longitudinal vibration frequencies for the unconstrained pipeline. Check that your analytical result for ω_1 agrees with equation (10.93).

10.12   Based on the equation of longitudinal vibrations for the pipeline derived in Problem 10.11, calculate the free vibration frequencies for a uniform pipeline clamped at each end. Compare your value of ω_1 to ω_ℓ of equation (10.93), and discuss briefly the reasons for the difference in the two results.

10.13   The horizontal cross brace modeled in Figure 10.4c is located at $z = h$ in water of depth d. The beam is subjected to transverse wave excitation that is modeled by linear wave theory. For $S_\eta(\omega)$, assume that the Pierson-Moskowitz wave spectrum applies, where the significant wave height is 10 m. Choose the transfer function $G(\omega)$ given by equation (4.30). See *Example Problem 10.1* for the explicit forms of frequencies and mode shapes. Write a computer program to calculate the midspan values of the displacement response spectrum, the variance of v and the rms value of v. Specify carefully all input data. Test the program by choosing a realistic numerical example. If $S_\eta(\omega)$ is Gaussian, how are these results interpreted in terms of expected peak displacements?

10.14   For the horizontal beam of Figure 10.4b, derive an expression for the spectral density of the bending moment in terms of the spectral density of its displacement given by equation (10.120). Where along its length would the peak rms value of the bending moment occur?

REFERENCES

Barr, R. A., and Johnson, V. E., Evaluation of Analytical and Experimental Methods for Determining OTEC Plant Dynamics and CWP Loads, *Proceedings of the Offshore Technology Conference*, 1979.

Borgman, L. E., Ocean Wave Simulation for Engineering Design, *Journal of the Waterways and Harbors Division*, ASCE **95** (4), November 1969.

Carnahan, B., Luther, H. A., and Wilkes, J. O., *Applied Numerical Methods*, Vol. 2, Wiley, New York, 1964.

Clough, R. W., and Penzien, J., *Dynamics of Structures*, second ed., McGraw-Hill, New York, 1993.

Gould, L. P., and Abu-Sitta, S. H., *Dynamic Response of Structures to Wind and Earthquake Loading*, Wiley, New York, 1980.

Green, W. L., Gray, D. W., Landers, E. A., and Calkins, D. E., Hybrid OTEC Plants–An Assessment of Feasibility, *Proceedings of the Offshore Technology Conference*, 1980.

Griffin, A. and Mortaloni, L.R., Baseline Designs for Three OTEC CW Pipes, *Proceedings of the Offshore Technology Conf*erence, 1980.

Hogben, N., Wave Loading on Structures, *Proceedings of the Conference on Behavior of Offshore Structures, BOSS,* Trodheim, 1976.

Hogben, N., and Standing, R. G., Experience in Computing Wave Loads on Large Bodies, OTC 2819, *Proceedings of the Offshore Technology Conference*, 1975.

Hove, D., and Grote, P. B., OTEC CWP Baseline Designs, *Proceedings of the Offshore Technology Conference*, 1980.

Lubkin, I., and Stokes, J. J., Stability of Columns and Strings under Periodic Forces, *Quarterly of Applied Mathematics*, $1(1)$, 1943.

McGuiness, T., Griffin, A., and Hove, D., Preliminary Design of OTEC CW Pipes, SB-1, *Proceedings of the Offshore Technology Conference*, 1979.

O'Brien, J. T., and Muga, B. J., Sea Tests on a Spread-Moored Landing Craft, *Proceedings of the 8th Conference on Coastal Engineering*, Lisbon, Portugal, 1964.

Pandey, P., Dynamic Analysis of Pipelines for Ocean Thermal Energy Conversion, Master's Thesis, Department of Civil Engineering, Duke University, Durham, NC, 1980.

Scotti, R., and Galef, A., An Assessment of the Reliability of the OTEC CWP Design Analyses, ID-1, *Proceedings of the Offshore Technology Conference*, 1980.

Timoshenko, S., and Goodier, J. N., *Theory of Elasticity*, McGraw-Hill, New York, 1951.

Trogdon, S. A., Wilson, J. F., and Munson, B. R., Dynamics of Flexible Cables under Combined Vortex and Parametric Excitation, *Journal of Dynamic Systems, Measurement and Control*, 1976.

Whitney, A. K., and Chung, J. S., Vibrations of Long Marine Pipes due to Vortex Shedding, *Journal of Energy Resources Technology*, 1981.

Wilson, J. F., Muga, B. J., and Pandey, P., Dynamics during Deployment of Pipes for Ocean Thermal Energy Conversion, *Proceedings of the 1st Offshore Mechanics, Arctic Engineering, Deepsea Systems Symposium*, New Orleans, 1982.

11

Behavior of Piles Supporting Offshore Structures

Lymon C. Reese

As discussed in detail in other chapters, the behavior of structures under dynamic loading depends on the geometry and mass of the particular unit, the nature of embedment into soil, and the relevant characteristics of the soil itself. Each of these features is discussed herein.

Offshore structures maintain their stability under loading due to the mass of the structure, a *gravity* foundation, or due to embedded piles. The piles, of course, can be considered as an extended portion of the structure, or the piles can be analyzed separately. In the latter case, a free body is selected at the pile heads or at the top of a pile cap and methods of analysis achieve compatibility between the piles and the superstructure. In the case of a gravity foundation, the mass of the structure is sufficient that environmental loads do not cause a lift off or excessive soil deformation. Whatever the concept of foundation design, the engineer must make a study of the soil supporting the structure. The nature of the response of soil under loading is discussed and methods of obtaining the needed properties of the soil at offshore locations are presented briefly.

With regard to the loading of a pile, four types can occur: short-term static, cyclic, dynamic, and sustained. Static loading can be assumed in many designs with an appropriate factor of safety. However, cyclic loading must be addressed in detail for lateral loading. Dynamic loading is discussed briefly in the last section. Sustained loading usually needs addressing only if the supporting soil is a soft, saturated clay. Methods of analyzing piles under axial and lateral loading are presented because pile-supported structures dominate those at offshore locations. The analytical methods relate specifically to short-term and cyclic loadings. An example problem involving a single pile and an actual soil profile from the Gulf of Mexico is solved by two analytical methods and the results are compared.

11.1   CHARACTERISTICS OF SOIL AND RESPONSE TO EXTERNAL LOADING

An investigation of soil characteristics is a necessary initial step preceding design. Prior to taking boring equipment to the site, the engineer will investigate the engineering geology, will gain as much information as possible from previous borings, and will develop a preliminary plan about the number of borings, their depth, and investigative techniques. The provision of a marine vessel necessary to sustain the soil-boring equipment is costly so engineers give attention to all details that can facilitate and limit the time of the soil investigation. Seismology is sometimes used to augment the soil sampling and testing.

For the Gulf of Mexico, Fisk (1956) presented an important overview that has been valuable in giving engineers a general understanding of the near surface geology. An example of a feature requiring careful attention is the existence of stream beds that were later filled with soft sediments, leading to the possibility that supporting piles could vary significantly in length across the platform.

McClelland (1956) published the results of a series of soil investigations in the Gulf of Mexico that also gave valuable general information. While such results are insufficient for design, useful information is given on the detailed planning for an efficient deployment of the marine equipment.

While indirect methods of characterizing soils are being used, such as the instrumented cone, many engineers prefer to have tube samples for examination and testing. The common procedure is to lower to the sea floor a motion-compensated drill string with an interior diameter to allow sampling tubes to be lowered on a wire line. A weight is dropped with a wire line to acquire samples of cohesive soil and to get the number of blows to drive the sampler a given distance into granular soil. Hvorslev (1949) made a comprehensive study of soil sampling and noted that the most favorable way to sample clays is with a steady, continuous push of the tube. Emrich (1971) made a study of the reduction in strength of clays caused by driving the sampler and found the loss to be significant.

Special tools, operated hydraulically, can be latched to the bottom of the drill string for performing special tests, such as vane shear, cone penetration, or pressuremeter. Nevertheless, the sampling by dropping a weight remains popular because of the speed of the work. When a sample is retrieved, the drill string is advanced by rotary drilling to the next depth.

The engineer may make a visual classification and perform some tests in the field, such as using a miniature vane on samples of clay, in order to determine the required depth of the boring so piles can be designed to sustain the expected loads. Tests in the laboratory, including grain-size distribution and Atterberg limits, allow the soil to be classified accurately. Water content and unit weight is obtained for all samples. For cohesive soils, a determination of the shear strength and sensitivity is necessary, along with some information on stiffness, if possible. Consolidation characteristics are normally a secondary consideration. For cohesionless soils, undisturbed samples are usually unavailable and an estimation of the relative density and friction angle is usually made from the blow counts found in the field.

11.2   DESIGN OF SINGLE PILES UNDER AXIAL LOADING

Single piles are found in onshore structures and occasionally in an offshore structure. But most offshore structures are founded on groups of piles. If the piles are four or five diameters apart, pile-soil-piles interaction is negligible and the piles can be designed as if they were isolated. If the spacing is closer, an efficiency factor is employed to reduce the capacity of piles in the group. The mechanics of the transfer of load from the pile to the soil for the single pile can be modified to include a reduction coefficient.

A rational model for the pile-soil system is shown in Figure 11.1 (Reese and Van Impe, 2001). The pile is shown as a spring to reflect that all piles, regardless of material, will deform under load. This spring has a constant stiffness with depth but its stiffness can vary with length without causing analytical difficulty.

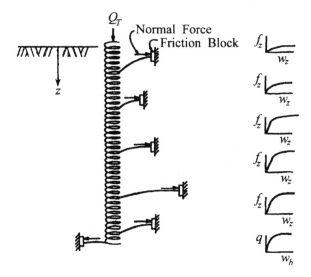

Figure 11.1   Mechanical model of axially loaded deep foundation.

The soil is represented by a series of mechanisms attached to the spring, varying in character with depth, to show that the transfer of load from the pile to the soil depends on the relative movement between pile and soil. The springs (load transfer functions) for resistance along the sides of the pile are definitely nonlinear with relative deflection the unit load transfer is given as f_z versus the relative movement w_z , where the z indicates depth below the ground surface. The load-transfer function for the base of the pile is given by a curve where w_b represents the downward movement of the pile and q represents the unit end bearing. The model can be used for computing the capacity of a pile due to uplift where q is assumed to be zero.

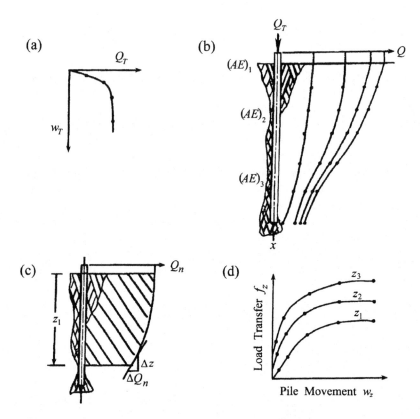

Figure 11.2   Illustration of procedure for development of curves showing load-transfer versus pile movement.

The technique for obtaining load-transfer curves is shown in Figure 11.2, which is representative of experimental results from an instrumented pile under axial loading Q_T. Figure 11.2a shows the load Q_T versus settlement w_T for the top of the pile. Figure 11.2b shows the distribution of load Q at depth x, where each curve represents a different value of Q_T. Here, the points are the measurements obtained from transducers placed in the pile. A section of the load distribution curve for the nth loading, say to the depth z_1, is shown in Figure 11.2c. Starting from the top of the pile, numerical integration of the area under the curve, when divided by the appropriate value of AE, will yield the elastic shortening of the pile to the point z_1. The elastic shortening can be subtracted from the observed settlement w_T for the nth load to obtain the relative movement between the pile and the soil at the point z_1. The value of $\Delta Q_n / \Delta z$ is found by differentiating the curve at point z_1. Failure is assumed to occur at the pile-soil interface, even though some experiments reveal a layer of clay at the surface of piles that have been recovered. Therefore, the value of f_z is found by dividing $\Delta Q_n / \Delta z$ by the circumference of the pile C . Performing similar computations for other points along the pile will yield a family of curves such as shown in Figure 11.2d.

For a particular applied load, values of settlement w_b at the base of the pile can be found by procedures given above. The corresponding values of unit end bearing load q can be found by dividing the load at the base of the pile by the area of the base. The full q/w_b curve can be obtained by analyzing all of the loadings, provided the pile settles sufficiently. If a number of such experiments in a variety of soils for a variety of piles are performed, correlations can be developed between soil properties and the load-transfer curves.

Referring to Figure 11.2a, the load test should be performed to cause plunging, where additional deflection results in no increase in load. The result will be the development of the load-transfer curves to ultimate values, as shown by the curves for the depths z_1, z_2, and z_3. The ultimate values of f_{ult} can be correlated with the shear strength of the soil in order to develop equations for computing the ultimate load in skin friction Q_s. If the end bearing has reached an ultimate value, the ultimate value of q can be correlated with the shear strength at the base of the pile and yields the ultimate load in end bearing Q_b. The capacity of the pile under axial load Q is $Q_s + Q_b$.

The procedure for computing the capacity and settlement of a pile in clay, found offshore at many locations, is illustrated in the paragraphs below. Let

$$Q_s = \int_0^L f_{ult} C \, dz \tag{11.1}$$

and

$$Q_b = \frac{1}{2} q_u N_c A_b \tag{11.2}$$

where L is the penetration of the pile; $f_{ult} = q_u \alpha_z / 2$; q_u is the unconfined compressive strength of the clay and assumed to be equal to twice the undrained shear strength c_u; N_c is the bearing-capacity factor which is taken as 9.0 for all except very short piles; and A_b is the area of the base of the pile. The value of α_z can be interpolated from the following list: $\alpha_z = 1.0$ at $q_u = 0$ tons/ft^2; 0.9 at 0.75; 0.8 at 1.12; 0.7 at 1.45; 0.6 at 1.82; 0.5 at 2.36, and 0.43 at 3.0. To simplify calculations, these numbers represent *average* values, selected from a curve with a wide range (see page 288 of Peck et al., 1974; Tomlinson, 1980). For the behavior of piles in clay under axial loading, the selection of the value of α_z varies among authors, and even the method of computing the axial capacity varies among investigators (American Petroleum Institute, 1993). The lack of agreement among investigators for piles driven into clay, and even more disparity for piles driven into sand, is due to the scarcity of high quality experimental data. The models for ultimate capacity and for settlement will remain useful even as more data become available.

The pile selected for analysis is an open steel pipe with a diameter b of 36 in. and assumed to have been driven to a penetration of 140 ft into clay at an offshore location with the following properties: 0 to 50 ft, $q_u = 1.0$ tons/ft^2; 50 to 100 ft, $q_u = 1.8$ tons/ft^2; 100 to 175 ft, $q_u = 2.5$ tons/ft^2. Interpolation of values of α_z from the list in the previous paragraph yielded the following values for the three strata, starting with the top: 0.83, 0.61, and 0.48. With these numerical values, equation (11.1) becomes

$$Q_s = 3\pi \left[(50)(1/2)(0.83) + (50)(1.8/2)(0.61) + (40)(2.5/2)(0.48)\right] = 680 \text{ tons}$$

The reasonable assumption is made that the pile plugged at some point during driving and thus the end bearing can be computed as if the pile were solid. The following result is found by substituting values into equation (11.2).

$$Q_b = (2.5/2)(9.0)\pi(1.5)^2 = 80 \text{ tons}$$

Thus, the total load the pile can sustain was computed to be 760 tons or 1520 kips. The safe load would be found by using an appropriate factor of safety.

The load-settlement curve for the pile can be computed by implementing load-transfer curves. With regard to the load transfer in skin friction for piles in clay, Coyle and Reese (1966) examined experimental data and proposed the results shown in Table 11.1.

Table 11.1   Pile Load-Settlement Data

f/f_{ult}	Pile Movement, in.
0	0
0.18	0.01
0.38	0.02
0.79	0.04
0.97	0.06
1.00	0.08
0.97	0.12
0.93	0.16
0.93	0.20
0.93	>0.2

With regard to the load transfer in end bearing for piles in clay, the work of Skempton (1951) is used, where the end bearing of a plate loaded in clay is shown to correlate with the laboratory stress-strain curve. He noted that the settlement w_b at one-half the ultimate unit end bearing of the base is equal to

$$\frac{w_{ult}}{2} = 2b\varepsilon_{50} \tag{11.3}$$

In the absence of a laboratory stress-strain curve for the soil, the following values of ε_{50} can be taken as a function of the unconfined compressive strength (consistency) of the clay: soft (<0.5 tons/ft^2) 0.0.2; medium (0.5 to 1.0 tons/ft^2) 0.01; and stiff (>1.0 tons/ft^2) 0.005. For the example problem, the value of q_u at the base of the pile was 2.5 tsf, so the value of ε_{50} was selected as 0.005, with $w_{ult}/2$ equal to 0.36 in. Numerous stress-strain curves for soil have been plotted on log-log paper and found to be a straight line, many with a slope of about

0.5 up to the failure stress. Therefore, the other points on the load settlement curve for the base can be computed by the following equation:

$$Q_b = \varsigma \sqrt{w_b} \qquad (11.4)$$

Based on $Q_b = 40$ tons for $w_b = 0.36$ in., the coefficient ς was computed as 66.7 tons/$\sqrt{in}$.. With this coefficient, sets (Q_b tons , w_b in.) can be computed from this equation.

The data are at hand for computing the axial load-settlement curve for the pile. An examination of the values of load-transfer in side resistance shows that the curves indicate movement-softening. Depending on the characteristics of the soil at the site, the engineer might assume that movement-softening does not occur and modify the curves somewhat. Comparison of the curves for side resistance end bearing reveals that the side resistance is generated with much smaller pile movements than the end bearing. Therefore, to illustrate the computation procedure, a movement of the base of the pile is selected as 0.05 in., giving a load in end bearing of 15 tons or 30 kips using equation (11.4).

With regard to the computed load transfer in side resistance, the accurate approach is to select small increments of length, perhaps as small as one or two feet. However, the procedure can be shown by taking increments along the pile equal to the three strata of soil; therefore, the first increment is from the tip of the pile at 1680 in. to the top of the lower stratum at 1200 in.

For the first trial, the assumption is made that the movement at the midheight of the stratum is the same as at the base: 0.05 in. Using the data on load transfer in side resistance and in movement to achieve the value of relative load transfer (ignoring movement softening), the value of unit load transfer was 7.33 lb/in.^{2}. The load transfer in side resistance between 1680 in. and 1200 in. was computed to be 398 kips. Thus, the load at the bottom of the 480 in. section was 30 kips and the top was 428 kips. The elastic shortening from the tip to the midheight of the section, at 1440 in., using elementary mechanics was computed to be 0.0091 in., yielding a midheight movement of the lower section of 0.0591 in. rather than 0.05 in. Employing the new midheight movement, the load transfer in the lower section was computed to be 435 kips, compared to the 398 kips for the first trial. Another iteration was done and the midheight movement of the lower section was found to be 0.0614 in., which yielded a revised load transfer for the lower section of 440 kips, compared to the 435 kips for the previous iteration. Convergence was assumed, which gave a load at 1200 in. of 461 kips and a computed movement of 0.0863 in.

Employing the above procedure, computations were done for the top two strata of soil, and loads and deformations were accumulated, with the computed top load of 1369 kips and a top movement of 0.4423 in. The procedure could be continued for other values of tip movement in order to obtain the full curve for load versus settlement for the top of the pile. Shown in Figure 11.3 is a computer-generated top load versus settlement curve. As can be seen, the ultimate load is 1520 kips. This plot is very instructive. Assuming that the load-transfer curves are acceptable, the engineer can see readily the settlement necessary to develop any portion of the load transfer, which is valuable information in selecting the safe load on the pile.

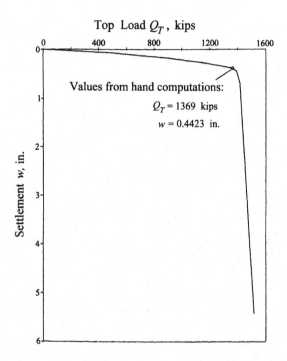

Figure 11.3   Computed curve of axial load versus settlement for a pile with a length of 140 ft driven into clay soils.

Plotted in Figure 11.3 is a point computed by hand. As may be seen, the crude partitioning of the pile that was employed yielded a relatively good agreement with the computer solution for the assumed top settlement of 0.05 in. Agreement may not have been so favorable for other selections of top movement used for obtaining the complete load settlement curve, distributions of load, and movement along the length of the pile.

11.3   DESIGN OF SINGLE PILES UNDER LATERAL LOADING

A problem that was recognized by the early designers of offshore structures was the computation of deflection and bending stresses in the foundation piles. The problem was vexing because the static equations were insufficient to solve it. Whereas axially loaded piles can be designed by simple static methods, the design of laterally loaded piles must address the interaction of the soil and the structure.

A research program was undertaken, supported principally by the petroleum industry, and the method described in this chapter was developed. Although additional research is needed, particularly in improving the prediction of soil response, the method presents an acceptable approach to a complex problem and is currently in use in the design of foundations for offshore platforms (American Petroleum Institute, 1993; Det Norske Veritas, 1977).

Pile Model and Method of Solution

The model for the problem of the pile under lateral loading is shown in Figure 11.4 where the response of the soil is described in terms of p-y curves which relate the soil resistance to the pile deflection at various depths below the ground surface. A set of typical p-y curves is shown in the figure. In general, these curves are nonlinear and depend on several parameters, including depth, shearing strength of the soil, and number of load cycles.

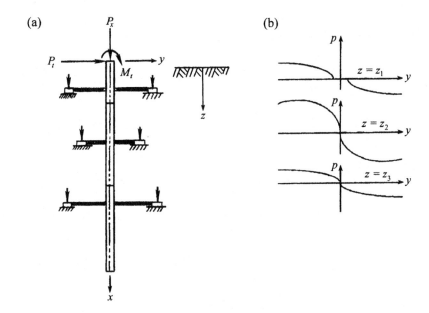

Figure 11.4   Model for pile under lateral loading showing typical p-y curves.

The solution of a soil-structure interaction problem requires the satisfaction of the conditions of equilibrium and of compatibility. The governing differential equation, derived by Hetenyi (1946), is

$$EI\frac{d^4y}{dx^4} + P_x\frac{d^2y}{dx^2} - p = 0 \tag{11.5}$$

where EI is the flexural rigidity of the pile, y is the pile deflection at position x along its length, P_x is the axial load, and p is the soil reaction force per unit length.

Equation (11.5) can be solved using a digital computer (Reese and Van Impe, 2001, p. 29); however, nondimensional methods can sometimes be employed to yield an acceptable solution for cases where EI is constant and there is no axial load. Both methods of solution give all the necessary design information including the moment, deflection, and shear at desired lengths along the pile. The methods described herein have received wide acceptance and are used in many design offices around the world.

Response of Soil

For convenience in solving equation (11.5), a secant modulus of soil reaction, E_{py}, can be used, which is defined by

$$E_{py} = \frac{p}{y} \tag{11.6}$$

The value of p from the last equation can be substituted into equation (11.5) and a solution obtained for the values of y with respect to points along the pile. Because E_{py} is a nonlinear function, equation (11.5) can be solved by iteration using procedures developed for piles in a variety of soils and rock (Reese and Van Impe, 2001, p. 49). The recommendations are based principally on the results of full-scale experiments which are augmented with theory to the extent possible. Reese and Van Impe (2001, p. 259) showed the comparison of results from experiments with results from analysis for a sizable number of cases. The validity of the analytical method has been well established within a reasonable degree of accuracy.

Consider solutions for lateral loading for pile, embedded in a soft clay below the water surface, a condition encountered frequently at offshore locations. Matlock (1970) presented procedures for developing p-y curves for soft clays below the water surface for two loading conditions: short-term static and cyclic. Those procedures, somewhat simplified, are now summarized.

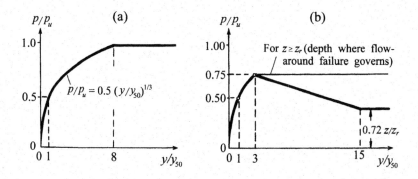

Figure 11.5   Characteristic shapes of the p-y curves for soft clay below water surface: (a) static loading; (b) cyclic loading (Matlock, 1970).

Short-Term Static Loading

The following step-by-step procedure for short-term static loading is based on the curve shown in Figure 11.5a.

1. Obtain the best possible estimate of the variation of shear strength and effective unit weight with depth. Also obtain the value of ε_{50}, the stain corresponding to one-half the maximum principal stress difference. If no values are available, use a typical value given in Table 11.2 (Skempton, 1951).

Table 11.2   Typical Values of ε_{50}

Consistency of Clay	ε_{50}
Soft	0.020
Medium	0.010
Stiff	0.005

2. Compute the ultimate soil resistance per unit length of shaft, p_u, using the smaller of the values given by the following two equations:

$$p_u = \left(3 + \frac{\gamma'}{c}z + \frac{0.5}{b}z\right)cb \qquad (11.7)$$

$$p_u = 9cb \qquad (11.8)$$

where γ' is the average effective (submerged) unit weight from ground surface to the p-y curve, x is the depth from ground surface to the p-y curve, c is the undrained shear strength of the undisturbed clay soil at depth x, and b is the pile width. Compute at each depth where a p-y curve is desired, based on the shear strength at that depth.

3. Compute the deflection, y_{50}, at one-half the ultimate soil resistance. Use the following equation:

$$y_{50} = 2.5\varepsilon_{50}b \qquad (11.9)$$

4. Compute points describing the p-y curve from equation (11.10), or

$$\frac{p}{p_u} = 0.5\left(\frac{y}{y_{50}}\right)^{1/3} \qquad (11.10)$$

The value of p remains constant beyond $y = 8y_{50}$.

Cyclic Loading

The following step-by-step procedure for cyclic loading is based on the curve shown in Figure 11.5b.

1. Construct the p-y curve in the same manner as for short-term static loading, for values of p less than $0.72p$.

2. Equate equations (11.7) and (11.8) and compute the depth $z = z_r$ where the transition occurs. If the unit weight and shear strength are constant in the upper zone, then

$$z_r = \frac{6cb}{\gamma'b + 0.5c} \tag{11.11}$$

3. If the depth to the p-y curves is greater than or equal to z_r, then the value of p is equal to $0.72p$ for all values of y greater than $3y_{50}$.

4. If the depth to the p-y curve is less than or equal to z_r, then the value of p decreases from $0.72p_u$ at $y = 3y_{50}$ to the value given by the following expression at $y = 15y_{50}$.

$$p = 0.72\, p_u\, (z/z_r) \tag{11.12}$$

The value of p remains constant beyond $y = 15y_{50}$.

Examining the parameters used in the equations for the analysis of lateral loading of piles reveals that the engineer can describe the pile with good accuracy, but describing the response of the soil depends on the quality of the p-y curves. Therefore, field experiments may be dictated in some instances.

Scour

At most of the places where offshore structures are installed, there is a probability that a certain amount of scour or erosion of the surface soil will occur. Much can be said about the use of available technology to predict the amount of scour that will occur at a given site; however, the number of parameters involved in making a prediction and the variability of those parameters are such that scour predictions are usually inexact. Some characteristics of scour are summarized as follows:

1. Scour is usually more severe close to the piles, but the mudline may be lowered over the entire area occupied by an offshore, pile-supported structure.

2. Scour is usually minor at sites where the surface soils are clays; however, some designers assume that the mudline may be lowered 3 to 5 ft (1 to 1.5 m) at such sites.

3. If the surface soil consists of fine sands, the erosion of the mudline can be severe. Some means of preventing such scour is usually adopted and installed at the time when the platform is being erected.

4. Predictions of the amount of scour are usually made on the basis of observations at other structures in the vicinity. If no data are available, a program for making periodic observations of the amount of scour is usually established.

Solutions Using Nondimensional Parameters

An iterative procedure using nondimensional coefficients is recommended for solving equation (11.5) for cases where there is no axial load and where the pile stiffness is constant (see Reese and Van Impe, 2001, p.35). A brief description of the solution procedure is given below for three sets of boundary conditions at the top of the pile: (1) pile head free to rotate, (2) pile head fixed against rotation, and (3) pile head restrained against rotation. These boundary conditions, along with the sign convention used in the following solutions, are shown in Figure 11.6. In these solutions, the reaction modulus for the soil is referenced to the ground or mudline surface by the symbol z; however, the symbol x is used here because the origin for the nondimensional solutions is the same for the ground surface and the top of the pile.

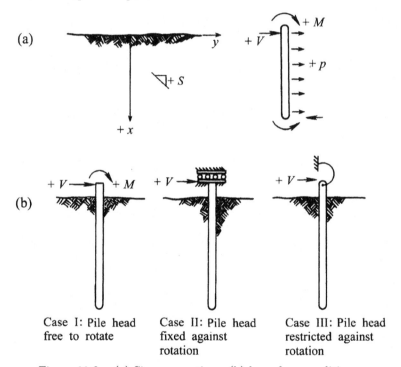

Case I: Pile head free to rotate

Case II: Pile head fixed against rotation

Case III: Pile head restricted against rotation

Figure 11.6   (a) Sign conventions; (b) boundary conditions.

Case I: Pile Head Free to Rotate

1. Construct p-y curves at various depths by procedures recommended earlier in this chapter. The chosen spacing between p-y curves should be closer near the ground surface than it is near the bottom of the pile.

2. Assume a numerical value for T, the relative stiffness factor, defined by

$$T = \left(\frac{EI}{k_{py}}\right)^{0.2}$$

(11.13)

where EI is the flexural rigidity of the pile and k_{py} is a constant relating the secant modulus of soil reaction to depth ($E_{py} = k_{py}x$).

3. Compute the depth coefficient Z_{max} as follows:

$$Z_{max} = \frac{x_{max}}{T} \tag{11.14}$$

4. Compute the deflection y at each depth along the pile where a p-y curve is available. Use the following equation:

$$y = A_y \frac{P_t T^3}{EI} + B_y \frac{M_t T^2}{EI} \tag{11.15}$$

Here A_y = deflection coefficient, found in Figure 11.7a; P_t = shear at the top of pile; T = relative stiffness factor; B_y = deflection coefficient, found in Figure 11.7b; and M_t = moment at the top of the pile. The particular curves to be employed in getting the A_y and B_y coefficients depend on the value of Z_{max} computed in Step 3.

(a)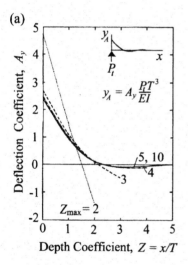

(b)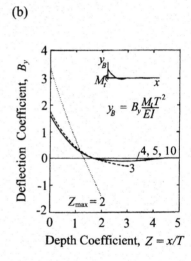

Figure 11.7   Case I: Pile deflection produced by: (a) lateral load at the mudline; (b) moment at the mudline.

5. Select from a p-y curve the value of soil resistance p that corresponds to the pile deflection value y at the depth of the p-y curves. Repeat this procedure for every p-y curve that is available.

6. Compute a secant modulus of soil reaction E_{py} using equation (11.6). Plot the E_{py} values versus depth.

7. From the E_{py} versus depth plot of Step 6, compute the constant, k_{py}, which relates E_{py} to depth ($k_{py} = E_{py}x$), giving more weight to the E_{py} values near the ground surface.

8. Compute a value of the relative stiffness factor T from the value of k_{py} found in Step 7. Compare this value of T to the values of T assumed in Step 2. Repeat Steps 2 through 8 using the new value of T each time until the assumed value of T equals the calculated value of T.

9. When the iterative procedure has been completed, the values of deflection along the pile are known from Step 4 of the final iteration. Values of soil reaction are computed from the basic expression: $p = E_{py}y$. Values of slope, moment, and shear along the pile can be found by using the following equations:

$$S = A_s \frac{P_t T^2}{EI} + B_s \frac{M_t T}{EI} \tag{11.16}$$

$$M = A_m P_t T + B_m M_t \tag{11.17}$$

$$V = A_v P_t + B_v \frac{M_t}{T} \tag{11.18}$$

The appropriate coefficients to be used in these equations can be obtained from Figures 11.8 through 11.10.

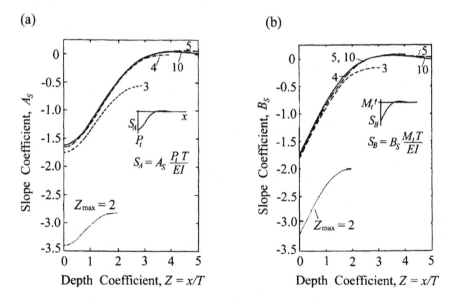

Figure 11.8 Case I: Slope of pile caused by: (a) lateral load at mudline; (b) moment at mudline.

Case II: Pile Head Fixed Against Rotation

When a pile is fixed firmly to a very stiff superstructure, the pile head can be considered to be fixed against rotation. The further assumption is made that any rotation of the superstructure is negligible.

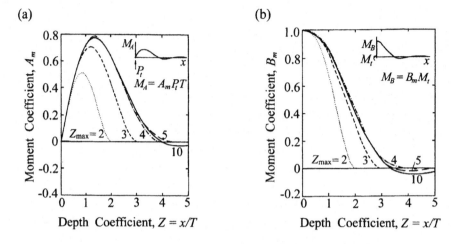

Figure 11.9   Case I: Pile bending moment produced by: (a) lateral load at the mudline; (b) moment applied at the mudline.

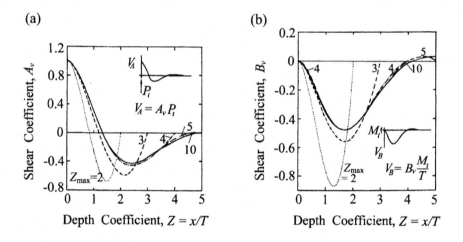

Figure 11.10   Case I: Deflection of pile fixed against rotation at mudline.

1. Perform Steps 1-3 of the solution procedure for free head piles, Case I.

2. Compute the deflection y at each depth along the pile where a p-y curve is available by using the following equation:

$$y_F = F_y \frac{P_t T^3}{EI} \qquad (11.19)$$

The deflection coefficients F_y can be found by entering Figure 11.11 with values of Z, using the appropriate curve according to the value Z_{max}.

3. Proceed in steps similar to those for the free head case.

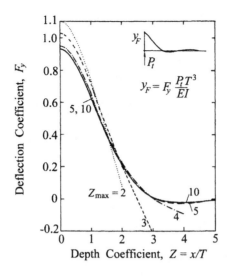

Figure 11.11   Case II: Deflection of pile fixed against rotation at the mudline.

4. Compute M_t, the moment at the top of the pile, from the following equation:

$$M_t = F_{Mt}P_tT \qquad (11.20)$$

The value of F_{Mt} can be found by entering Table 11.3 with the appropriate value of Z_{max}.

Table 11.3   Moment Coefficients at Top of Pile for the Fixed Head Case

Z_{max}	F_{Mt}
2	−1.06
3	−0.97
4	−0.93
5 and above	−0.93

5. Compute values of slope, moment, shear, and soil reaction along the pile by following the procedure for the free head pile.

Case III: Pile Head Partially Restrained Against Rotation

This case can be used to obtain a solution when the rotational restraint is known for the pile on entering the superstructure. Such a case is illustrated by the example that follows.

1. Perform Steps 1-3 of the solution procedure for free head piles, Case I.

2. Obtain k_θ, the value of the spring stiffness of the pile-superstructure system. This spring stiffness is defined as

$$k_\theta = \frac{M_t}{S_t} \qquad (11.21)$$

where M_t is the moment at the top of the pile and S_t denotes the slope there. (Note that S_t sometimes denotes soil sensitivity, but in context, there should be no confusion of symbols.)

3. Compute S_t, the slope at the top of pile, as follows:

$$S_t = A_{st} \frac{P_t T^2}{EI} + B_{st} \frac{M_t T}{EI} \qquad (11.22)$$

Here A_{st} is the slope coefficient A_s found in Figure 11.8a; and B_{st} is the slope coefficient B_s found in Figure 11.8b.

4. Solve equations (11.21) and (11.22) for M_t, the moment at the top of the pile.

5. Perform Steps 4-9 of the solution procedure for free head piles, Case I.

This completes the solution of the laterally loaded pile problem for three sets of boundary conditions. The solution gives values of deflection, slope, moment, shear, and soil reaction as a function of depth. The example problem in the next section illustrates the use of this method.

The nondimensional method as presented above has the following limitations: the influence of an axial load was not considered; the pile must have a constant value of EI; and perhaps of most importance, the soil must be of one type and preferably have a shear strength that increases linearly with depth from zero at the mudline. In spite of these limitations, the nondimensional method can give good answers to a considerable share of cases of lateral loading of piles. Furthermore, this hand solution can reveal explicitly the influence of various parameters.

Computer-Aided Solutions

Computer codes have been written to eliminate the limitations in the nondimensional method by solving equation (11.5) using finite-difference techniques. If a pile is divided into n increments of length h, then $n + 1$ equations of the following form can be written:

$$\frac{EI}{h^4}(y_{m-2} - 4y_{m-1} + 6y_m - 4y_{m+1} + y_{m+2})$$

$$+ \frac{P_x}{h^2}(y_{m-1} - 2y_m + y_{m+1}) - E_{py} y_m = 0 \qquad (11.23)$$

Two imaginary points must be introduced at the top of the pile and two at the bottom, leading to $n + 5$ unknown deflections. Two boundary equations at the bottom of the pile and two at the top can be added to the $n + 1$ standard equations. Solutions of these $n + 5$ simultaneous equations yield the deflections along the length of the pile, from which slope, bending moment, shear, and soil reaction may be readily computed. Equation (11.23) is written for a constant EI; however, the equation can be rewritten to account for a change in pile stiffness. This expanded equation presents little additional difficulty in writing the computer program.

A number of methods can be used to solve the simultaneous, finite difference equations. A method of Gaussian elimination discussed by Reese and Van Impe (2001) was found to be efficient. The mesh size h and the number of significant figures in the computation procedure must be controlled in order to achieve acceptable accuracy. Reese and Van Impe (2001) present several case studies.

The definite advantages of a computer-aided solution are summarized:

1. Changes in flexural stiffness of the pile can be introduced at any depth.
2. The pile length can be changed as desired.
3. The p-y data can be introduced in several ways.
4. The E_{py} value can be changed from point to point as dictated by the soil response.
5. An axial load can be specified and accounted for in obtaining the shape of the deflected pile.

11.4 AN EXAMPLE: LATERAL PILE LOADING

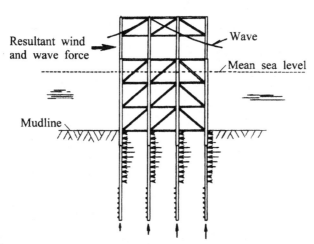

Figure 11.12   A plane truss or bent from an offshore platform.

An example problem is now solved using the nondimensional coefficients, and this solution is compared with results from a finite-difference solution. Many features of the example are consistent with current practice, except that no attempt has been made to establish exact compatibility between the structure and pile.

Problem Description

The problem is illustrated in Figure 11.12, which shows a plane truss or bent from an offshore platform. A template or *jacket,* which consists of a welded pipe framework, is first set on the ocean floor. In this particular case the legs of the jacket are assumed to penetrate into the soft surface soil. The penetration is 140 in.; mud sills are set below the bottom bracing to keep the jacket from sinking further.

After the jacket is set, the piles are stabbed into the jacket legs and driven to a predetermined penetration. The annular space between the pile and the jacket is then grouted. (Some designers prefer not to use a jacket-leg extension and not to grout the annular space in order to reduce the bending moment at the bottom panel points. In this case, the wall thickness of the pile must be greater than in the case being considered.)

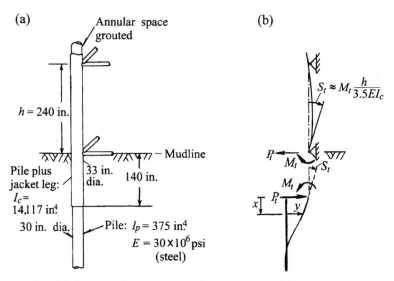

Figure 11.13   (a) Section of jacket showing pile and superstructure at mudline; (b) interaction between pile and jacket leg with assumed equation for rotational restraint.

Shown in Figure 11.13 are a pile and a portion of the superstructure. Two boundary conditions are indicated: the lateral load P_t and the rotational restraint k_θ. As indicated in the figure, the restraint against rotation at the pile head is provided by the superstructure and can be computed using equation (11.21), or

$$k_\theta = \frac{M_t}{S_t} \simeq \frac{3.5EI_c}{h} = \frac{(3.5)(30 \times 10^6)(14{,}117)}{240} = 6.18 \times 10^9 \text{ in.-lb/rad}$$

(Note that h in the above computations is the panel length and not the increment length as defined earlier.) A constant pile stiffness EI must be employed in the

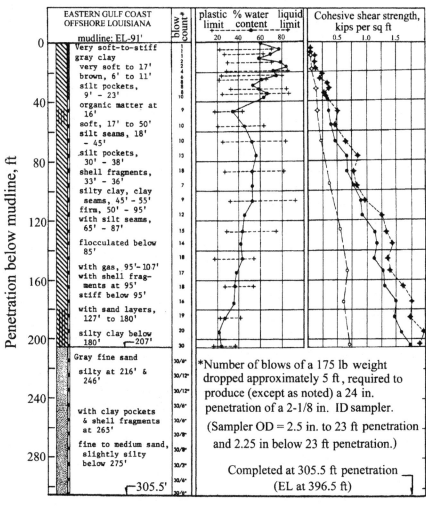

Legend for Shear Strength Plot

• unconfined compression ▲ unconsoliated-undrained triaxial
✦ miniature vane ■ consolidated-undrained triaxial

Figure 11.14   Soil profile with properties for the example problem.

nondimensional solutions, and the stiffness of the upper portion of the pile is selected. Experience has shown that the behavior of the upper portion of a pile under lateral loading has a significant effect on the result of a solution.

Solutions Using Nondimensional Parameters

The p-y curves are nonlinear, which leads to a nonlinear solution with respect to the lateral load P_t; therefore, a series of solutions should be generated for a range of values of P_t. An approach that is useful is to increase the P_t by multiplying the service load by the factor of safety and to check the behavior of the pile under the increased load. However, it is wise to employ loads below and above the design load because nature of the p-y curves is such that in some cases a slight increase in lateral load can cause a large increase in deflection and bending moment.

The soil conditions employed in this analysis are shown in Figure 11.14. The soil conditions are representative of those in the eastern Gulf of Mexico and offshore from the Louisiana coast. The water depth is 91 ft. The undrained shear strength and the water content vary with depth in a manner like that of a normally consolidated clay. The soil profile was simplified somewhat, and the following values of undrained shear strength and submerged unit weight were selected for the analysis:

$$c = 0 \text{ at } x = 0; \quad c = 12.15 \text{ lb/in.}^2 \text{ at } x = 2400 \text{ in.}$$
$$\gamma' = 0.020 \text{ lb/in.}^3 \text{ at } x = 0 ; \quad \gamma' = 0.036 \text{ lb/in.}^3 \text{ at } x = 2400 \text{ in.}$$

In the absence of stress-strain curves, the value of ε_{50} was assumed to be 0.02.

In view of the soil conditions, the method used to compute the p-y curves was for soft clay below the water surface, as detailed earlier. It is assumed that the loading is cyclic because the maximum lateral load on an offshore platform occurs during a storm. The p-y curves were computed for the following depths in inches: 0, 50, 100, 200, 300, 400, 500, 700, and 900. In generating nondimensional solutions, the curves are spaced more closely near the mudline. As can be seen later, two or three additional curves between the mudline and a depth of 300 in. would have been helpful.

The first step in computing the p-y curves is to compute the ultimate soil resistance p_u using the smaller of the values from equations (11.7) and (11.8). These values are shown in Table 11.4, along with values of $z_r = x_r$ computed from equation (11.11).

Shown in Figure 11.15 are computer-generated p-y curves computed for various depths up to 900 in., all based on a pile diameter of 33 in. The rotational restraint at the pile head is assumed to be constant for all of the lateral loads.

The soil profile of Figure 11.14 shows relatively soft clay to a considerable depth. No computations are shown here for the pile penetration that is required to sustain the expected axial loading; however, it is likely that the piles would tip in the sand deposit. As noted earlier, when a pile under lateral loading reaches a length where it can be termed as *long* pile, any additional length has no effect on its behavior. For the following solutions, a pile length of 1500 in. was selected.

Table 11.4   Computed Values of p_u and $z_r = x_r$

Depth z, in.	Submerged Unit Weight γ', lb/in.^{3}	Undrained Shear Strength c, lb/in.^{2}	Ultimate Soil Resistance p_u, lb/in.	z_r, in.
0	0.0200	0	0	–
50	0.0203	0.253	64.87	62.90
100	0.0207	0.506	143.70	107.03
200	0.0213	1.013	300.86	165.85
300	0.0220	1.519	451.14	202.47
400	0.0227	2.025	601.43	227.61
500	0.0233	2.531	751.71	246.33
700	0.0277	3.544	1052.57	271.23
900	0.0260	4.556	1353.13	287.66

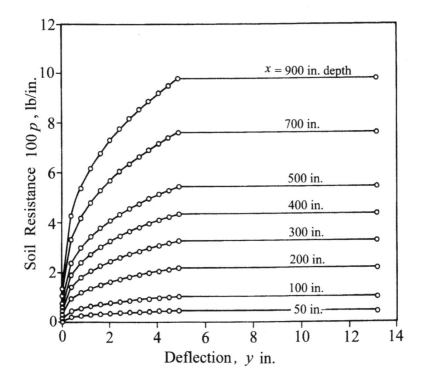

Figure 11.15 p-y curves for the example problem.

COMPUTATION SHEET

Let $P_t = 60,000$ lb   Try $T = 200$ in.

$Z_{\max} = 1,500/200 = 7.5 > 5$ Use *long* pile coefficients

From equation (11.21): $S_t = M_t/6.18 \times 10^9$

From equation (11.22): $S_t = (-1.623)(60,000)(200)^2/(4.23 \times 10^{11})$

$$-(1.75)(200)M_t/(4.23 \times 10^{11})$$

Solve the above two equations simultaneously for M_t.

$$68.447M_t = -3.8952 \times 10^9 - 350M_t; \qquad M_t = -9.31 \times 10^6 \text{ in.-lb}$$

From equation (11.15): $S_t = A_y(60,000)(200)^2/4.23 \times 10^{11}$

$$+B_y(-9.31) \times 10^6(200)^2/(4.23 \times 10^{11})$$

$$y = 1,135A_y - 0.880B_y$$

Table 11.5   Computed Results for the First Trial

x in.	Z	A_y in.	y_A in.	B_y	y_s in.	y in.	p lb/in.	E_{py} lb/in.^{2}
0	0	2.435	2.763	1.623	-1.428	1.335	0	0
50	0.25	2.032	2.306	1.210	-1.065	1.241	29.5	24
100	0.50	1.644	1.866	0.873	-0.768	1.098	62.5	57
200	1.00	0.962	1.092	0.364	-0.320	0.772	116.2	151
300	1.50	0.463	0.526	0.070	-0.062	0.464	146.7	316
400	2.00	0.142	0.161	0.007	-0.006	0.155	134.5	868
500	2.50	-0.025	-0.028	0.100	0.088	0.060	73.6	1227
700	3.50	-0.060	-0.068	0.066	0.053	-0.015	25.8	1720
900	4.50	-0.020	-0.022	0	0	0.022	48.6	2209
$k_{py} = 320/395 = 0.81$ lb/in.^{3}				$T = (4.23 \times 10^{11}/0.81)^{0.2} = 221$ in.				

The Computation Sheet and Table 11.5 show the first trial for a lateral load $P_t = 60,000$ lb. A relative stiffness factor T of 200 in. was selected. A value of the pile-head moment M_t was computed and the nondimensional expressions were used to compute a lateral deflection y at the depth of each of the p-y curves. The p-y curves of Figure 11.15 were entered with the computed deflection to obtain a value of p, with which the values of soil modulus E_{py} were computed.

The values of soil modulus are plotted in Figure 11.16, and the best straight line passing through zero is fitted through the plotted points. Somewhat more weight is given to the points near the mudline in fitting the line. The slope of the line is the value of k_{py}, the variation of the soil modulus with depth. The computed values of k_{py} and T are shown at the bottom of Table 11.5. Note that the first trial value of $T = 200$ in. yielded a computed value of $T = 221$ in.

Second and third trials were made following this same computation procedure, and the resulting values of E_{py} are plotted in Figure 11.16. A second trial value of T of 250 in. yielded a computed value of 244 in. A plot of the tried values of T against the computed values that were obtained is shown in Figure 11.17. As shown in this figure, T will converge to a value of about 239 in. When the third trial was made with a tried value of T of 239 in., convergence was indeed found to be this value.

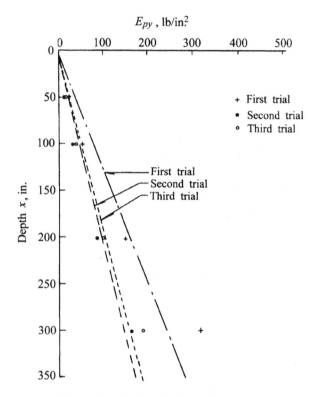

Figure 11.16   Trial plots of E_{py} values.

Using the same procedures demonstrated for $P_t = 60$ kips, values of T were found for values of P_t of 20, 100, and 140 kips. With these values of T, the nondimensional curves and equations (11.15)-(11.18), together with $p = E_{py}y$, the deflection, slope, moment, shear, and soil resistance (y, S, M, V, and p) can be computed for all depths.

Computer-Aided Solutions

As a means of evaluating the usefulness of the nondimensional method, even for the case where the soil properties were ideally suited to $E_{py} = k_{py}y$, computer solutions were obtained for the nonlinear differential equation using finite difference techniques. The following changes were made: an axial load of 500 kips was assumed, the pile stiffness and diameter were allowed to be reduced below a depth of 140 in., and the p-y curves were computed from the soil properties that existed at each depth. Computations were made for values of P_t in increments of 20 kips up to a value of 160 kips.

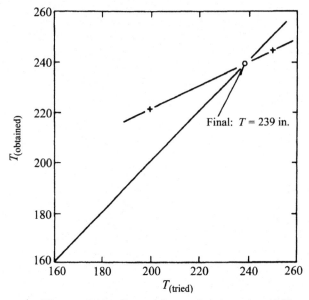

Figure 11.17   Interpolation for the value of T.

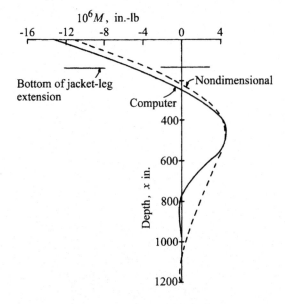

Figure 11.18   Bending moment versus depth for two methods of solution.

The bending moment along the length of a pile is of major concern for an offshore structure and pile deflection in most cases is less important. Figure 11.18 shows these computed values of bending moments with the two procedures for a lateral load of 60 kips. There is reasonable agreement in the results of the two methods, with the computer results showing somewhat more bending moment, perhaps because the use of the axial load of 500 kips. The position of the jacket leg extension is indicated, which allows the moment at the top of the 30 in. section to be found. The maximum values of combined stress in the 30 in. section and the 33 in. section were computed, assuming that the jacket leg extension was fully grouted so that the axial load at the mudline was taken by both the 33 in. and the 30 in. sections; and it was found that the most critical stress occurred at the top to the pile or where the pile joined the jacket.

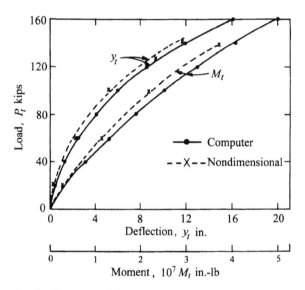

Figure 11.19   Deflection and bending moment at the pile head based on two computation methods.

The computed values of pile deflection and maximum bending moment for the top of the pile for the two methods of computation are shown in Figure 11.19. The computed values agree reasonably well, although the results from the computer are somewhat greater. If mild steel with a yield strength of 36 ksi is used in the design, the computed lateral load to cause a plastic hinge at the top of the pile can be found. The axial stress f_a can be computed as follows for the areas of 51.05 in.^{2} and 69.92 in.^{2} of the respective sections of 33 × 1/2 in. and the 30 × 3/4 in. Thus

$$f_a = (500)/(51.05 + 69.92) = 4.13 \text{ ksi}$$

The allowable bending stress would then be 31.83 ksi. The section modulus of the section (14,117/16.5) yields a bending moment at failure of 27,233 in.-kip. By entering Figure 11.19 with the bending moment, a lateral load of about 105 kips is found. The computed pile-head deflection at that load was about 6.7 in. A portion of the lateral load could be taken by dynamic loading and some load factor, according to the design criteria, to find the loading under service conditions.

Conclusions

In addition to the comments made in the presentation about the importance of obtaining the best possible prediction of p-y curves, the following points can be made:

1. The nondimensional solution can yield solutions that agree fairly well with the computer solution if the soil profile shows favorable values of shear strength with depth (sands and normally consolidated clays).

2. The nondimensional solution can (a) serve as a check to computer solutions and (b) reveal clearly the nature of the lateral-load problem and the importance of various parameters.

3. The plot of bending moment with respect to depth shown in Figure 11.18 clearly demonstrates the importance of analyzing the piles and superstructure as a unit to ensure the best estimate of rotational restraint at the top of the pile.

4. The engineer can achieve the most efficient design of the piles by taking into account the length of the extension of the jacket, the details of grouting between pile and jacket, the wall thickness of the pile with respect to depth, and careful estimation of loadings on a platform as a function of time.

11.5   RESPONSE OF PILES TO DYNAMIC LOADING

Earthquakes

Several steps are taken in order to design the foundation for a pile-supported structure in a seismic region. The location of the fault with respect to the structure must be found and the expected magnitude of the event must be selected. Then, the characteristics of the soils and rocks at the site must be considered in order to perform microzonation. With such data at hand, the time-dependent, free-field motion of the supporting soils and rocks at the building site can be computed or estimated. The engineer can then decide if there is a chance of liquefaction of any loose granular soil below the water table. If liquefaction appears to be likely for the selected earthquake, steps must be taken to improve the supporting soil or to design a structure that remains stable, even though some of the supporting soil liquefies.

If the free-field motion of the soils at the site are known as a function of depth, a fully rational solution can be undertaken which will require extensive and complicated computations. The validity of the results of such computations will be difficult to ascertain because of the lack of experimental data on the detailed performance of a variety of structures in a variety of seismic events.

A procedure sometimes used for the design of the pile foundation is to select a horizontal acceleration, which is some function of the acceleration of gravity. The acceleration can be used to obtain a pseudo-horizontal load, which is some function of the mass of the superstructure. A judgment about the safety of the foundation can be made on the basis of the pseudo-horizontal load.

The American Petroleum Institute (1993) requires that structures be designed to be safe in seismic zones. The evaluation of the area where the structure is to be placed with respect to seismic activity is required. Designs may be made using procedures from dynamic analysis such as response-spectrum analysis or time-history analysis.

Zeevaert (1983) studied data from records of seismographs during two significant earthquakes. The instruments were located at sites in the Valley of Mexico where deep beds of soft soils are present. In addition to the valuable records on accelerations, Zeevaert had extensive information on structures that were not damaged during the earthquakes or damaged to various degrees of severity. He studied the records of acceleration in detail and developed pseudo-acceleration spectra as a function of various amounts of damping. The characteristics of the soil were studied experimentally and analytically. Zeevaert developed a prediction of ground motion as a function of depth and extended the work into predicting the bending moment in a pile at a specific location during the earthquake selected for design. A photograph was included to show the rupture of a reinforced-concrete pile due to high bending moments. With regard to piles broken by an earthquake, Professor Ishahara at a breakfast in Tokyo years ago told the writer that he had not seen broken piles except due to liquefaction (Ishahara, 1977).

Time-Dependent Loading Above the Mudline

A completely different approach to the design of the piles is needed for the case where a forcing function is applied to the superstructure. The loading may come from machinery on the platform deck, from waves, and possibly from ship impact. The problem can be solved in either the time domain or in the frequency domain. The responses of a single pile or group of piles to dynamic loading in the frequency domain have been analyzed by a number of engineers. For example, Ensoft, Inc. (1999) has prepared a computer code that yields the dynamic stiffness of a pile-supported foundation, which is defined as the ratio between the applied force and the resulting displacement under the steady state vibration at a given frequency. Specifically, the stiffness can be related to lateral-load versus deflection, axial-load versus deflection, and moment versus rotation, and the configurations of the piles. The close spacing of piles is taken into account by pile-soil-pile interaction.

The Ensoft code is based principally on papers by Kausel (1974), Blaney, et al. (1975), Poulos (1971), and Roesset and Kausel (1975). The principal analytical techniques employed are the consistent boundary matrix method and the finite element method. This code requires the entry of the geometry and material characteristics of the piles and geometry and material characteristics of the layers of soil. For the soil, the shear-wave velocity, Poisson's ratio, mass density, and damping ratio are required. In addition, for pile groups the mass of the pile cap, a damping ratio, and a lateral stiffness can be input. With this code, the engineer can obtain a family of nonlinear curves showing pile stiffness as a function of frequency. For a single pile or a pile group, the curves will show the horizontal stiffness the vertical stiffness. For the pile group, the curves will also present the rocking stiffness in the x and y directions. With such sets of curves at hand, the engineer can proceed to analyze the dynamic response of the superstructure.

PROBLEMS

11.1   Compute a point on the curve of axial load versus settlement for the example in the text using a base settlement of 0.02 in. (instead of 0.05 in.). Compare the result with the computer solution in Figure 11.3.

11.2   Refer to the American Petroleum Institute's (1993) Recommended Practice and compute the allowable load that can be sustained by the axially loaded pile in the example.

11.3   Given a normally consolidated clay with an undrained shear strength that increased from zero at the mudline to a value of 1250 psf at a depth of 100 ft, an average submerged unit weight of 45 pcf, and an ε_{50} of 0.02, make necessary computations and plot the p-y curves for both static and cyclic loading for depths of 0 ft, 6 ft, and 12 ft for a pile with a diameter of 24 in. and a wall thickness of 1.0 in. The pile is assumed to be at an offshore location; select an appropriate value of the submerged unit weight in your computations.

11.4   Assume a fixed-head, open, steel-pipe pile with a diameter of 30 in. and a wall thickness of 0.75 in. and assume further that the pile behaves as a *long* pile. Compute the magnitude of the lateral load at the mudline to yield a maximum bending moment of 20 ksi if $E_{py} = k_{py}x$, where $k_{py} = 12$ lb/in.^{3}. Assume no axial load and a constant EI with depth. How long should the pile be? Use the nondimensional method.

11.5   Repeat Problem 11.4 with the assumption that the head of the pile is free to rotate.

11.6   For Problems 11.4 and 11.5, find the rotational restraint such that the maximum negative moment at the pile head and the maximum positive moment are equal. Use the lateral load found in Problem 11.5.

11.7   Use data from Problem 11.3 and assume a lateral load is applied at 10 ft above the mudline. Solve for the energy from a docking boat that can be sustained by the pile if the maximum allowable pile bending stress is 30 ksi. Use the nondimensional method of analysis and assume the number of repetitions is small so that the p-y curves for static loading are appropriate. (Hints: The lateral load to cause the maximum allowable bending stress is about 40 kips. Because several computations are required, the work may be divided among several students.)

REFERENCES

American Petroleum Institute, Recommended Practice for Planning, Designing and Constructing Fixed Offshore Platforms — Working Stress Design, *API Recommended Practice, RP2A-WSD,* twentieth ed., 1993.

Blaney, G. W., Kausel, E., and Roesset, J. M., Dynamic Stiffness of Piles, *Proceedings of the 2nd International Conference on Numerical Methods of Geomechanics,* Blacksburg, VA, 1975.

Coyle, H. M., and Reese, L. C., Load Transfer for Axially Loaded Piles in Clay, 1-26, *Journal of the Soil Mechanics and Foundations Division ASCE* **92**, SM2, 1966.

Det Norske Veritas, *Rules for the Design, Construction and Inspection of Offshore Structures,* Veritsveien 1, 1322 Hovek, Norway, 1977.

Ensoft, Inc., *Analysis of a Group of Piles Subjected to Axial and Lateral Loading,* Austin, TX, 1999.

Emrich, W. J., Sampling of Soils and Rock, *Performance Study of Soil Sampler for Deep Penetration Marine Borings, ASTM STP 483,* 1971.

Fisk, H. N., Nearsurface Sediments on the Continental Shelf Off Louisiana, *Proceedings of the 8th Texas Conference on Soil Mechanics and Foundation Engineering,* Bureau of Engineering Research, University of Texas, Austin, 1956.

Hetenyi, M., *Beams on Elastic Foundation,* University of Michigan Press, Ann Arbor, 1946.

Hvorslev, M. J., *Subsurface Exploration and Sampling of Soils for Civil Engineering Purposes,* Waterways Experiment Station, Corps of Engineers, Vicksburg, MI, 1949.

Ishahara, K., private communication, 1977.

Kausel, E., *Forced Vibrations of Circular Foundations on Layered Media,* Research Report R74-11, Civil Engineering Dept., Massachusetts Institute of Technology, Cambridge, 1974.

Matlock, H., Correlations for Design of Laterally Loaded Piles in Soft Clay, OTC 1204, *Proceedings of the Offshore Technology Conference,* 1970.

McClelland, B., Engineering Properties of Soils on the Continental Shelf of the Gulf of Mexico, *Proceedings of the 8th Texas Conference on Soil Mechanics and Foundation Engineering,* Bureau of Engineering Research, University of Texas, Austin, 1956.

Peck, R. B., Hanson, W. E. and Thornburn, T. H., *Foundation Engineering,* second ed., Wiley, New York, 1974.

Poulos, H. G., Behavior of Laterally Loaded Piles: Pile Groups, *Journal of the Soil Mechanics and Foundations Division ASCE* **97**, SM5, 1971.

Reese, L. C., and Van Impe, W., *Single Piles and Pile Groups Under Lateral Loading*, Balkema, Rotterdam, 2001.

Reese, L. C., and Wang, S. T., *TZPile for Windows*, Ensoft, Inc., Austin, TX 1996.

Roesset, J. M., and Kausel, E., Dynamic Soil Structure Interaction, *2nd International Conference on Numerical Methods of Geomechanics*, Blacksburg, VA, 1975.

Skempton, A. W., The Bearing Capacity of Clay, *Proceedings of the Building Research Congress, Division 1,* London, 1951.

Tomlinson, M. J., *Foundation Design and Construction*, fourth ed., Pitman, London, 1980.

Zeevaert, L., *Foundation Engineering for Difficult Subsoil Conditions,* second ed., Van Nostrand Reinhold, New York, 1983.

CONVERSION TABLE

Customary U.S. (English) Units to SI (Metric) Units

Quantity	Conversion	Multiplier
Acceleration	in./sec$^2 \longrightarrow$ m/s^2	2.54×10^{-2}
	ft/sec$^2 \longrightarrow$ m/s^2	3.048×10^{-1}
Area	in.$^2 \longrightarrow$ m^2	6.4516×10^{-4}
	ft$^2 \longrightarrow$ m^2	9.2903×10^{-2}
Area Moment of Inertia	in.$^4 \longrightarrow$ m^4	4.1623×10^{-7}
Density	lb($avoir.$)/in.$^3 \longrightarrow$ kg/m^3	2.7680×10^4
	lb($avoir.$)/ft^3 (pcf)$\longrightarrow$ kg/m^3	1.6018×10
Energy	ft-lb $\longrightarrow$ J	1.3558
	Btu $\longrightarrow$ J	1.0551×10^3
Force	lb $\longrightarrow$ N	4.4482
	kip (1000 lb) $\longrightarrow$ N	4.4482×10^3
	ton (2000 lb) $\longrightarrow$ N	8.8964×10^3
Length	in. $\longrightarrow$ m	2.54×10^{-2}
	ft $\longrightarrow$ m	3.048×10^{-1}
	mi $\longrightarrow$ m	1.6093×10^3
	nautical mi $\longrightarrow$ m	1.852×10^3
Mass	lb-sec^2/ft (slug) $\longrightarrow$ kg	1.4594×10
	lb ($avoir.$) $\longrightarrow$ kg	4.5359×10^{-1}
Mass Moment of Inertia	lb-ft-sec^2 (slug-ft^2) $\longrightarrow$ kg·m^2	1.3558
Moment	in.-lb $\longrightarrow$ N·m	1.1298×10^{-1}
	ft-lb $\longrightarrow$ N·m	1.3558
Momentum, linear	lb-sec $\longrightarrow$ kg·m/s	4.4482
Pressure	lb/in.^{2} (psi) $\longrightarrow$ Pa	6.8948×10^3
	lb/ft^2 (psf) $\longrightarrow$ Pa	4.7880×10
	tons/ft^2 (tsf) $\longrightarrow$ Pa	9.5760×10^4
Spring Constant, linear	lb/ft $\longrightarrow$ N/m	2.1016×10^3
Spring Constant, angular	ft-lb/rad $\longrightarrow$ N·m/rad	1.3558
Velocity	ft/sec $\longrightarrow$ m/s	3.048×10^{-1}
	mi/hr $\longrightarrow$ m/s	4.4704×10^{-1}
	nautical mi/hr (knot) $\longrightarrow$ m/s	5.1444×10^{-1}
Volume	in.$^3 \longrightarrow$ m^3	1.6387×10^{-5}
	ft$^3 \longrightarrow$ m^3	2.8317×10^{-2}

Index

323

CPSIA information can be obtained at www.ICGtesting.com
Printed in the USA
BVOW08*0714280515

401263BV00018B/109/P